INTRODUCTION TO

# INTRODUCTION TO MAGNETIC RESONANCE

## WITH APPLICATIONS TO CHEMISTRY AND CHEMICAL PHYSICS

## A. Carrington, F.R.S.

PROFESSOR OF CHEMISTRY, UNIVERSITY OF SOUTHAMPTON

## A. D. McLachlan

RESEARCH FELLOW, MRC LABORATORY OF MOLECULAR
BIOLOGY, CAMBRIDGE

LONDON
CHAPMAN AND HALL
A HALSTED PRESS BOOK
JOHN WILEY & SONS, INC., NEW YORK

First published 1967
by Harper & Row, Inc.

First published as a Science Paperback 1979
by Chapman and Hall Ltd
11 New Fetter Lane, London EC4P 4EE

© 1967, 1979 A. Carrington and A. D. McLachlan

Printed in Great Britain by
J. W. Arrowsmith Ltd., Bristol

ISBN 0 412 21700 7

Distributed in the U.S.A. by Halsted Press,
a Division of John Wiley & Sons, Inc., New York

# ACKNOWLEDGMENTS

We wish to thank both publishers and authors of a number of papers for permission to quote copyright material in figures and tables, as listed below.

Fig. 3.2, Pake, *J. Chem. Phys.* **16,** 327 (1948) p. 331; Table 3.1, Dailey and Townes, *J. Chem. Phys.* **23,** 118 (1955) p. 121; Fig. 6.6, Chestnut and Sloan, *J. Chem. Phys.* **33,** 637 (1960) p. 638; Fig. 11.11, Rogers and Pake, *J. Chem. Phys.* **33,** 1107 (1960) p. 1107; Fig. 13.1, Phillips, Looney, and Ikeda, *J. Chem. Phys.* **27,** 1435 (1957) p. 1435; Fig. 13.2, Bloembergen and Morgan, *J. Chem. Phys.* **34,** 842 (1961) p. 846; reprinted from the Journal of Chemical Physics by permission of the American Institute of Physics.

Figs. 11.9 and 11.10, Bloembergen, Purcell, and Pound, *Phys. Rev.* **73,** 679 (1948) pp. 705, 706; Fig. 13.7, Bloom and Shoolery, *Phys. Rev.* **97,** 1261 (1955) p. 1265; reprinted from the Physical Review by permission of the American Institute of Physics.

Fig. 3.3, Deeley and Richards, *J. Chem. Soc.* 3697 (1954) p. 3699; by permission of the Chemical Society, London.

Fig. 12.2, Looney, Phillips, and Reilly, *J. Amer. Chem. Soc.* **79,** 6136 (1957) p. 6138; Fig. 12.5, De Boer and Mackor, *J. Amer. Chem. Soc.* **86,** 1513 (1964) p. 1515; by permission of the copyright owners, the American Chemical Society.

Fig. 7.6, Horsfield, Morton, and Whiffen, *Mol. Phys.* **4,** 425 (1961) p. 426; Formula of triptycene, Van Der Waals and De Groot, *Mol. Phys.* **6,** 545 (1963) p. 545; Fig. 12.3, Pople, *Mol. Phys.* **1,** 168 (1958) p. 173; by permission of Taylor and Francis Ltd.

# PREFACE

The development of magnetic resonance spectroscopy ranks among the most important advances in chemical physics during the past two decades. Few other spectroscopic methods offer such a direct and detailed insight into events at the molecular level, and virtually every branch of chemistry has been given a new impetus. The organic chemist has, in nuclear resonance, an analytical tool of remarkable power. Applications in structural inorganic chemistry, less spectacular as yet, are nevertheless of growing importance, and much of our detailed knowledge of the electronic structure of transition metal ions derives from electron resonance studies. Free radical chemistry is already developing with increasing vigor. Photochemistry, radiation chemistry, the study of rapid equilibrium processes—these and many other applications suggest that few modern chemists can afford to be without a grasp of the essentials of magnetic resonance.

The first draft of this book was based on a course of sixteen lectures delivered to final year undergraduate students of chemistry in Cambridge. Our aim is to present the important principles clearly but succinctly and at the same time to illustrate the scope and applications of the method with key examples taken from current chemical research. The book has deliberately been kept short and free of unnecessary detail, but we have tried to give a well-balanced and coherent account of the subject which accurately reflects the importance of each part of it, and is as up-to-date as possible. In order to leave space for essential principles we have limited the discussion of applications in chemistry to a minimum. The use of high-resolution nuclear resonance in organic structure analysis, for example, receives a disproportionately short discussion in relation to its practical importance. The reason for this is that the principles are few and simple, while the applications, which are numerous and often of great sophistication, are described in several excellent books to which we refer. Again we have not generally given references to the original papers which are important in the development of the subject, because these also can be traced by reference to more advanced books. Our chief object, therefore, is to introduce the essentials of both nuclear and electron spin resonance to chemists with no previous knowledge of the subject. Then understanding the principles and the possibilities of magnetic resonance, they can usefully turn to the more specialized books and research papers to follow up their own special interests.

The main problem in writing a book such as this is to decide on what level the basic phenomena are to be treated. A quantum-mechanical approach is clearly essential to do justice to fundamental ideas like energy-level diagrams, transition probabilities, and the spin Hamiltonian. It is also important to deduce the main results logically and step by step from well-known physical principles, rather than to produce a magic formula out of the air. Therefore, our aim has been to present the main theoretical ideas accurately but simply, so that the reader follows each step in the argument and understands where each result comes from. We have omitted extra details which complicate the theory without introducing important new principles. At some points a compromise is necessary. For example, we have not felt it desirable to include a derivation of the Fermi contact hyperfine interaction, since the elementary proofs are unconvincing. We also wish to stress features which are common to both nuclear and electron spin resonance. For example, the student who masters the theory of nuclear dipole-dipole coupling, essential to an understanding of broad-line nuclear resonance in solids, has also learned much of what lies behind electron resonance investigations of triplet states. Again, the indirect nuclear spin-spin couplings seen in high-resolution n.m.r. spectra depend on the same magnetic interaction as the isotropic hyperfine splittings in the e.s.r. spectra of free radicals. There may well be people who would prefer their first introduction to magnetic resonance to be taught in entirely non-mathematical terms without recourse to quantum mechanics. If so, this book is not for them. We believe that such an approach makes magnetic resonance a very dull subject with no intellectual challenge, and hardly measures up to the needs of a modern chemist.

Having said this, we believe that this book is at a level suitable mainly for first year graduate students, although we hope that parts of it will prove useful to undergraduates in their final year. The reader should have a sound knowledge of elementary quantum mechanics and theoretical chemistry and in particular should be conversant with angular momentum operators, the solution of eigenvalue-eigenvector problems, and the use of perturbation theory. He should also know something about molecular orbital theory, matrix algebra, and vectors and tensors. We have included several mathematical appendixes; these are deliberately kept brief and are designed to indicate to the student what he should know and understand. They are not intended to be complete or self-contained.

The last four chapters, on transition metal ions, line widths, and spin relaxation, are necessarily more advanced than the others and could well be omitted at a first reading. The ideas of relaxation theory, in particular, require a sound knowledge of statistical mechanics and perturbation theory. The problems at the end of each chapter should be worked through. Many of them bring out important ideas which cannot be discussed in the body of the text. The reading lists generally refer to books or papers where the reader can find more detailed information about various topics. They are not necessarily the most important and original papers in the history of the subject, and we have not gone out of our way to compile any kind of roll of honor, including all of the famous physicists and chemists who have contributed to the development of magnetic resonance, from Abragam and Bloch through to Wasserman and Weissman.

Many of our colleagues and friends have helped us directly or indirectly, and we are grateful for their encouragement. Our principal debt is to Dr. Ruth Lynden-Bell who read the entire manuscript, made innumerable valuable suggestions, and saved us from committing a number of errors to print. Dr. Edel Wasserman, Dr. John Weil, and Dr. Anthony Stone have each helped us to improve parts of the book. We are most grateful to Professor John D. Roberts who made it possible for one of us to spend a year at the California Institute of Technology and provided all kinds of assistance in the writing and preparation of our manuscript. It is a pleasure to acknowledge the skillful and intelligent help of Mrs. Ruth Hanson in the long task of typing and preparing the manuscript from our early drafts. We also thank our publishers, particularly Walter E. Sears and Joan C. Paddock, for their keen encouragement and patient support.

Finally we dedicate our book to Professor H. C. Longuet-Higgins, F.R.S. He has taught us much and helped to shape our ideas in the years that we have spent at Cambridge. His acute insight into theoretical problems and his keen interest in magnetic resonance have been a constant source of inspiration and guidance, and our debt to him is enormous. We hope that his friends will recognize something of his spirit in the pages of this book.

We shall be glad to hear from anyone who has comments or suggestions which may help to improve the book. They will be considered carefully.

ALAN CARRINGTON
ANDREW MCLACHLAN

*Cambridge, England*

# PREFACE TO THE PAPERBACK EDITION

This edition of our book is unaltered from the first edition, originally published by Harper & Row in 1967. It was originally our intention to describe the important principles of magnetic resonance, and to illustrate the methods with suitably chosen examples. We did not intend to provide an exhaustive coverage of the applications of magnetic resonance in chemistry, physics and biology, and such a task would now, in 1978, be monumental. Happily the principles have not changed and if our book is now outdated this is mainly because of what it does not contain, rather than through inadequacies of the existing text. The most important advances since 1967 have been in nuclear magnetic resonance, particularly the development of Fourier transform and high frequency technique. The study of solids and of large molecules has been revolutionized by these new methods.

Developments in electron spin resonance have been less spectacular. The extension of magnetic resonance methods, particularly in conjunction with infra-red lasers to study gaseous free radicals, has been noteworthy. Multiple resonance experiments have also become more common, particularly in the study of condensed matter.

We are grateful to the many friends and colleagues who have read and used our book and who have made valuable comments about it.

ALAN CARRINGTON
ANDREW MCLACHLAN

*Southampton and Cambridge*
*July 1978*

# CONTENTS

# SHORT BIBLIOGRAPHY

We shall often refer the reader to the following books for further details about various aspects of magnetic resonance and quantum mechanics.

Abragam, A., *The Principles of Nuclear Magnetism* (Oxford: Clarendon Press, 1961). An advanced monograph for experts. The authoritative work on nuclear spin relaxation, line shapes, and saturation effects. Mainly written for physicists.

Andrew, E. R., *Nuclear Magnetic Resonance* (Cambridge: Cambridge University Press, 1955). An early book on n.m.r. with good chapters on nuclear resonance in solids and in metals.

Ballhausen, C., *Introduction to Ligand Field Theory* (New York: McGraw-Hill Book Co., Inc., 1962). A detailed monograph on the theory of transition metal ions. Contains much information about electron resonance studies and many illustrative calculations.

Bloembergen, N., *Nuclear Magnetic Relaxation* (New York: W. A. Benjamin, Inc., 1961). A reprint of a famous thesis which gives the first presentation of the theoretical principles of nuclear spin relaxation processes in solution.

Coulson, C. A., *Valence,* Second Edition (Oxford: Oxford University Press, 1961). Excellent descriptive text on valence theory.

Eyring, H., Walter, J., and Kimball, G. E., *Quantum Chemistry* (New York: John Wiley & Sons, Inc., 1944). A most useful concise introduction to molecular quantum mechanics.

Griffith, J. S., *The Theory of Transition Metal Ions* (Cambridge: Cambridge University Press, 1961). The standard work on this topic. The emphasis is on powerful and general theoretical ideas, and there is an excellent account of the spin Hamiltonian.

Landau, L. D., and Lifshitz, E. M., *Quantum Mechanics* (Oxford: Pergamon Press, 1959). Advanced text. Specially good for angular momentum.

Orgel, L. E., *An Introduction to Transition Metal Chemistry* (New York: John Wiley & Sons, Inc., 1960). Short descriptive text on transition metal ions.

Pake, G. E., *Paramagnetic Resonance* (New York: W. A. Benjamin, Inc., 1962). An introductory text on electron spin resonance, with special emphasis on transition metal ions and the theory of spin relaxation.

Pauling, L., and Wilson, E. B., *Introduction to Quantum Mechanics* (New York: McGraw-Hill Book Co., Inc., 1935.) Text on quantum mechanics. Recommended for angular momentum, perturbation theory, and valence theory.

Pople, J. A., Schneider, W. G., and Bernstein, H. J., *High Resolution Nuclear Magnetic Resonance* (New York: McGraw-Hill Book Co., Inc., 1959). The standard work on n.m.r. in organic chemistry. Highly recommended.

Ramsey, N. F., *Nuclear Moments* (New York: John Wiley & Sons, Inc., 1953). Detailed monograph on the theory of nuclear properties and magnetic interactions. Written for physicists.

Slichter, C. P., *Principles of Magnetic Resonance* (New York: Harper & Row, Publishers, Inc., 1963). An excellent exposition of the theoretical principles at a moderately advanced level.

# TABLE OF ATOMIC AND MAGNETIC CONSTANTS

### General constants

| | |
|---|---|
| Velocity of light | $c = 2.99793 \times 10^{10}$ cm/sec |
| Electron mass | $m = 9.1083 \times 10^{-28}$ gm |
| Electron charge | $e = 4.80286 \times 10^{-10}$ e. s. u. |
| Planck's constant | $h = 1.05443 \times 10^{-27}$ erg sec |
| Boltzmann's constant | $k = 1.38044 \times 10^{-16}$ erg/deg |
| Avogadro's number | $N = 6.0249 \times 10^{23}$ atoms/mole |
| Proton/electron mass ratio | $M/m = 1836.12$ |

### Magnetic constants

| | |
|---|---|
| Bohr magneton | $eh/2mc = 0.92731 \times 10^{-20}$ erg/gauss |
| Nuclear magneton | $eh/2Mc = 0.50504 \times 10^{-23}$ erg/gauss |
| $g$-value of free electron | $g_e = 2.002322$ |
| $g$-value of proton | $g_H = 5.585486$ |
| Electron-proton moment ratio | $g_e\beta/g_H\beta_N = 658.229$ |
| Proton-deuteron moment ratio | $g_D/g_H = 0.15351$ |
| Gyromagnetic ratio of the proton | $\gamma_H = 2.67530 \times 10^4$ radians/sec gauss |
| Hyperfine splitting of the H atom | $a_H = 1420.4058$ Mc/s |
| | $= 506.82$ gauss |

### Magnetic resonance

Resonance frequency of electron at 10,000 gauss = 28,025 Mc/s ($g = 2.002322$).
Wavelength for e.s.r. at 10,000 gauss: $\lambda = 1.0697$ cm.
Electron density at the nucleus of H atom $|\psi(0)|^2 = 2.14813 \times 10^{24}$ electrons/cm³.

### Energy conversion table

| | | cm$^{-1}$ | erg | electron volt | °K | Mc/s |
|---|---|---|---|---|---|---|
| 1 cm$^{-1}$ | $=$ | 1 | $1.9862 \times 10^{-16}$ | $1.2398 \times 10^{-4}$ | 1.4388 | $2.9979 \times 10^4$ |
| 1 erg | $=$ | $5.0348 \times 10^{15}$ | 1 | $6.2420 \times 10^{11}$ | $7.2440 \times 10^{15}$ | $1.5094 \times 10^{20}$ |
| 1 eV | $=$ | 8066.0 | $1.6020 \times 10^{-12}$ | 1 | 11,605 | $2.4181 \times 10^8$ |
| $kT$ at 1°K | $=$ | 0.69504 | $1.3805 \times 10^{-16}$ | $8.6169 \times 10^{-5}$ | 1 | $2.0837 \times 10^3$ |
| 1 Mc/s | $=$ | $3.3356 \times 10^{-5}$ | $6.6252 \times 10^{-21}$ | $4.1355 \times 10^{-9}$ | 1 | $4.7993 \times 10^{-5}$ |

# CHAPTER 1

# PRINCIPLES OF MAGNETIC RESONANCE

## 1·1 INTRODUCTION

The principal ideas behind magnetic resonance are common to both electron spin resonance (e.s.r.) and nuclear magnetic resonance (n.m.r.), but there are differences in the magnitudes and signs of the magnetic interactions involved, which naturally lead to divergences in the experimental techniques employed. In this introductory chapter we will describe the basic experiments and consider some of the necessary conditions for these experiments to be successful.

## 1·2 THE N.M.R. EXPERIMENT

All nuclei with odd mass number possess the property of spin; the spin angular momentum vector, which we denote by the symbol $I\hbar$ is measured in units of $\hbar$, where $\hbar$ is Planck's constant divided by $2\pi$. The value of the spin, $I$, is an odd integral multiple of $1/2$. Nuclei with even isotope number may be either spinless if the nuclear charge is even, or possess an integral spin $I$ with value 1, 2, 3, etc. The chemist is mostly concerned with the simplest nuclei having spin $1/2$, and of these the proton is by far the most important.

The possession of both spin and charge confers on the nucleus a magnetic moment $\mathbf{\mu}_N$ which is proportional to the magnitude of the spin, that is,

$$\mathbf{\mu}_N = \gamma_N \hbar \mathbf{I} = g_N \beta_N \mathbf{I} \tag{1}$$

$\gamma_N$ is called the magnetogyric ratio of the nucleus and is measured in radians·sec$^{-1}$·gauss$^{-1}$. Alternatively, as we see from Eq. (1), the magnetic moment may be expressed in terms of a dimensionless constant $g_N$ called the nuclear $g$ factor, and the nuclear magneton $\beta_N$. $\beta_N$ is equal to $e\hbar/2Mc$ where $e$ and $M$ are, respectively, the charge and mass of the proton and $c$ is the velocity of light. $g_N$ and $\mathbf{I}$ are the quantities whose different values distinguish one nucleus from another, so far as magnetic resonance is concerned. Sometimes the nuclear moment in nuclear magneton units is expressed in terms of the scalar *magnetic moment* $\mu_N$ which is defined to be equal to $g_N I$.

The magnetic moments and spins of some important nuclei are given in Table 1.1. Also tabulated is the electric quadrupole moment $Q$, which we shall introduce in Section 1.6.

TABLE 1.1. Nuclear Moments and Spins

| Nucleus | $I$ | $g_N$ | $\gamma_N$ (radians sec$^{-1}\cdot$gauss$^{-1}$) | Nucleus | $I$ | $g_N$ | $\gamma_N$ (radians sec$^{-1}\cdot$gauss$^{-1}$) | $Q$ (10$^{-24}$ cm$^2$) |
|---|---|---|---|---|---|---|---|---|
| H$^1$ | 1/2 | 5.585 | 26,753 | D$^2$ | 1 | 0.857 | 4,107 | 0.00274 |
| C$^{13}$ | 1/2 | 1.405 | 6,728 | Li$^7$ | 3/2 | 2.171 | 10,398 | 0.02 |
| N$^{15}$ | 1/2 | −0.567 | −2,712 | N$^{14}$ | 1 | 0.403 | 1,934 | 0.02 |
| F$^{19}$ | 1/2 | 5.257 | 25,179 | O$^{17}$ | 5/2 | −0.757 | −3,628 | −0.0265 |
| Si$^{29}$ | 1/2 | −1.111 | −5,319 | Na$^{23}$ | 3/2 | 1.478 | 7,081 | +1.00 or |
| P$^{31}$ | 1/2 | 2.263 | 10,840 | | | | | −0.836$^a$ |
| | | | | S$^{33}$ | 3/2 | 0.429 | 2,054 | −0.064 |
| | | | | Cl$^{35}$ | 3/2 | 0.548 | 2,624 | −0.079 |
| | | | | Cl$^{37}$ | 3/2 | 0.456 | 2,184 | −0.062 |
| | | | | K$^{39}$ | 3/2 | 0.261 | 1,250 | 0.113 |

NOTE: Nuclei with no spin: C$^{12}$ O$^{16}$ O$^{18}$ Si$^{28}$ S$^{32}$ Ca$^{40}$
$^a$ The quadrupole moment of Na$^{23}$ is uncertain.

Quantum theory demands that the allowable nuclear spin states are quantized; the component $m_I$ of the nuclear spin vector in any given direction can only take up one of a set of discrete values which are $+I, (I-1), \cdots -I$. $m_I$ is called the nuclear spin quantum number. For the proton with $\mathbf{I} = 1/2$, $m_I$ may only take the values $+1/2$ or $-1/2$. If we apply a steady magnetic field $\mathbf{H}$ on the proton, there is an interaction between the field and the magnetic moment $\boldsymbol{\mu}_N$, which may be represented in terms of the Hamiltonian

$$\mathcal{H} = -\boldsymbol{\mu}_N \cdot \mathbf{H} \qquad (2)$$

The precise meaning of this equation will be discussed in the next chapter. For the moment we merely note that if the direction of the magnetic field is defined to be the $z$ direction, the interaction may be rewritten

$$\mathcal{H} = -\gamma \hbar H I_z = -g_N \beta_N H I_z \qquad (3)$$

where $I_z$, the allowed component of the nuclear spin in the $z$ direction, has the value $+1/2$ or $-1/2$. These possibilities are summarized in terms of the energy level diagram shown in Fig. 1.1. The nuclear magneton $\beta_N$ is positive, and $g_N$ is positive for the proton. Thus the lower level with $m_I = +1/2$ corresponds to the situation in which the magnetic field $H$ and the nuclear moment are parallel; in the upper level they are antiparallel. In subsequent discussions we shall denote the state for which $m_I = +1/2$ by the symbol $|\alpha\rangle$ and that with $m_I = -1/2$ by $|\beta\rangle$.

In a macroscopic assembly of protons subjected to an external field $H$ we expect to find some protons with $\alpha$ spin and some with $\beta$ spin. We might further expect the distribution of spins over the two possible states to be governed by the Boltzmann law, so that the number of $\beta$ spins, $N_\beta$, divided by the number of $\alpha$ spins, $N_\alpha$, is equal to $e^{-\Delta E/kT}$ where $\Delta E$, the energy difference, is equal to $g_N \beta_N H$. We shall see subsequently that the establishment and maintenance of thermal equilibrium raises subtle problems of the utmost importance; for the moment we take it for granted that thermal equilibrium exists so that slightly more protons have $\alpha$ spin than $\beta$.

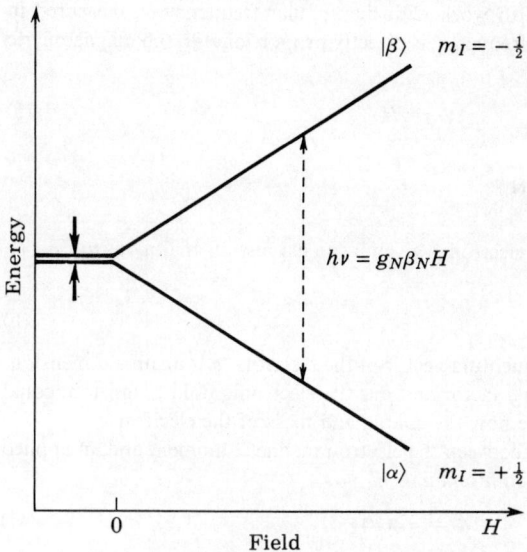

FIG. 1.1. Proton spin levels in a magnetic field.

In order to induce transitions between the two nuclear spin levels, an oscillating electromagnetic field is now applied to the system. Absorption of energy occurs provided the *magnetic* vector of the oscillating field is *perpendicular* to the steady field $H$ (we shall see why in Chapter 2) and provided the frequency $v$ of the oscillating field satisfies the resonance condition,

$$hv = g_N \beta_N H \tag{4}$$

The oscillating field is equally likely to produce transitions up from $\alpha$ to $\beta$ or down from $\beta$ to $\alpha$, and we can only detect absorption of energy from the radiation field if there are more $\alpha$ spins than $\beta$ spins.

Equation (4) indicates that one could look for a nuclear resonance absorption by varying either the magnetic field $H$ or the frequency $v$. In practice one usually works with a fixed frequency $v$, sweeping the magnetic field through the resonance value and obtaining a spectrum in which the absorption of energy is plotted as a function of magnetic field strength.

Because of variations in the nuclear $g$ factor and spin, different nuclei require widely different values of $H$ and $v$ for resonance to occur. In Table 1.2 we list the conditions required for four important nuclei, each with spin 1/2.

TABLE 1.2.   The Resonance Conditions

| Nucleus | Frequency $v$ in Mc/s in a Constant Field of 10,000 Gauss | Field $H$ at a Constant Frequency of 20 Mc/s |
|---------|:---:|:---:|
| $H^1$ | 42.577 | 4,697 |
| $F^{19}$ | 40.055 | 4,993 |
| $C^{13}$ | 10.705 | 18,633 |
| $P^{31}$ | 17.235 | 11,604 |

Sometimes it is convenient to work with the angular frequency $\omega$, measured in radians/sec. Then the resonant frequency is directly proportional to the magnetogyric ratio, with

$$\omega = \gamma H \tag{5}$$

## 1·3 THE E.S.R. EXPERIMENT

The magnetic moment $\mu_e$ of the electron is given by an expression analogous to Eq. (1), i.e.,

$$\mu_e = -g\beta \mathbf{S} \tag{6}$$

Here $\hbar \mathbf{S}$ is the spin angular momentum vector of the electron. $g$ is again a dimensionless constant called the electron $g$ factor and $\beta$ is the electronic Bohr magneton, equal to $e\hbar/2mc$, where $-e$ and $m$ are now the charge and mass of the electron.

As before, the interaction between the electron magnetic moment and an applied field $\mathbf{H}$ is represented by the Hamiltonian

$$\mathcal{H} = -\mu_e \cdot \mathbf{H} \tag{7}$$

which becomes

$$\mathcal{H} = g\beta H S_z \tag{8}$$

if the applied magnetic field is in the $z$ direction. Note that because of the negative charge of the electron $\mu_e$ in Eq. (6) is negative, and also that we have defined $\beta$ to be positive.

Again because $S = 1/2$ for the electron, there are two allowed orientations of the spin, parallel or antiparallel to $H_z$, and these are illustrated in Fig. 1.2. Note that for an electron the lowest state (which we shall again denote by the symbol $|\beta\rangle$) has $m_S = -1/2$, in contrast to the proton. The difference arises simply because of the different signs of $\mu_N$ and $\mu_e$.

Application of an oscillating field perpendicular to $H$ induces transitions provided the frequency $v$ is such that the resonance condition

$$hv = g\beta H \tag{9}$$

is satisfied. However we find that for a field of 10,000 gauss the resonance frequency of a free electron ($g = 2.002322$) is 28,026 Mc/s, approximately a thousand times larger than is required for any nuclear resonance. Radiation of this frequency has a wavelength of approximately 1 cm. As with n.m.r., experiments are nearly always carried out with a fixed frequency, and two frequencies are commonly used. An $X$-band spectrometer operates at about 3 cm wavelength with $v \simeq 9,500$ Mc/s, the field for resonance of a free electron then being about 3,400 gauss; with a $Q$-band spectrometer, $v \simeq 35,000$ Mc/s and $H$ is 12,500 gauss. The shorter wavelength $Q$-band spectrometers are commonly employed for studies at very low temperatures; otherwise $X$-band is usually more convenient, particularly for liquid phase studies.

## 1·4 THERMAL EQUILIBRIUM AND SPIN RELAXATION

We have already commented that resonance absorption can be detected only if there is a population difference between the two spin levels and we must now examine this

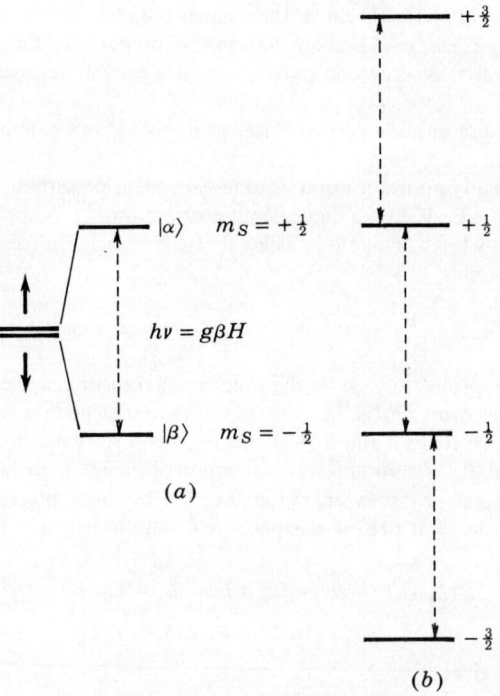

FIG. 1.2. Electron spin levels in a magnetic field. (a) Spin of 1/2. (b) 3 electrons with total spin 3/2.

assertion more carefully. The argument is identical for electron and nuclear spins; we will talk in terms of nuclear levels.

We will suppose that a macroscopic specimen containing $N$ spins has been placed in a steady magnetic field and that there are $N_\alpha$ spins in the lower state and $N_\beta$ in the upper. In thermal equilibrium there is a slight excess of spins in the $\alpha$ state, which gives rise to a small temperature-dependent paramagnetism. The ratio of $\alpha$ to $\beta$ spins is

$$\frac{N_\alpha}{N_\beta} = e^{g_N\beta_N H/kT} \tag{10}$$

by Boltzmann's law; and it follows, of course, that $N_\alpha + N_\beta = N$, the total number of spins. At ordinary temperatures, $g_N\beta_N H \ll kT$, and the Boltzmann factor in Eq. (10) is approximately $[1 + (g_N\beta_N H/kT)]$. The populations of the spin states are almost equal, and a simple calculation shows that on the average $\frac{1}{2}N[1 + (g_N\beta_N H/2kT)]$ nuclei have a spin $\alpha$, while $\frac{1}{2}N[1 - (g_N\beta_N H/2kT)]$ have spin $\beta$. The component of the magnetic moment resolved along the field direction is either $+\frac{1}{2}g_N\beta_N$ or $-\frac{1}{2}g_N\beta_N$ and so the whole sample in thermal equilibrium acquires a magnetic moment $M$ equal to $Ng_N^2\beta_N^2 H/4kT$. Since $M = \chi H$, where $\chi$ is the bulk magnetic susceptibility, we see that nuclei with spin 1/2 have $\chi = Ng_N^2\beta_N^2/4kT$. In general, the susceptibility of $N$ nuclei with spin $I$ is given by the equation

$$\chi = N\frac{g_N^2\beta_N^2 I(I+1)}{3kT} \tag{11}$$

For protons in a field of 10,000 gauss, the magnetic energy $g_N\beta_N H$ is only $10^{-3}$ cm$^{-1}$, whereas the thermal energy $kT$ is about 200 cm$^{-1}$. Consequently the populations of the two spin states differ by only one part in $10^5$ and nuclear resonance absorption is relatively weak.

We now enquire into the consequences of applying an oscillating magnetic field to the spin system.

First we note an important result of time-dependent perturbation theory which is proved in Appendix E. If a time-dependent perturbation $V(t)$ is applied to any system with discrete energy levels, the rate at which it causes transitions from level $a$ to level $b$ is given by the formula

$$P_{ab} = \frac{2\pi}{\hbar} |\langle b|V|a\rangle|^2 \delta(E_b - E_a - h\nu) \tag{12}$$

$P_{ab}$, the transition probability, gives the number of transitions per second, while the so-called matrix element $\langle b|V|a\rangle$ expresses the requirement that the perturbation $V$ couples or mixes the states $a$ and $b$. $\delta$ stands for the Dirac delta function. It imposes the condition that $P_{ab}$ is zero unless the quantum of energy $h\nu$ is exactly equal to the energy difference $E_b - E_a$ between the states. To be more precise, the $\delta$ function represents a line shape of perfect sharpness and infinite height, with a total area of unity.

$$\delta(E_b - E_a - h\nu) = \infty \text{ if } h\nu = E_b - E_a$$

$$= 0 \quad \text{if } h\nu \neq E_b - E_a \tag{13}$$

while integration gives

$$\int_{-\infty}^{+\infty} \delta(x)dx = 1 \tag{14}$$

Of course, a world in which absorption lines were $\delta$ functions would be a spectroscopist's heaven, and we shall see in due course how Eq. (12) has to be modified to take account of finite line widths. We shall also give a more explicit definition to the perturbation $V(t)$, which in our case is the rf field. However the result which is important for our immediate needs is that

$$|\langle b|V|a\rangle|^2 = |\langle a|V|b\rangle|^2 \tag{15}$$

so that the probabilities of upward and downward transitions are equal. The transitions caused by the oscillating field are called "stimulated transitions," as opposed to the spontaneous transitions which may also occur for other reasons. For spins of 1/2 let us denote the stimulated transition probabilities $P_{\alpha\beta}$ and $P_{\beta\alpha}$ by $P$ as shown in Fig. 1.3(a). Then the rate of change of population of the $|\alpha\rangle$ state is given by the equation

$$\frac{dN_\alpha}{dt} = N_\beta P_{\beta\alpha} - N_\alpha P_{\alpha\beta}$$

$$= P(N_\beta - N_\alpha) \tag{16}$$

We now introduce the new variable, the population difference $n = N_\alpha - N_\beta$ and express $N_\alpha$ and $N_\beta$ in terms of $N$ and $n$. Clearly

$$N_\alpha = \frac{1}{2}(N + n) \qquad N_\beta = \frac{1}{2}(N - n) \tag{17}$$

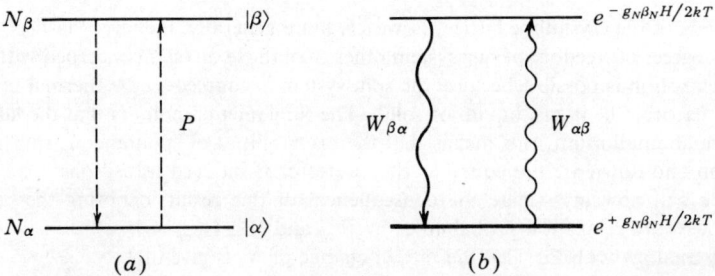

FIG. 1.3. (a) Stimulated transitions for a nuclear spin $1/2$. (b) Spontaneous transitions and Boltzmann factors.

Substituting (17) in (16) we obtain

$$\frac{dN_\alpha}{dt} = \left(\frac{1}{2}\right)\frac{dn}{dt} = -Pn \tag{18}$$

so that

$$\frac{dn}{dt} = -2Pn \tag{19}$$

The solution of this simple differential equation for the rate of change of the population difference is

$$n = n(0)e^{-2Pt} \tag{20}$$

where $n(0)$ is the difference at time $t = 0$. Finally we shall be interested in the rate of absorption of energy from the radiation field, $dE/dt$, which is given by the equation

$$\frac{dE}{dt} = N_\alpha P_{\alpha\beta}(E_\beta - E_\alpha) + N_\beta P_{\beta\alpha}(E_\alpha - E_\beta) = nP\Delta E \tag{21}$$

Now consider the rather startling implications of Eqs. (20) and (21). Equation (20) states that although we may start with a population difference $n(0)$, application of the resonant rf field results in exponential decay of the population difference and eventually the levels will be equally populated. This state of affairs is described as *saturation*. Equation (21) tells us that absorption of energy from the radiation field occurs only if $n$ is finite; the net result of (20) and (21) is therefore the depressing news that our resonance absorption line will ultimately disappear.

Clearly some other factor must be involved, and the reader may have already noted that our argument started with the apparently unjustified assumption that $N_\alpha > N_\beta$. Since $N_\alpha$ is equal to $N_\beta$ in the absence of an external magnetic field, the establishment of thermal equilibrium between the two states after application of $H$ must inevitably require that there are interactions between the nuclei and their surroundings which cause the spin orientation to change, while the excess magnetic energy is transferred to other degrees of freedom. This process, of nonradiative transitions between the two states $|\alpha\rangle$ and $|\beta\rangle$, is called *spin-lattice relaxation*. The name was originally used because nuclear spins in a solid transfer energy $\Delta E$ to other degrees of freedom, including the vibrations of the crystal lattice, each time a spin turns over.

The term "lattice" is a convenient if somewhat loose description of the physical systems which concern us. In the case of solids, relaxation processes do indeed involve vibrations of the crystalline lattice; however, more generally, the term "lattice" refers to the degrees of freedom of our system other than those directly concerned with spin. Spin relaxation is possible because the spin system is coupled to the thermal motions of the "lattice," be it gas, liquid, or solid. The fundamental point is that the lattice is at thermal equilibrium; this means that the probabilities of spontaneous spin transitions up and down are *not* equal, as they were for rf induced transitions.

We will now investigate the consequences of this result, denoting the upward and downward relaxation probabilities by $W_{\alpha\beta}$ and $W_{\beta\alpha}$ ($W_{\alpha\beta} \neq W_{\beta\alpha}$).

By analogy with Eq. (16) the rate of change of $N_\alpha$ is given by

$$\frac{dN_\alpha}{dt} = N_\beta W_{\beta\alpha} - N_\alpha W_{\alpha\beta} \tag{22}$$

At thermal equilibrium $dN_\alpha/dt = 0$, and denoting the equilibrium populations by $N_\alpha^0$ and $N_\beta^0$ we see that $N_\beta^0/N_\alpha^0 = W_{\alpha\beta}/W_{\beta\alpha}$. The populations follow from Boltzmann's law (10), and so the ratio of the two transition probabilities must also be equal to $e^{-\Delta E/kT}$. Expressing $N_\alpha$ and $N_\beta$ in terms of $N$ and $n$ as before we obtain

$$\frac{dn}{dt} = -n(W_{\beta\alpha} + W_{\alpha\beta}) + N(W_{\beta\alpha} - W_{\alpha\beta}) \tag{23}$$

This may be rewritten as

$$\frac{dn}{dt} = -\frac{(n - n_0)}{T_1} \tag{24}$$

in which $n_0$, the population difference at thermal equilibrium, is equal to $N[(W_{\beta\alpha} - W_{\alpha\beta})/(W_{\beta\alpha} + W_{\alpha\beta})]$ and $1/T_1$ is equal to $(W_{\alpha\beta} + W_{\beta\alpha})$. $T_1$ thus has the dimensions of time and is called the "spin-lattice relaxation time." It is a measure of the time taken for energy to be transferred to other degrees of freedom, that is, for the spin system to approach thermal equilibrium; large values of $T_1$ (minutes or even hours for some nuclei) indicate very slow relaxation.

We may now obtain a rather more complete description of the spin system by combining Eq. (19), which represents the effect of the rf field, with Eq. (24), representing the relaxation effect. We obtain

$$\frac{dn}{dt} = -2Pn - \frac{(n - n_0)}{T_1} \tag{25}$$

so that at equilibrium ($dn/dt = 0$) in the presence of an rf field the population difference $n$ is given by

$$n = \frac{n_0}{(1 + 2PT_1)} \tag{26}$$

Equation (26) combined with (21) gives a final expression for the rate of absorption of rf energy

$$\frac{dE}{dt} = nP\Delta E = n_0\Delta E \frac{P}{(1 + 2PT_1)} \tag{27}$$

Equation (27) indicates that provided $2PT_1 \ll 1$, it is relatively easy to avoid saturation. We shall see later that $P$ is proportional to the square of the rf field amplitude, so that one normally operates with low powers to avoid saturation. In certain special studies however, one saturates deliberately; we will encounter these cases in Chapter 13.

# 1·5 THE RESONANCE LINE SHAPE

It is now possible to say something about the width and shape of the resonance absorption line, which certainly cannot be represented by a $\delta$ function. First it is clear that, because of the spin relaxation, the spin states have a finite lifetime. The resulting line broadening can be estimated from the uncertainty relation $\Delta v \Delta t \approx 1$, and we find that the line width due to spin-lattice relaxation will be of the order of $1/T_1$. Spin-lattice relaxation is by no means the only mechanism of line broadening however. Many other processes which occur in both solids and liquids have the effect of varying the *relative energies* of the spin levels, rather than their lifetimes. Such processes are characterized by a relaxation time $T_2$, often called the spin-spin relaxation time but more satisfactorily, the transverse relaxation time.

These remarks may seem to suggest that there is little connection between $T_1$ and $T_2$. On the contrary these quantities are closely related because both modes of relaxation depend on the same interactions between the spins and their surroundings. Thus those interactions which lead to a finite lifetime for the spin states may also modulate the energy levels. As we shall see later, in Chapter 11, the total line width $1/T_2$ generally consists of two contributions. One, equal to $1/2T_1$, comes from spin-lattice relaxation, while another part called $1/T'_2$ comes from the energy fluctuations.

The true form of a broadened line is described empirically by a shape function $g(v)$ which shows how the energy absorption varies near resonance. That is, Eq. (12) for the stimulated transitions becomes

$$P(v) = \frac{2\pi}{\hbar} |\langle b|V|a\rangle|^2 g(v) \tag{28}$$

with $g(v)$ instead of the $\delta$ function. The shape function is normalized to unit area so that

$$\int_0^\infty g(v)dv = 1 \tag{29}$$

More commonly the shape function is defined on an angular frequency scale, $g(\omega) = g(v)/2\pi$. Magnetic resonance lines in solution almost always show the so-called Lorentz line shape

$$g(\omega) = \frac{T_2}{\pi} \frac{1}{1 + T_2^2(\omega - \omega_0)^2} \tag{30}$$

characteristic of a damped oscillatory motion. The resonance has a sharp peak at $\omega = \omega_0$ and the width between the points where the absorption has half its maximum height is $2/T_2$, as shown in Fig. 1.4(a). The next most common shape [Fig. 1.4(b)] is a Gaussian curve

$$g(\omega) = \frac{T_2}{\sqrt{2\pi}} e^{-\frac{1}{2}T_2^2(\omega - \omega_0)^2} \tag{31}$$

which often occurs with n.m.r. in crystalline solids. N.M.R. spectrometers usually record the absorption directly, but in e.s.r. the instrument is designed to draw the slope $dg/d\omega$ of the absorption line shape. The e.s.r. spectrum of a radical in solution

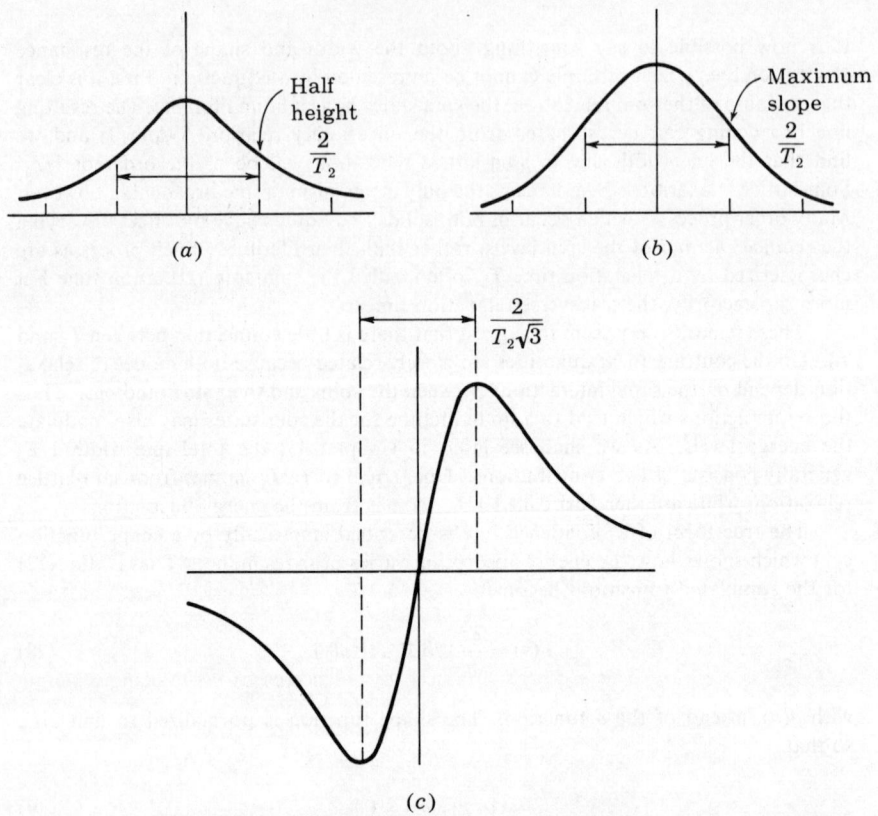

FIG. 1.4. Line shapes (a) Lorentz line; (b) Gaussian line; (c) derivative of the Lorentz line.

with a Lorentzian line then has the characteristic form shown in Fig. 1.4(c). The two peaks of the derivative curve correspond to points of maximum slope in the absorption, and the separation between them is $2/T_2\sqrt{3}$, or $2/T_2$ for a Gaussian line. The main difference between the two types of line shape (30) and (31) is that a Lorentzian drops off very much more slowly in the outer wings of the line.

We should note that these expressions give the shape of the *unsaturated* line, and depend only on the transverse relaxation time $T_2$. In a strong rf field the denominator of Eq. (27), often referred to as the *saturation factor*, is modified by the effects of line broadening and becomes $[1 + 2P(v)T_1]$. Thus we see that the possibility of power saturation depends on both the lifetimes and energy fluctuations of the spin levels.

## 1·6   MAGNETIC INTERACTIONS

It is of interest to look ahead and consider why the physics discussed in this chapter is of importance to chemists. In our discussions of the n.m.r. and e.s.r. experiments we were referring to bare nuclei and isolated electrons. The chemist is concerned with nuclei in molecules and we shall find, for example, that for protons in molecules the resonance magnetic field in Eq. (2) must be replaced by an effective field whose value is slightly different from $H$ and varies according to the chemical environment of the proton. The chemist is likewise concerned with electrons with unpaired spins in molecules, and here we shall find that the effective $g$ value varies according to the molecular environment of the unpaired electrons.

These are by no means the only complications. Molecules possessing more than one magnetic nucleus frequently give n.m.r. spectra showing that the magnetic moments of different nuclei interact with each other, the nature of the interaction again being characteristic of the molecular environment of the nuclei. Similarly in molecules possessing one unpaired electron and magnetic nuclei there are often electron-nuclear interactions which can, in principle, be studied by either n.m.r. or e.s.r. techniques. Molecules with two or more unpaired electrons give e.s.r. spectra with complications due to electron-electron interactions.

Some nuclei with spins of 1 or more also possess an electric quadrupole moment $eQ$ which occurs because the distribution of electric charge density $\rho(r)$ inside the nucleus is ellipsoidal rather than spherical. It is defined by the equation

$$eQ = \int \rho(r)(3z^2 - r^2)dv \qquad (32)$$

Here the $z$ direction is parallel to the spin vector $\mathbf{I}$ and the integral extends over the interior of the nucleus. $Q$ has the dimensions of an area, and is of the order of $10^{-24}$ cm$^2$. Nuclei with quadrupole moments tend to orient themselves in the strongly nonuniform electric fields produced by valence electrons in a molecule and this produces various complications in magnetic resonance spectra.

Finally we have already seen that the width of a resonance absorption line is dependent upon fluctuations in the energy and lifetime of the spin states, and we may anticipate that fluctuations in molecular environment will have profound effects on the appearance of magnetic resonance spectra.

One or more of these magnetic interactions is present in nearly every molecule and their study and elucidation, the subject of this book, represents one of the major advances in chemical physics during the past fifteen years.

## PROBLEMS

1. In Bohr's theory of the hydrogen atom, the lowest orbit has an electronic orbital angular momentum of $\hbar$ about the $z$ axis. Show that the orbital magnetic moment is equal to $-e\hbar/2mc$.

2. Compute the ratio $N_\alpha/N_\beta$ for electron spins in a field of 10,000 gauss at a temperature of 300°K.

3. Compute the paramagnetic susceptibility of a gram-molecule of protons at 1°K. How low a temperature is needed to align 75% of the spins in a field of 10,000 gauss?

4. The derivative X-band e.s.r. spectrum of a free radical in solution with $g = 2$ is found to have a peak-to-peak separation of 0.05 gauss. What is the relaxation time $T_2$?

5. In an n.m.r. experiment at 40 Mc/s, the sample consists of 0.1 cc of water at room temperature, and the spin-lattice relaxation time is 2.3 sec. The rf power is adjusted to make the saturation factor 1/2. At what rate is energy absorbed by the spins?

## SUGGESTIONS FOR FURTHER READING*

Ramsey: Chapters 1 and 2. Nuclear magnetic moments and magnetic interactions.

Slichter: Section 1.3. Fuller treatment of transition probabilities and relaxation.

Bloembergen: Chapters 1, 2, and 4. Theory of resonance absorption, $T_1$ and $T_2$.

Eyring, Walter, and Kimball: Pages 107–116, 124–127, and 347–350. Emission and absorption of electromagnetic radiation. Quantum mechanics of electron spin. Spin paramagnetism.

---

* Author's name only indicates that the book is listed in the Bibliography, pp. xvii–xviii.

CHAPTER 2

# MAGNETIC RESONANCE SPECTRA OF THE HYDROGEN AND HELIUM ATOMS

## 2·1 INTRODUCTION

In this chapter we shall derive expressions for the energies of the spin levels of the hydrogen atom in a magnetic field; we will then proceed to examine the possible spin transitions and hence calculate the details of the e.s.r. and n.m.r. spectra. Even this relatively simple system, consisting of one proton and one electron, possesses most of the complications which we shall meet later; the techniques which we employ here will be used many times throughout this book.

The solution of the problem involves several distinct stages. First we know that there are two possible orientations of the electron spin and two orientations of the proton spin, so that four distinct combinations can arise. We must formulate appropriate wave functions to describe these four arrangements, which are degenerate in the absence of any magnetic interactions. Next the interactions between the magnetic field **H** and the electron spin **S**, and between **H** and the nuclear spin **I**, are expressed in the form of operators which are added to the Hamiltonian and modify the energies of the spin states. We then have to introduce the coupling of the electron and nuclear spins and see how it further modifies the energies of the spin states and even, to some extent, mixes them together. Finally, we will use perturbation theory to calculate the effects of this so-called "hyperfine coupling" when the atom is in a strong magnetic field.

This calculation falls into two distinct parts. The first task is to find the forms of the stationary state spin wave functions and their energies. The final step is to determine which transitions may be induced by an oscillating magnetic field, and to derive expressions for their relative intensities.

# 2·2 THE MAGNETIC HAMILTONIAN

The spin Hamiltonian for the hydrogen atom is much simpler than most because the atom is spherical, and molecules show additional anisotropic effects which are introduced at the end of this Chapter. Here we commence by describing the three main magnetic interactions.

### 2.2.1 The Zeeman Energy

The nucleus and the electron both interact with the steady magnetic field giving an energy

$$\mathscr{H}_0 = g\beta H S_z - g_N \beta_N H I_z \tag{1}$$

As we have seen in Chapter 1, the nuclear energy is much smaller than the electronic part. $g$ and $g_N$ are the $g$ factors for the electron and the nucleus, and for the present we assume that $g$ takes the value 2.00232 characteristic of a free electron. Later we shall find that small corrections have to be made to both the electronic and nuclear Zeeman energies, even in a spherical atom.

### 2.2.2 The Isotropic Hyperfine Coupling

The magnetic moments of electrons and nuclei are coupled *via* the so-called *contact interaction*. This interaction, first introduced by Fermi to account for hyperfine structure in atomic spectra, represents the energy of the nuclear moment in the magnetic field produced at the nucleus by electric currents, which are associated with the spinning electron. It has the form

$$\mathscr{H}_1 = a\mathbf{I}\cdot\mathbf{S} = a(I_x S_x + I_y S_y + I_z S_z) \tag{2}$$

The coupling constant $a$ is proportional to the squared amplitude of the electronic wave function at the nucleus. $a$ has the dimensions of energy,

$$a = \frac{8\pi}{3} g\beta g_N \beta_N |\psi(0)|^2 \tag{3}$$

or it may also be expressed as a frequency $a/h$. An alternative notation is to use the Dirac $\delta$ function $\delta(\mathbf{r})$ and represent $\mathscr{H}_1$ by the operator

$$\mathscr{H}_1 = \frac{8\pi}{3} g\beta g_N \beta_N \delta(\mathbf{r}) \mathbf{I}\cdot\mathbf{S} \tag{4}$$

The $\delta$ function imposes the condition that $\mathbf{r} = 0$ when we integrate over the coordinates of the electron. Hence contact interaction can only occur when the electron has a finite probability density at the nucleus; since the $p$, $d$, $f$, or higher orbitals have nodes at the nucleus the electron must have some $s$-orbital character.

In the hydrogen atom the unpaired electron occupies the $1s$ orbital

$$\psi = \frac{1}{\sqrt{\pi a_0^3}} e^{-r/a_0}$$

$$a_0 = \hbar^2/me^2 \tag{5}$$

where $a_0$ is the Bohr radius, 0.52918 Ångstroms. Substitution of this wave function into Eq. (3) gives a predicted hyperfine frequency $a/h$ of 1422.74 Mc/s; however the experimental value of 1420.406 Mc/s is significantly smaller because of various corrections to Fermi's theory.

### 2.2.3 The Dipolar Interaction

There is also a magnetic coupling between the magnetic moments of the electron and nucleus which is entirely analogous to the classical dipolar coupling between two bar magnets. The classical interaction energy $E$ between two magnetic moments $\mu_e$ and $\mu_N$ is given by

$$E = \frac{\mu_e \cdot \mu_N}{r^3} - \frac{3(\mu_e \cdot r)(\mu_N \cdot r)}{r^5} \tag{6}$$

where $r$ is the radius vector from $\mu_e$ to $\mu_N$ and $r$ is the distance between the two moments. The quantum mechanical version of (6) is obtained simply by substituting

$$\mu_e = -g\beta S$$
$$\mu_N = g_N \beta_N I \tag{7}$$

which yields the dipolar interaction Hamiltonian

$$\mathcal{H}_2 = -g\beta g_N \beta_N \left( \frac{I \cdot S}{r^3} - \frac{3(I \cdot r)(S \cdot r)}{r^5} \right) \tag{8}$$

The expression (8) has to be averaged over the entire probability distribution $|\psi(r)|^2$ of the odd electron. For reasons which are discussed later the dipolar Hamiltonian $\mathcal{H}_2$ averages out to zero whenever the electron cloud is spherical. Hence the magnetic interactions in the hydrogen atom are necessarily isotropic.

In summary, then, the complete Hamiltonian is

$$\mathcal{H} = g\beta HS_z - g_N \beta_N HI_z + aS \cdot I \tag{9}$$

Eq. (9) is called the spin Hamiltonian since it operates on wave functions expressed in terms of spin variables only. In a sense it is an empirical expression; the hyperfine term $aS \cdot I$ represents an isotropic coupling but does not imply anything about the mechanism of the coupling. It is the task of the experimentalist to determine the appropriate spin Hamiltonian and to measure the values of the parameters. The theorist (who may, of course, also have made the measurements!) has the task of interpreting these parameters.

## 2.3 PERTURBATION THEORY

Before dealing with the specific problem of the hydrogen atom we will describe the general approach to be used. The three magnetic interactions $H \cdot S$, $H \cdot I$, and $S \cdot I$ which make up the operator $\mathcal{H}$ are divided into two distinct parts, $\mathcal{H}_0$ and $\mathcal{H}_1$. $\mathcal{H}_0$ is regarded as the main part of the Hamiltonian and $\mathcal{H}_1$ as a small perturbation (further details of perturbation theory are given in Appendix B). The spin wave functions are then expressed as linear combinations of four basis functions $\phi_1$, $\phi_2$, $\phi_3$, $\phi_4$ which are deliberately chosen to be eigenfunctions of $\mathcal{H}_0$ with the "zero-order" unperturbed energy values $\varepsilon_1$, $\varepsilon_2$, $\varepsilon_3$, $\varepsilon_4$.

The perturbation $\mathscr{H}_1$ yields modified wave functions and energies of the form

$$\psi_n = \phi_n - \sum_{m \neq n} \frac{\langle m|\mathscr{H}_1|n\rangle}{\varepsilon_m - \varepsilon_n} \phi_m \tag{10}$$

$$E_n = \varepsilon_n + \langle n|\mathscr{H}_1|n\rangle - \sum_{m \neq n} \frac{\langle m|\mathscr{H}_1|n\rangle\langle n|\mathscr{H}_1|m\rangle}{\varepsilon_m - \varepsilon_n} \tag{11}$$

The two extra terms on the right-hand side of Eq. (11) are described as the first-order and second-order corrections to the energy. $\langle n|\mathscr{H}_1|n\rangle$ and $\langle m|\mathscr{H}_1|n\rangle$ are the matrix elements of $\mathscr{H}_1$ (see Appendix A), the former representing the fact that $\mathscr{H}_1$ shifts the energy of $\phi_n$ without changing its form, the latter indicating that $\mathscr{H}_1$ also mixes $\phi_n$ with other functions $\phi_m$.

For the hydrogen atom $\mathscr{H}_0$ is the Zeeman energy (1) and $\mathscr{H}_1$ is the contact interaction (2). Finally we ought to convince ourselves that $\mathscr{H}_1$ is indeed small enough to be a genuine perturbation. In a field of 10,000 gauss the hyperfine frequency 1420.4 Mc/s is considerably smaller than the free electron e.s.r. frequency of 28,026 Mc/s. This is actually sufficient to justify the perturbation technique since the quantity which has to be small [see Eq. (32)] is the ratio of the hyperfine frequency to the sum of the e.s.r. and n.m.r. frequencies.

## 2·4    THE BASIC SPIN FUNCTIONS AND ZERO-ORDER ENERGIES

In Chapter 1 we saw that the electron has a spin vector $\mathbf{S}$ equal to $1/2$ and that there are two allowed components of the spin along any chosen direction which we shall call the $z$ axis. The two possible *spin functions* will now be denoted by the symbols $|\alpha_e\rangle$ and $|\beta_e\rangle$ with spin quantum numbers $m_S = +1/2$ and $-1/2$, respectively. These statements are expressible in terms of the operator equations

$$S_z|\alpha_e\rangle = +\tfrac{1}{2}|\alpha_e\rangle$$
$$S_z|\beta_e\rangle = -\tfrac{1}{2}|\beta_e\rangle \tag{12}$$

A much fuller discussion of these relations is given in Appendix C. Similar equations also apply to the proton spin

$$I_z|\alpha_N\rangle = +\tfrac{1}{2}|\alpha_N\rangle$$
$$I_z|\beta_N\rangle = -\tfrac{1}{2}|\beta_N\rangle \tag{13}$$

Having defined more clearly what is meant by the statement that a particle has $\alpha$ or $\beta$ spin we can proceed to write down the starting wave functions. Each orientation of the electron spin can be associated with either of the two proton spin orientations so that appropriate basis functions are written as products of an electron and a nuclear part

$$\phi_1 = |\alpha_e\alpha_N\rangle \qquad \phi_2 = |\alpha_e\beta_N\rangle \qquad \phi_3 = |\beta_e\alpha_N\rangle \qquad \phi_4 = |\beta_e\beta_N\rangle \tag{14}$$

These four states are initially degenerate (neglecting $\mathscr{H}_1$), but the effect of applying the steady magnetic field, represented by the Hamiltonian $\mathscr{H}_0$, is to separate the energy levels. The functions (14) are all eigenfunctions of $\mathcal{H}_0$ and the matrix of $\mathcal{H}_0$ is diagonal. For example

$$
\begin{aligned}
\mathcal{H}_0|\alpha_e\beta_N\rangle &= g\beta H S_z|\alpha_e\rangle \cdot |\beta_N\rangle - g_N\beta_N H|\alpha_e\rangle \cdot I_z|\beta_N\rangle \\
&= \tfrac{1}{2}g\beta H|\alpha_e\rangle \, |\beta_N\rangle + \tfrac{1}{2}g_N\beta_N H|\alpha_e\rangle \, |\beta_N\rangle \\
&= (\tfrac{1}{2}g\beta H + \tfrac{1}{2}g_N\beta_N H)|\alpha_e\beta_N\rangle
\end{aligned} \tag{15}
$$

The zero-order energy values for the four states are then

$$
\begin{aligned}
\varepsilon_1 &= \tfrac{1}{2}g\beta H - \tfrac{1}{2}g_N\beta_N H && |\alpha_e\alpha_N\rangle \\
\varepsilon_2 &= \tfrac{1}{2}g\beta H + \tfrac{1}{2}g_N\beta_N H && |\alpha_e\beta_N\rangle \\
\varepsilon_3 &= -\tfrac{1}{2}g\beta H - \tfrac{1}{2}g_N\beta_N H && |\beta_e\alpha_N\rangle \\
\varepsilon_4 &= -\tfrac{1}{2}g\beta H + \tfrac{1}{2}g_N\beta_N H && |\beta_e\beta_N\rangle
\end{aligned} \tag{16}
$$

The results are also shown in the form of an energy level diagram in Fig. 2.1.

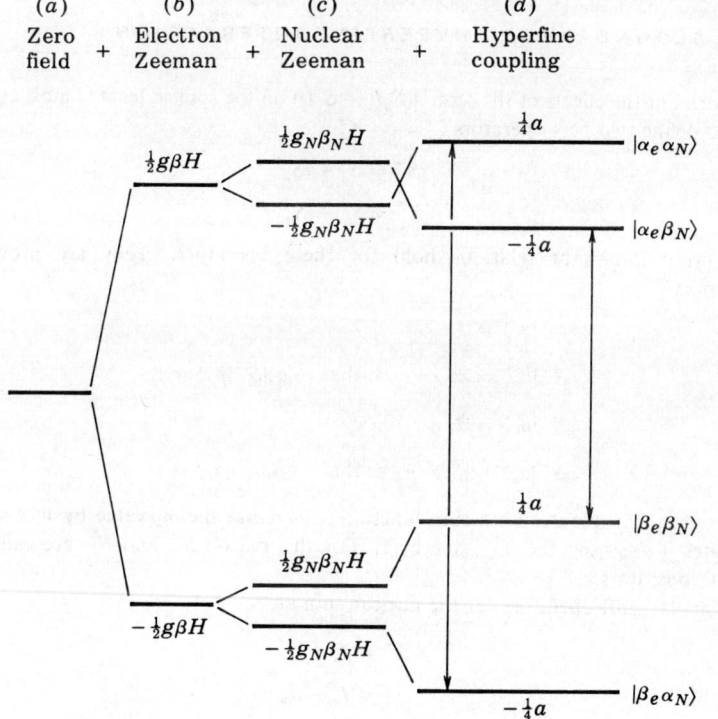

FIG. 2.1. First-order spin energy levels of the hydrogen atom, and the allowed e.s.r. transitions.

$2 \cdot 5$ **THE FIRST-ORDER HYPERFINE ENERGIES**

We now consider the effect of $\mathscr{H}_1$ acting as a perturbation on the zero-order wave functions (14). As already indicated $\mathscr{H}_1$ may be expanded as follows.

$$\mathscr{H}_1 = a\mathbf{S} \cdot \mathbf{I} = a[S_z I_z + S_x I_x + S_y I_y] \tag{17}$$

We have not considered the operators $S_x$, $S_y$, $I_x$, $I_y$; for the moment we merely comment that they do not affect the energies of $\phi_1$ and $\phi_4$, but do mix $\phi_2$ and $\phi_3$ together. The net effect of the term $a(S_x I_x + S_y I_y)$ is thus to produce a second-order change in the energies which will be calculated in Section 2.6. But first we restrict ourselves to a first-order treatment.

The operator $S_z I_z$ simply multiplies each spin function by a factor of $(\pm 1/2)$ for the electron and $(\pm 1/2)$ for the nucleus, giving a diagonal matrix for $aI_z S_z$.

$$\langle \alpha_e \alpha_N | a S_z I_z | \alpha_e \alpha_N \rangle = \tfrac{1}{4}a, \qquad \langle \alpha_e \beta_N | a S_z I_z | \alpha_e \beta_N \rangle = -\tfrac{1}{4}a$$

$$\langle \beta_e \alpha_N | a S_z I_z | \beta_e \alpha_N \rangle = -\tfrac{1}{4}a, \qquad \langle \beta_e \beta_N | a S_z I_z | \beta_e \beta_N \rangle = \tfrac{1}{4}a \tag{18}$$

These are just the first-order energy shifts in (11), and assuming $a$ to be positive the new energies of the spin levels are as shown in Fig. 2.1.

$2 \cdot 6$ **THE SECOND-ORDER HYPERFINE INTERACTION**

To work out the effects of the term $a(S_x I_x + S_y I_y)$ on the spin states it is most convenient to define two new operators,

$$S^+ = S_x + iS_y$$

$$S^- = S_x - iS_y \tag{19}$$

Four very important relations hold for these operators. They are proved in Appendix C.

$$S^+ |\alpha_e\rangle = 0 \tag{20}$$

$$S^+ |\beta_e\rangle = |\alpha_e\rangle \qquad \text{so that} \quad \langle \alpha_e | S^+ | \beta_e \rangle = 1 \tag{21}$$

$$S^- |\beta_e\rangle = 0 \tag{22}$$

$$S^- |\alpha_e\rangle = |\beta_e\rangle \qquad \text{so that} \quad \langle \beta_e | S^- | \alpha_e \rangle = 1 \tag{23}$$

Thus when $S^+$ operates on a spin function it *increases* the $m_S$ value by 1; when $S^-$ operates it *decreases* the $m_S$ value by 1. For this reason $S^+$ and $S^-$ are called the "shift" operators.

Similar shift operators for the nuclear spin are

$$I^+ = I_x + iI_y$$

$$I^- = I_x - iI_y \tag{24}$$

These operate on the nuclear spin functions in precisely the same way:

$$I^+ |\alpha_N\rangle = 0, \qquad I^+ |\beta_N\rangle = |\alpha_N\rangle, \qquad I^- |\alpha_N\rangle = |\beta_N\rangle, \qquad I^- |\beta_N\rangle = 0 \tag{25}$$

We can now rewrite the operator $S_x I_x + S_y I_y$ in terms of the shift operators. First (19) and (24) yield

$$S^+ I^- = (S_x I_x + S_y I_y) + i(S_y I_x - S_x I_y) \tag{26}$$

$$S^- I^+ = (S_x I_x + S_y I_y) - i(S_y I_x - S_x I_y) \tag{27}$$

so that

$$S_x I_x + S_y I_y = \tfrac{1}{2}(S^+ I^- + S^- I^+) \tag{28}$$

From Eqs. (20–23) and (25) it is clear that the operator $S^+ I^-$ gives zero unless it acts on the state $|\beta_e \alpha_N\rangle$ shifting it to $|\alpha_e \beta_N\rangle$. Similarly $S^- I^+$ annihilates all states except $|\alpha_e \beta_N\rangle$. The only nonvanishing matrix elements of the two operators are

$$\langle \alpha_e \beta_N | S^+ I^- | \beta_e \alpha_N \rangle = 1$$
$$\langle \beta_e \alpha_N | S^- I^+ | \alpha_e \beta_N \rangle = 1 \tag{29}$$

This indeed demonstrates that $\phi_1$ and $\phi_4$ are unaffected but $\phi_2$ and $\phi_3$ are mixed together. Combining (29) and (18) we are at last able to write down the complete matrix of the operator $\mathscr{H}_1$.

$$a\mathbf{I}\cdot\mathbf{S} = \frac{1}{4}a \begin{bmatrix} 1 & 0 & 0 & 0 \\ 0 & -1 & 2 & 0 \\ 0 & 2 & -1 & 0 \\ 0 & 0 & 0 & 1 \end{bmatrix} \begin{matrix} \alpha_e \alpha_N \\ \alpha_e \beta_N \\ \beta_e \alpha_N \\ \beta_e \beta_N \end{matrix} \tag{30}$$

The two modified wave functions, obtained through use of the perturbation formula (10), are

$$\psi_2 = |\alpha_e \beta_N\rangle + \frac{a}{2(g\beta H + g_N \beta_N H)} |\beta_e \alpha_N\rangle$$

$$\psi_3 = |\beta_e \alpha_N\rangle - \frac{a}{2(g\beta H + g_N \beta_N H)} |\alpha_e \beta_N\rangle \tag{31}$$

and their energies, which are shown schematically in Fig. 2.2 are

$$E_2 = \left(\frac{1}{2}g\beta H + \frac{1}{2}g_N \beta_N H\right) - \frac{1}{4}a + \frac{a^2}{4(g\beta H + g_N \beta_N H)}$$

$$E_3 = -\left(\frac{1}{2}g\beta H + \frac{1}{2}g_N \beta_N H\right) - \frac{1}{4}a - \frac{a^2}{4(g\beta H + g_N \beta_N H)} \tag{32}$$

Perturbation theory has been used here to illustrate the technique which is generally used to analyze electron resonance spectra, but the spin energy levels of the hydrogen atom can be solved exactly by finding the eigenvalues of the complete energy matrix. To save space we use the temporary abbreviations $\Delta_e$ and $\Delta_N$ for $g\beta H$ and $g_N \beta_H H$. The secular determinant (defined in Appendix A) is then

|  | $|\alpha\alpha\rangle$ | $|\alpha\beta\rangle$ | $|\beta\alpha\rangle$ | $|\beta\beta\rangle$ |
|---|---|---|---|---|
| $\langle\alpha\alpha|$ | $\tfrac{1}{2}(\Delta_e-\Delta_N)+\tfrac{1}{4}a-E$ | 0 | 0 | 0 |
| $\langle\alpha\beta|$ | 0 | $\tfrac{1}{2}(\Delta_e+\Delta_N)-\tfrac{1}{4}a-E$ | $\tfrac{1}{2}a$ | 0 |
| $\langle\beta\alpha|$ | 0 | $\tfrac{1}{2}a$ | $-\tfrac{1}{2}(\Delta_e+\Delta_N)-\tfrac{1}{4}a-E$ | 0 |
| $\langle\beta\beta|$ | 0 | 0 | 0 | $-\tfrac{1}{2}(\Delta_e-\Delta_N)+\tfrac{1}{4}a-E$ |

$$\tag{33}$$

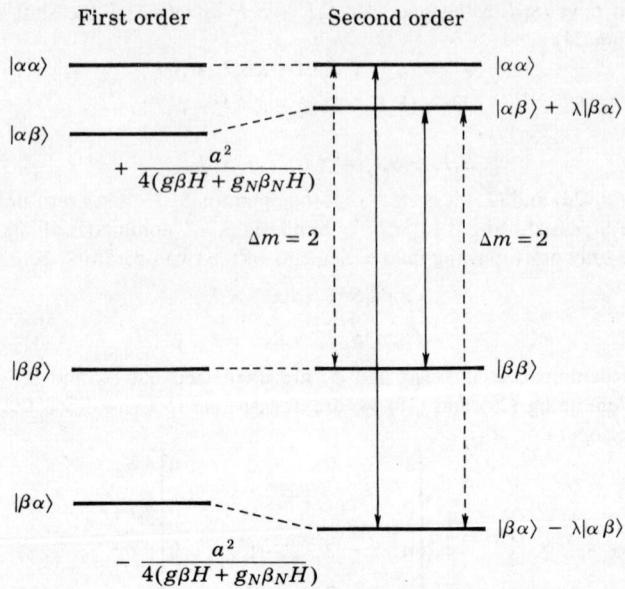

FIG. 2.2. Second-order hyperfine energy levels of the hydrogen atom. The vertical arrows indicate the "allowed" and "forbidden" electron resonance transitions.

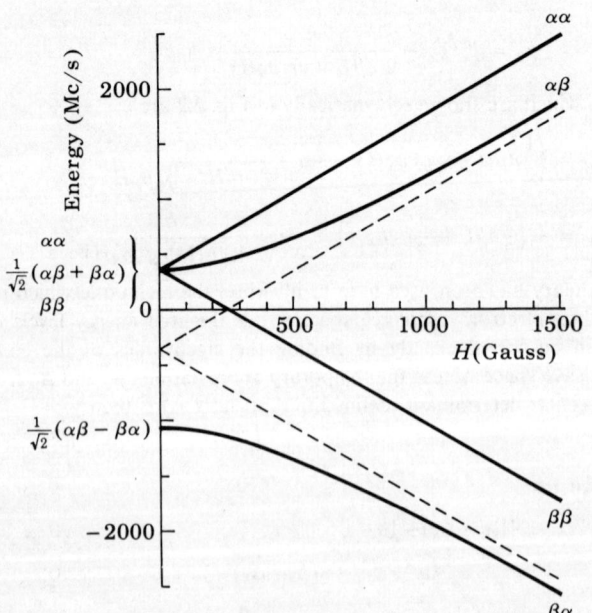

FIG. 2.3. Exact energy levels of the hydrogen atom in varying magnetic fields.

Fig. 2.3 shows the behavior of the exact energy levels as a function of the magnetic field $H$.

## 2.7 THE FIRST-ORDER E.S.R. SPECTRUM

To a first approximation the spin states and energies of the hydrogen atom are those given in Eqs. (16) and (18), or in Fig. 2.1. Suppose an oscillating magnetic field of strength $2H_1 \cos \omega t$ now acts on the atom (the factor of 2 is conventional). If the angular frequency $\omega$ is correct for resonance several kinds of spin transitions may be induced. Transitions like $\alpha_e \alpha_N \rightarrow \beta_e \alpha_N$ where only the electron spin changes are called e.s.r. transitions, while those like $\alpha_e \alpha_N \rightarrow \alpha_e \beta_N$ which involve the nucleus alone are n.m.r. ones. Finally the third kind of transition $\alpha_e \beta_N \rightarrow \beta_e \alpha_N$ where both spins change has a very low probability, and is known as a "forbidden transition."

An oscillating $H_1$ field along the $z$ direction merely modulates the energy levels of the spin system and causes no energy absorption. Therefore it must be applied across the steady field $H$ in the $x$ direction. The resulting time-dependent perturbation on the atom is

$$V(t) = 2(g\beta H_1 S_x - g_N \beta_N H_1 I_x) \cos \omega t$$

$$= 2V \cos \omega t \tag{34}$$

and then the transition probability from state $n$ to state $m$ is equal to

$$P_{nm} = \frac{2\pi}{\hbar^2} |\langle n|V|m \rangle|^2 \delta(\omega_{mn} - \omega) \tag{35}$$

as we have seen in Chapter 1 and Appendix E. The problem reduces to one of finding whether the matrix element of $V$ between each pair of states vanishes or not.

As the electron resonance transitions are caused by the effect of $H_1$ on the *electron* spins we can omit the nuclear spin operator $I_x$ from (34), and write

$$P_{nm} = \frac{2\pi}{\hbar^2} g^2 \beta^2 H_1^2 |\langle n|S_x|m \rangle|^2 \delta(\omega_{mn} - \omega) \tag{36}$$

It is convenient to use the shift operators again

$$S_x = \tfrac{1}{2}(S^+ + S^-) \tag{37}$$

Then a typical matrix element is calculated as follows

$$\langle \alpha_e \alpha_N |S_x| \beta_e \alpha_N \rangle = \langle \alpha_e| \tfrac{1}{2}(S^+ + S^-)|\beta_e \rangle \langle \alpha_N | \alpha_N \rangle$$

$$= \tfrac{1}{2}\langle \alpha_e |S^+| \beta_e \rangle = \tfrac{1}{2} \tag{38}$$

and the transition probability is equal to

$$P = \frac{\pi}{2\hbar^2} g^2 \beta^2 H_1^2 \, \delta(\omega_{mn} - \omega) \tag{39}$$

Notice that since the $S_x$ operator only acts on the electron spin the nuclear spin parts in (38) can be factored out. The orthogonality relations $\langle \alpha_N | \alpha_N \rangle = \langle \beta_N | \beta_N \rangle = 1$ and $\langle \alpha_N | \beta_N \rangle = \langle \beta_N | \alpha_N \rangle = 0$ show that the nuclear spin cannot be changed by $S_x$. At this stage it is useful to simplify (39) by introducing the magnetogyric ratio $\gamma$ instead of $g\beta$ and the general line shape function $g(\omega)$ of Section 1.5 instead of the $\delta$ function. The probability is then

$$P = \tfrac{1}{2}\pi\gamma^2 H_1^2 g(\omega) \qquad (\alpha_e\alpha_N \leftrightarrow \beta_e\alpha_N) \tag{40}$$

The only other e.s.r. transition is from $\alpha_e\beta_N$ to $\beta_e\beta_N$ and the required matrix element of $S_x$ is the same as (38)

$$\langle\alpha_e\beta_N|S_x|\beta_e\beta_N\rangle = \tfrac{1}{2}\langle\alpha_e|S^+|\beta_e\rangle = \tfrac{1}{2} \tag{41}$$

The other matrix elements of $S_x$ vanish, and the corresponding transitions are forbidden in e.s.r. To sum up, we have found that the spin quantum numbers in allowed e.s.r. transitions obey the selection rules

$$\Delta m_S = \pm 1 \qquad \Delta m_I = 0 \tag{42}$$

The transitions $\alpha_e\alpha_N \leftrightarrow \beta_e\alpha_N$ and $\alpha_e\beta_N \leftrightarrow \beta_e\beta_N$ have equal probabilities and hence equal intensities (neglecting the slight effect due to thermal population differences). The frequencies of the transitions are as follows:

$$hv_a = (\tfrac{1}{2}g\beta H - \tfrac{1}{2}g_N\beta_N H + \tfrac{1}{4}a) - (-\tfrac{1}{2}g\beta H - \tfrac{1}{2}g_N\beta_N H - \tfrac{1}{4}a)$$

$$= g\beta H + \tfrac{1}{2}a \tag{43}$$

$$hv_b = (\tfrac{1}{2}g\beta H + \tfrac{1}{2}g_N\beta_N H - \tfrac{1}{4}a) - (-\tfrac{1}{2}g\beta H + \tfrac{1}{2}g_N\beta_N H + \tfrac{1}{4}a)$$

$$= g\beta H - \tfrac{1}{2}a \tag{44}$$

The first-order e.s.r. spectrum therefore consists of two equally intense lines separated by $a$, the hyperfine coupling constant.

## 2·8 THE SECOND-ORDER E.S.R. SPECTRUM— FORBIDDEN TRANSITIONS

We will now see how the conclusions in Section 2.7 are affected by the second-order modifications (31) and (32) to the energies and wave functions. First we note that the allowed e.s.r. transitions occur at slightly different energies but the separation between them is unaltered and still gives directly the value of $a$. Their intensities are reduced slightly because of small changes in the spin wave functions.

The most interesting new feature is that the transition which we described as $\alpha_e\beta_N \leftrightarrow \beta_e\alpha_N$, strictly forbidden in first-order, is now weakly allowed provided the oscillating magnetic field is polarized *parallel* to the static field, rather than perpendicular. To compute the transition probabilities we use the wave functions (31), and denote the mixing coefficient $a/2(g\beta H + g_N\beta_N H)$ by $\lambda$. The perturbation $2H_1g\beta S_z \cos\omega t$ has a matrix element

$$\langle\psi_2|S_z|\psi_3\rangle = \langle\alpha_e\beta_N + \lambda\beta_e\alpha_N|S_z|\beta_e\alpha_N - \lambda\alpha_e\beta_N\rangle$$

$$= \langle\alpha_e\beta_N|S_z|\beta_e\alpha_N\rangle + \lambda\langle\beta_e\alpha_N|S_z|\beta_e\alpha_N\rangle - \lambda\langle\alpha_e\beta_N|S_z|\alpha_e\beta_N\rangle - \lambda^2\langle\beta_e\alpha_N|S_z|\alpha_e\beta_N\rangle$$

$$= 0 - \tfrac{1}{2}\lambda - \tfrac{1}{2}\lambda - 0$$

$$= -\lambda \tag{45}$$

The corresponding transition probability is

$$P = 2\lambda^2\pi\gamma^2 H_1^2 g(\omega) \tag{46}$$

$\lambda$ is, of course, very small, and at high fields $\lambda^2$ would be negligible. At low fields however, the transition should be detectable. Strictly speaking Eq. (45) is only valid for small values of $\lambda$ since the wave functions (31) should each contain a normalization factor of $(1 + \lambda^2)^{-1/2}$.

In the first-order treatment we saw that each level could be characterized by values of the quantum numbers $m_S$ and $m_I$, and the first-order e.s.r. transitions obeyed the selection rule $\Delta m_S = \pm 1$, $\Delta m_I = 0$. If we retain this scheme of level classification, we describe the second-order transition as a "forbidden" $\Delta m_S = 1$, $\Delta m_I = 1$ line. In reality, the $m_S$ and $m_I$ values of the exact wave functions are not sharply defined. Each state acquires some of the properties of the one it mixes with, and the forbidden transition borrows intensity from the allowed ones.

## 2·9  THE ZERO-FIELD LEVELS OF HYDROGEN

In the absence of an external field the dominant term in the energy is the isotropic hyperfine interaction $a\mathbf{I}\cdot\mathbf{S}$, which couples the spin vectors $\mathbf{I}$ and $\mathbf{S}$ into a resultant angular momentum

$$\mathbf{F} = (\mathbf{I} + \mathbf{S}) \tag{47}$$

By diagonalizing the matrix (30) we find that there are two types of state (see Appendix C). The "singlet" state with $F = 0$ has the wave function

$$|S\rangle = \frac{1}{\sqrt{2}} \{|\alpha_e \beta_N\rangle - |\beta_e \alpha_N\rangle\} \qquad E = -\frac{3}{4} a \tag{48}$$

while the "triplet" energy level with $F = 1$ consists of three states

$$\left. \begin{aligned} |T_{+1}\rangle &= |\alpha_e \alpha_N\rangle \\ |T_0\rangle &= \frac{1}{\sqrt{2}} \{|\alpha_e \beta_N\rangle + |\beta_e \alpha_N\rangle\} \\ |T_{-1}\rangle &= |\beta_e \beta_N\rangle \end{aligned} \right\} \qquad E = \frac{1}{4} a \tag{49}$$

These functions are distinguished from one another by their values of $F_z$, or $(m_I + m_S)$, which may be 1, 0, and $-1$. The e.s.r. spectrum consists of a single line at 1420 Mc/s which is due to strongly allowed singlet-triplet transitions.

## 2·10  THE N.M.R. SPECTRUM OF THE HELIUM ATOM

The hydrogen atom and all paramagnetic molecules present unusual problems so far as n.m.r. is concerned, because of the effects of the electron spin. This is because the electron spin relaxes, in some cases quite rapidly. The hyperfine coupling not only makes the n.m.r. frequency differ enormously from the value for a free proton, but may also broaden the n.m.r. lines. As a result the energies of the nuclear spin levels vary with time and n.m.r. lines in liquids and solids often become too broad to be detectable. We shall say more about these matters in Chapter 12, but now we turn instead to the n.m.r. spectrum of the helium atom, which is much more typical.

The pair of electron spins are coupled into a singlet state with no magnetic moment, so only the nucleus need be considered. The normal He$^4$ nucleus has no spin, but there is a rare isotope $He^3$ with an abundance of only about $10^{-4}$ % which has a spin of 1/2. The spin has two states $|\alpha_N\rangle$ and $|\beta_N\rangle$ in a magnetic field, and energy absorption by a bare nucleus will occur at the resonance frequency $\omega_N = \gamma_N H$.

A transition probability $P$ is calculated in precisely the same way as for an electron spin. The oscillating field $2H_1 \cos \omega t$ produces a perturbation

$$V(t) = 2\gamma_N \hbar H_1 I_x \cos \omega t \tag{50}$$

and we obtain the transition probability

$$P = 2\pi \gamma_N^2 H_1^2 |\langle \alpha_N | I_x | \beta_N \rangle|^2 g(\omega)$$
$$= 2\pi \gamma_N^2 H_1^2 |\langle \alpha_N | \tfrac{1}{2}(I^+ + I^-)|\beta_N\rangle|^2 g(\omega)$$
$$= \tfrac{1}{2}\pi \gamma_N^2 H_1^2 g(\omega) \tag{51}$$

The most significant feature of Eq. (51) is that $\gamma_N$ for all nuclei is much smaller than the electron gyromagnetic ratio $\gamma$. As a result the transition probability $P$ is a factor of about $10^{-5}$ smaller for n.m.r. transitions as compared with e.s.r.

# 2·11 CHEMICAL SHIELDING

The observed n.m.r. frequency of $He^3$ and other nuclei in atoms and molecules differs slightly from the theoretical value $g_N \beta_N H$ for a free nucleus. This is called the screening effect or the "chemical shift" and it arises from the magnetic effect of induced electronic currents set up in the atom by the external magnetic field $H$. The currents produce an additional field at the nucleus which opposes $H$ and has a strength proportional to $H$. The total effective magnetic field which acts on the nuclear moment can therefore be written

$$H_{eff} = (1 - \sigma)H \tag{52}$$

where $\sigma$ is a small number of order $10^{-6}$, called the screening constant. The nuclear Zeeman energy must now be altered to

$$\mathscr{H}_0 = -g_N \beta_N (1 - \sigma) H I_z \tag{53}$$

to allow for screening, and in the atom magnetic resonance transitions at a fixed frequency $v$ now appear at a higher field $H$ than they do for the free nucleus.

The theory of the chemical shift in a closed shell atom is very simple. In the external magnetic field the entire spherical electron cloud precesses about the field direction as if it were a rotating rigid sphere of electricity, and its angular velocity is $\omega = eH/2mc$. This is illustrated in Fig. 2.4(a). Since an electron at a distance $r$ from the nucleus [Fig. 2.4(b)] moves with the velocity $v = (\omega \times r)$ it produces a secondary magnetic field

$$\mathbf{H'} = -\frac{e}{c}\frac{\mathbf{r} \times \mathbf{v}}{r^3} = -\frac{e^2}{2mc^2}\frac{\mathbf{r} \times (\mathbf{H} \times \mathbf{r})}{r^3} \tag{54}$$

On the average the electrons are distributed over the atom with a probability density $\rho(r)$ and the average secondary field is

$$\mathbf{H'} = \frac{-e^2}{2mc^2}\int \frac{\mathbf{r} \times (\mathbf{H} \times \mathbf{r})}{r^3}\rho(r)dv \tag{55}$$

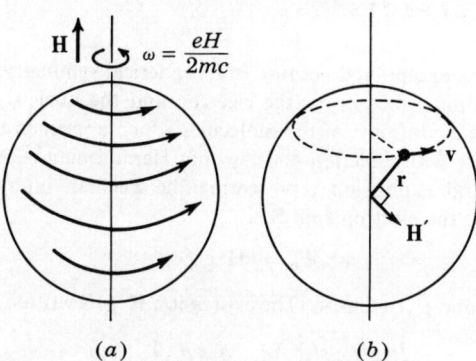

FIG. 2.4. Chemical shielding in the helium atom. (a) Precession of the electron cloud. (b) The screening magnetic field.

The $x$ and $y$ components of $\mathbf{H}'$ vanish, leaving a screening field $-\sigma H$ directed along the $z$ axis. We evaluate the vector product in (55) and obtain

$$\sigma = \frac{e^2}{2mc^2} \int \frac{(x^2 + y^2)}{r^3} \rho(r)dv = \frac{e^2}{3mc^2} \int \frac{\rho(r)}{r} \, dv \qquad (56)$$

This important relation is called the Lamb formula, and we see that $\sigma$ depends on the mean value of $1/r$ for all the electrons. For helium the integral in (56) works out at 3.377 in units of $1/a_0$ [Eq. (5)] and the calculated screening constant is

$$\sigma = 59.93 \times 10^{-6} \qquad (57)$$

or 59.93 parts per million. Larger atoms have much larger shifts. For example neon gives a value of about 547 ppm. We shall also see in Chapter 5 that the chemical shifts in molecules are considerably more complicated than they are in spherical closed shell atoms!

## 2·12  CORRECTIONS TO THE $g$ FACTOR

Similar effects operate in e.s.r. spectra, and even in a hydrogen atom the electron resonance frequency is not exactly equal to the theoretical value $hv = 2.00232\beta H$ for a free electron. In n.m.r. it is usual to take the line that $g_N$ is an inherent property of the nucleus, and any corrections to the Zeeman energy are then interpreted as screening effects just as we have done in Eq. (53). However e.s.r. spectroscopists prefer to describe changes in the resonance frequency by saying that the effective magnetic moment $\mu_e = -g\beta S$ of an electron can be different in different situations, so that the value of $g$ is not constant, but changes from one atom or molecule to another. The Zeeman energy is always written $\mathscr{H}_0 = g\beta \mathbf{H} \cdot \mathbf{S}$, but the $g$ value differs from the free-electron spin-only value (2.002322) because of spin orbit coupling, which endows the unpaired electron with a small orbital angular momentum and alters its effective magnetic moment.

## 2·13 ANISOTROPIC EFFECTS

The hydrogen atom is exceptional because of its spherical symmetry; the spin Hamiltonian (9) has isotropic $g$-factors for the electron and the nucleus and an isotropic hyperfine interaction $a$. In most of the molecules which appear in this book each of these quantities varies with direction and the spin Hamiltonian is anisotropic.

The most general expression representing the Zeeman interaction between a magnetic field $\mathbf{H}$ and the electron spin $\mathbf{S}$ is

$$\mathscr{H}_0 = \beta \mathbf{H} \cdot \mathbf{g} \cdot \mathbf{S} \tag{58}$$

$\mathbf{H}$ and $\mathbf{S}$ are vectors and $\mathbf{g}$ is a tensor. The expression $\mathbf{H} \cdot \mathbf{g} \cdot \mathbf{S}$ written in full becomes

$$[H_x, H_y, H_z] \begin{bmatrix} g_{xx} & g_{yx} & g_{zx} \\ g_{xy} & g_{yy} & g_{zy} \\ g_{xz} & g_{yz} & g_{zz} \end{bmatrix} \begin{bmatrix} S_x \\ S_y \\ S_z \end{bmatrix} \tag{59}$$

where $H_x, H_y, H_z, S_x, S_y, S_z$ are scalar components of the $\mathbf{H}$ and $\mathbf{S}$ vectors, defined in terms of an axis system $x, y, z$ fixed in the molecule. As will be seen in due course, the tensor elements are determined by studying an oriented specimen (single crystal) of the paramagnetic molecule. We will find that the $g$ tensor is almost always symmetric, i.e., $g_{xy} = g_{yx}$ and so on. If so, it follows that the $g$ tensor, having a symmetric matrix, can be diagonalized by means of an appropriate matrix transformation

$$\mathbf{L} \mathbf{g} \mathbf{L}^{-1} = \mathbf{g}(\text{diagonal}) \tag{60}$$

The transformation corresponds to a reorientation of the axes, and the transformation matrix $L$ defines the orientation of the new "principal axes" with respect to the old. Our readers who are new to tensor notation will find a short explanation in Appendix D. In the new axis system which makes the $g$ tensor diagonal the Zeeman Hamiltonian may be written

$$\mathscr{H}_0 = \beta[g_{xx}H_xS_x + g_{yy}H_yS_y + g_{zz}H_zS_z] \tag{61}$$

As an example, suppose we are interested in the e.s.r. spectrum of a bent radical like $NO_2$ trapped in a crystal (Fig. 2.5). Initially we choose an arbitrary set of axes $x, y, z$ fixed in the crystal. Usually this will prove to be a rather inconvenient axis system because the orientation of the molecules inside the crystal is not known and the $g$ tensor is not diagonal. However the molecular symmetry axes are clearly principal axes of the tensor; e.s.r. measurements can be used to find these axes, giving valuable clues as to the orientation of the radical.

There is one further point about the $g$ tensor which must be mentioned at this stage. The principal values of the $g$ tensor ($g_{xx}, g_{yy}$ and $g_{zz}$) in Eq. (61) can only be determined from solid state studies. In solution the paramagnetic molecules are tumbling rapidly and randomly and we can only measure the average $g$ value, which is equal to $\frac{1}{3}$ of the trace of the tensor, i.e., $\frac{1}{3}(g_{xx} + g_{yy} + g_{zz})$.

Most of the conclusions of the previous paragraphs apply to chemical shielding. The electronic currents induced in a molecule are anisotropic and the $x$ and $y$ components of the field $\mathbf{H}'$ in Eq. (54) no longer vanish. Hence $\mathbf{H}_{\text{eff}}$ is not parallel to $\mathbf{H}$ and the nuclear Zeeman energy is in general represented by

$$\mathscr{H}_0 = -g_N\beta_N\mathbf{H} \cdot (1 - \boldsymbol{\sigma}) \cdot \mathbf{I} \tag{62}$$

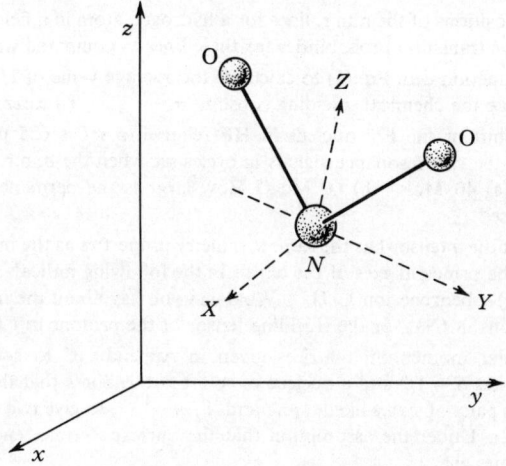

FIG. 2.5. Crystal axes and principal axes of the $g$ tensor for a bent radical like $NO_2$.

in which $\sigma$ is called the screening tensor. An n.m.r. spectrum in solution yields a scalar screening constant $\sigma$ which is one third of the trace of $\sigma$.

The reader will perhaps now expect hyperfine coupling to be described by means of a tensor and this is indeed the case, the Hamiltonian being written in the form

$$\mathscr{H}_1 = \mathbf{S} \cdot \mathbf{T} \cdot \mathbf{I} \tag{63}$$

where $\mathbf{S}$ and $\mathbf{I}$ are spin vectors and $\mathbf{T}$ is the hyperfine tensor. Here, too, a solution spectrum can only yield the value of $\frac{1}{3}$ Trace $\{\mathbf{T}\}$, a quantity which is denoted by the same symbol $a$ as in the hydrogen atom.

## PROBLEMS

1. Calculate the hyperfine splitting constant of the deuterium atom $(I = 1)$. Hence find the energy levels of the atom in zero magnetic field. (Hint: First read Appendix C.)
[Answer: $a = 218$ Mc/s. Energies $\frac{1}{2}a$, $-a$.]

2. What is the dipolar interaction energy between an electron and a proton spin 1 Ångstrom apart when both spins are in the $z$ direction and the vector $\mathbf{r}$ joining them makes an angle $\theta$ with the $z$ direction? Show that this energy vanishes on averaging over all possible directions of $\mathbf{r}$.

3. Calculate (a) the first-order, and (b) the second-order energies of the H atom in a field of 10,000 gauss (rounded off in Mc/s). Find the value of the parameter $\lambda$ in Eq. (45).
[Answer: (a) $14347, 13679, -14389, -13637$; (b) $14347, 13697, -14407, -13637$; $\lambda = 0.0253$]

4. Solve the energy matrix (33) exactly under the conditions of Problem 3. How good is perturbation theory? [Same as (b) above, Problem 3].

5. The rf field in an e.s.r. experiment is $(0.1) \cos \omega t$ gauss. Estimate the transition probability $P$ in Eq. (40). What can you say about the value of $T_1$ if the resonance does not saturate?

6. Hydrogen atoms in interstellar space emit radiation at 21 cm (1420 Mc/s) which is observed by radio astronomers. An rf field along the $x$ axis is applied to a hydrogen atom in a small steady magnetic field. Calculate the positions and relative intensities of the e.s.r. singlet-triplet and triplet-triplet transitions. Explain how you might detect small magnetic fields in the galaxy.

**7.** Work out the positions of the n.m.r. lines for a hydrogen atom in a field of 10,000 gauss. Calculate the relative transition probabilities for these lines as compared with the e.s.r. lines.

**8.** Use the wave function $\psi$ in Eq. (5) to calculate the average value of $1/r$ in the hydrogen atom. Hence estimate the chemical shielding constant $\sigma$.          [*Answer:*   $17.75 \times 10^{-6}$.]

**9.** The chemical shift of the $F^{19}$ nucleus in HF relative to $F_2$ is 625 ppm. What is the separation between the two resonance signals in cycles/sec when the n.m.r. apparatus works at a frequency of (a) 40 Mc/s? (b) 60 Mc/s? How large is the permanent magnetic field required in each case?

**10.** Assuming that the $g$ tensor has the same symmetry properties as the molecular structure as a whole, draw the principal axes of the tensor in the following radicals: $CH_3$, NO, $NH_2$, benzyl, $CH(COOH)_2$, benzene ion $C_6H_6^-$. What can you say about the hyperfine coupling tensors of the protons in $CH_3$, or the shielding tensor of the protons in $CH_4$?

**11.** Use the angular momentum matrices given in Appendix C to solve the following problem. An atom has $S = 1/2$ and a nucleus of spin $I > 1/2$. Show that the hyperfine interaction $a\mathbf{I} \cdot \mathbf{S}$ couples pairs of states like $|\alpha, m_I\rangle$ and $|\beta, m_I + 1\rangle$, to give two new states having $F_z = M = (m_I + 1/2)$. Under the assumption that the nuclear Zeeman energy is negligible, show that the energies are

$$E = -\tfrac{1}{4}a \pm \tfrac{1}{2}\{\Delta_e^2 + 2Ma\Delta_e + (I + \tfrac{1}{2})^2 a^2\}^{1/2}$$

What is the separation of the hyperfine levels in zero field?

## SUGGESTIONS FOR FURTHER READING

Slichter: Pages 84–89. A short proof of Fermi's formula.

Landau and Lifshitz: Pages 78–88 and 485–486. Angular momentum. Contact interaction.

Kopfermann: *Nuclear Moments* (New York: Academic Press Inc., 1958). Chapter 1. More detailed description of hyperfine energy levels.

Breit and Rabi: *Phys. Rev.*, **38**: 2082 (1932). Gives the full solution of Problem 11.

Lamb: *Phys. Rev.*, **60**: 817 (1941). The theory of chemical shielding in atoms.

Pipkin and Lambert: *Phys. Rev.*, **127**: 787 (1962). A very accurate determination of the hyperfine frequency of the hydrogen atom. Illustrates the high accuracy that is possible in resonance experiments.

# CHAPTER 3

# NUCLEAR RESONANCE
# IN SOLIDS

## 3·1 INTRODUCTION

The n.m.r. absorption lines from solids are rather broad and their study is commonly referred to as "broad line n.m.r." It would, however, be quite wrong to assume that broad line n.m.r. is just an inferior version of "high-resolution n.m.r." The n.m.r. spectra of liquids are indeed very sharp and as a result the very small but important electron-coupled spin-spin interactions and chemical shifts can be observed; *other* anisotropic interactions are lost, however, and it is the study of these interactions in the solid state which constitutes broad line n.m.r.

The main interaction between the spins of two nuclei is the dipole-dipole coupling between their magnetic moments. The magnitude of this interaction varies according to the orientation of the magnetic field with respect to the molecule and detailed study of this anisotropy enables the distance and relative orientation of the nuclei to be obtained with considerable precision. Thus the main applications of broad line n.m.r. are to the determination of internuclear distances and other crystal parameters. In this chapter we will discuss the basic theory underlying the measurements and illustrate the technique with a few suitable examples.

## 3·2 THE DIPOLAR COUPLING TENSOR

We start by considering the case of two identical protons which are separated by a distance $r$. Each proton has two possible spin orientations so that we commence the calculation with four basis functions, which we could write,

$$|\alpha_1\alpha_2\rangle, \quad |\alpha_1\beta_2\rangle, \quad |\beta_1\alpha_2\rangle, \quad |\beta_1\beta_2\rangle \tag{1}$$

where the subscripts distinguish the two protons. In fact we will find it more convenient to start with the functions,

$$
\left.
\begin{aligned}
|t_1\rangle &= |\alpha_1\alpha_2\rangle, \\
|t_0\rangle &= \frac{1}{\sqrt{2}}|\alpha_1\beta_2 + \beta_1\alpha_2\rangle, \\
|t_{-1}\rangle &= |\beta_1\beta_2\rangle,
\end{aligned}
\right\}
\qquad |s\rangle = \frac{1}{\sqrt{2}}|\alpha_1\beta_2 - \beta_1\alpha_2\rangle
\tag{2}
$$

$t_1$, $t_0$, $t_{-1}$ are the three components of a nuclear triplet state (total spin $I = 1$), while the function $s$ represents a singlet state ($I = 0$). We have already come across similar spin functions in Section 2.9.

The next problem is to formulate the appropriate spin Hamiltonian and we first define a coordinate system $x$, $y$, $z$, in terms of which the orientation of the vector $\mathbf{r}$ joining the two protons is defined. We will define the $z$ axis to be the direction of the magnetic field so that the nuclear Zeeman interaction will be written simply as

$$\mathscr{H}_0 = -g_N\beta_N H(I_{1z} + I_{2z}) \tag{3}$$

Strictly speaking the Hamiltonian should contain the effect of chemical shielding; however this effect is exceedingly small compared with the line width.

The dipolar interaction between the two proton magnetic moments is represented by the Hamiltonian

$$\mathscr{H}_D = g_N^2\beta_N^2\left\{\frac{\mathbf{I}_1\cdot\mathbf{I}_2}{r^3} - \frac{3(\mathbf{I}_1\cdot\mathbf{r})(\mathbf{I}_2\cdot\mathbf{r})}{r^5}\right\} \tag{4}$$

and the effects of $\mathscr{H}_D$ on the eigenstates and eigenvalues of $\mathscr{H}_0$ will be treated by first-order perturbation theory.

We now obtain the various vector products in (4) as follows:

$$\mathbf{I}_1\cdot\mathbf{I}_2 = I_{1x}I_{2x} + I_{1y}I_{2y} + I_{1z}I_{2z}$$

$$\mathbf{I}_1\cdot\mathbf{r} = I_{1x}x + I_{1y}y + I_{1z}z \tag{5}$$

$$\mathbf{I}_2\cdot\mathbf{r} = I_{2x}x + I_{2y}y + I_{2z}z$$

Hence Eq. (4) may be rewritten in the expanded form

$$\mathscr{H}_D = \frac{g_N^2\beta_N^2}{r^5}\{I_{1x}I_{2x}(r^2 - 3x^2) + I_{1y}I_{2y}(r^2 - 3y^2) + I_{1z}I_{2z}(r^2 - 3z^2)$$

$$-(I_{1x}I_{2y} + I_{1y}I_{2x})3xy - (I_{1y}I_{2z} + I_{1z}I_{2y})3yz - (I_{1z}I_{2x} + I_{1x}I_{2z})3zx\}. \tag{6}$$

The nature of the tensor coupling of spins, which was discussed at some length in Section 2.12 is nicely illustrated by the calculation just performed. Equation (6) can clearly be written in matrix form

$$\mathscr{H}_D/g_N^2\beta_N^2 = [I_{1x}, I_{1y}, I_{1z}]\begin{bmatrix} (r^2 - 3x^2)/r^5 & -3xy/r^5 & -3xz/r^5 \\ -3xy/r^5 & (r^2 - 3y^2)/r^5 & -3yz/r^5 \\ -3xz/r^5 & -3yz/r^5 & (r^2 - 3z^2)/r^5 \end{bmatrix}\begin{bmatrix} I_{2x} \\ I_{2y} \\ I_{2z} \end{bmatrix} \tag{7}$$

which is written in abbreviated form as

$$\mathscr{H}_D = g_N^2\beta_N^2\mathbf{I}_1\cdot\mathbf{D}\cdot\mathbf{I}_2 \tag{8}$$

where $\mathbf{D}$ is the dipolar coupling tensor. If the orientation of the $x$, $y$, $z$ axes with respect to $\mathbf{r}$ is chosen as in Fig. 3.1, the tensor $\mathbf{D}$ is clearly not diagonal. The transformation which diagonalizes it is, however, readily discovered. It is clear that the condition for $xy = yz = zx = 0$ is just that $\mathbf{r}$ points along one of the coordinate axes. Hence the tensor is diagonal in terms of a coordinate system in which the direction of the interproton vector is also the $z$ axis. The principal values of the tensor are then $D_{xx} = D_{yy} = 1/r^3$, $D_{zz} = -2/r^3$, so that the tensor is axially symmetric.

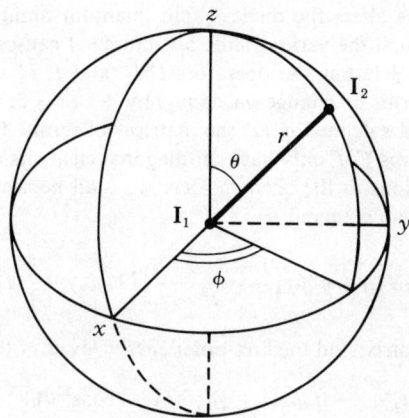

FIG. 3.1. Axis system and polar coordinates for the dipolar coupling of two protons.

The reader will also notice that the trace of the tensor is zero. This is a very important result because it means that the dipolar coupling does not contribute to the line positions and intensities of a liquid phase spectrum (see Appendix D). It does affect the widths of the lines, however, for we shall see that combination of the dipolar coupling and the random tumbling of the molecules in solution contributes to both relaxation times $T_1$ and $T_2$.

## 3·3 THE N.M.R. SPECTRUM OF TWO COUPLED PROTONS

In order to find the spin energy levels of our two protons we first transform from cartesian coordinates to the polar coordinates

$$x = r \sin \theta \cos \phi$$

$$y = r \sin \theta \sin \phi \qquad (9)$$

$$z = r \cos \theta$$

and rewrite the operators $I_x$ and $I_y$ in terms of the shift operators $I^+$ and $I^-$. There are six different kinds of term which are usually called $A$, $B$, $C$, $D$, $E$, $F$.

$$\mathscr{H}_D = \frac{g_N^2 \beta_N^2}{r^3} [A + B + C + D + E + F] \qquad (10)$$

$$
\left.
\begin{aligned}
A &= (1 - 3 \cos^2 \theta) I_{1z} I_{2z} \\
B &= -\tfrac{1}{4}(1 - 3 \cos^2 \theta)[I_1^+ I_2^- + I_1^- I_2^+] \\
C &= -\tfrac{3}{2} \sin \theta \cos \theta \, e^{-i\phi}[I_{1z} I_2^+ + I_1^+ I_{2z}] \\
D &= -\tfrac{3}{2} \sin \theta \cos \theta \, e^{+i\phi}[I_{1z} I_2^- + I_1^- I_{2z}] \\
E &= -\tfrac{3}{4} \sin^2 \theta \, e^{-2i\phi} I_1^+ I_2^+ \\
F &= -\tfrac{3}{4} \sin^2 \theta \, e^{+2i\phi} I_1^- I_2^-
\end{aligned}
\right\} \qquad (11)
$$

Each of these operators alters the nuclear spin quantum numbers $m_1$ and $m_2$ in a characteristic way through the various shift operators. $A$ causes no change; $B$ alters both nuclear spins by $\pm 1$, but the operators $I_1^+ I_2^-$ and $I_1^- I_2^+$ do not alter the sum $(m_1 + m_2)$; the other terms do change $(m_1 + m_2)$ by $\pm 1$ or $\pm 2$. In terms of the basis functions $t_1$, $t_0$, $t_{-1}$ and $s$ defined in (2) the matrices of terms $A$ and $B$ are diagonal, while the other four terms $C$–$F$ only have off-diagonal elements and hence they make second-order contributions to the energy. Here we shall neglect them entirely. Our effective spin Hamiltonian is therefore

$$\mathcal{H} = -g_N\beta_N H(I_{1z} + I_{2z}) + g_N^2\beta_N^2\left(\frac{1 - 3\cos^2\theta}{r^3}\right)\left[I_{1z}I_{2z} - \frac{1}{4}(I_1^+ I_2^- + I_1^- I_2^+)\right] \quad (12)$$

$\mathcal{H}$ is now a diagonal matrix, and the first-order energy levels of the states are

$$|t_1\rangle, \quad -g_N\beta_N H + \tfrac{1}{4}g_N^2\beta_N^2(1 - 3\cos^2\theta)/r^3$$

$$|t_0\rangle, \quad 0 \qquad - \tfrac{1}{2}g_N^2\beta_N^2(1 - 3\cos^2\theta)/r^3 \qquad (13)$$

$$|t_{-1}\rangle, \quad +g_N\beta_N H + \tfrac{1}{4}g_N^2\beta_N^2(1 - 3\cos^2\theta)/r^3$$

$$|s\rangle, \quad 0$$

It is readily found by the methods of Section 2.7 that the only transitions induced by an oscillating magnetic field are between $t_1$ and $t_0$ or between $t_0$ and $t_{-1}$. The energies of these transitions are

$$h\nu = g_N\beta_N H + \tfrac{3}{4}g_N^2\beta_N^2(1 - 3\cos^2\theta)/r^3 \qquad (t_0 \to t_{-1})$$

$$\qquad (14)$$

$$h\nu = g_N\beta_N H - \tfrac{3}{4}g_N^2\beta_N^2(1 - 3\cos^2\theta)/r^3 \qquad (t_1 \to t_0)$$

If we work with a fixed frequency $\nu$ for which the resonance of a *single proton* would occur at a field value $H^*$ given by the usual condition $h\nu = g_N\beta_N H^*$, the observed spectrum will consist of a pair of lines at field values

$$H = H^* \pm \tfrac{3}{4}g_N\beta_N(1 - 3\cos^2\theta)/r^3 \qquad (15)$$

The separation between the lines (in gauss) is thus equal to $\tfrac{3}{2}g_N\beta_N(1 - 3\cos^2\theta)/r^3$ where $\theta$, we recall, is the angle between the interproton vector and the magnetic field. From studies of the angular variation of the splitting it is thus possible to measure $1/r^3$ and determine the magnitude and direction of $\mathbf{r}$.

The first such application was to a single crystal of gypsum, $CaSO_4.2H_2O$. Each pair of protons in an $H_2O$ molecule is well separated from the other pairs so that the dipolar interaction between different pairs is very small. However the unit cell of the crystal contains two types of water molecule, distinguishable because their proton-proton vectors are not parallel. Hence for some orientations, four absorption lines are observed, rather than two. Figure 3.2 shows the absorption spectra obtained as the magnetic field is rotated in the crystallographic $xy$ plane. Note that when the field is along the $x$ direction only one line is observed, indicating that $3\cos^2\theta = 1$ for both types of proton pair, i.e., $\theta = 54°44'$. The separation constant $\tfrac{3}{2}g_N\beta_N/r^3$ was found to be 10.8 gauss, giving a distance of 1.58 Å between the protons in each water molecule.

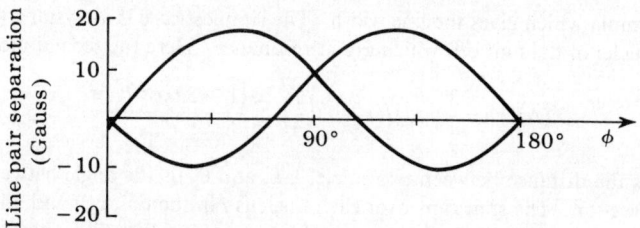

FIG. 3.2. Proton resonance spectrum of crystalline gypsum. The magnetic field is being rotated within the $xy$ plane of the crystal unit cell, perpendicular to the $z$ axis. The curves show the separation between the line pairs for the two kinds of water molecules. $\phi$ is the angle between the field and the $x$ axis.

## 3·4 THE SECOND MOMENT OF AN N.M.R. ABSORPTION LINE

Only in relatively rare cases does the crystal structure permit complete resolution of the dipole splitting; however it is still possible to obtain structural information from the width of the n.m.r. line by less direct methods.

Systems in which three or more protons within the same molecule have strong dipolar interactions would give very complex n.m.r. spectra if the lines could be resolved. However the intermolecular interactions with more distant nuclei contribute markedly to the line width and even a single crystal specimen gives only a broad resonance with little or no resolved structure. Despite these difficulties, valuable structural information can often be obtained through measurement of the *second moment* of the absorption line, which we shall now define.

Measurements are made at a fixed frequency by varying the magnetic field $H$. Suppose that $f(H)$ represents the normalized line shape as a function of the field value $H$. The center of the resonance line is the average magnetic field

$$H_{av} = \int_0^\infty H f(H) dH \tag{16}$$

The second moment is the mean square width $(\Delta H)^2$ measured from the center of the resonance line,

$$(\Delta H)^2 = \int_0^\infty (H - H_{av})^2 f(H) dH \tag{17}$$

It can be evaluated graphically from the observed line shape.

Quite often the broadened n.m.r. line has a Gaussian shape described by the formula

$$f(H) = \frac{1}{\Delta H \sqrt{2\pi}} e^{-\{(H - H_{av})^2 / 2(\Delta H)^2\}} \tag{18}$$

Let us now consider the case of a single crystal where a number of different magnetic dipolar interactions result in a broad resonance line with little or no resolved structure. The field $\Delta H$ is the root mean square value of the local magnetic fields produced at the nucleus $N$ by all the other nuclear dipoles in the crystal, averaged over all possible spin states of the other nuclei. There is a famous relation known as Van

Vleck's formula which gives the line width. The simplest case is a crystal containing $n$ identical nuclei in the unit cell which are at resonance. Here the second moment is

$$(\Delta H)^2 = \frac{3}{4} g_N^2 \beta_N^2 I(I+1) \left(\frac{1}{n}\right) \sum_{j,k} \frac{(1-3\cos^2\theta_{jk})^2}{r_{jk}^6} \tag{19a}$$

where $r_{jk}$ is the distance between two nuclei $j$, $k$, and $\theta_{jk}$ is the angle between $\mathbf{r}_{jk}$ and the field direction. The sum runs over each nucleus $j$ in the unit cell and all its neighbors $k$ within the crystal (including ones in the same unit cell!). The crystal may contain nuclei $N'$ of other atoms which are not at resonance simultaneously with the first kind. They also contribute to the second moment, but their effect is smaller by a factor of 4/9:

$$(\Delta H)^2 = \frac{1}{3} g_N'^2 \beta_N^2 I'(I'+1) \left(\frac{1}{n}\right) \sum_{j,f} \frac{(1-3\cos^2\theta_{jf})^2}{r_{jf}^6} \tag{19b}$$

Now the sum $f$ runs over all the nonresonant nuclei, and the complete second moment is the sum of (19a) and (19b). If a polycrystalline specimen is studied, then $(3\cos^2\theta - 1)^2$ is replaced by 4/5, its average value over all orientations. The final expression for the second moment thus becomes:

$$(\Delta H)^2 = \frac{3}{5} g_N^2 \beta_N^2 \left(\frac{1}{n}\right) I(I+1) \sum_{j,k} \frac{1}{r_{jk}^6}$$

$$+ \frac{4}{15} g_N'^2 \beta_N^2 \left(\frac{1}{n}\right) I'(I'+1) \sum_{j,f} \frac{1}{r_{jf}^6} \tag{20}$$

The theoretical calculations which lead to Van Vleck's formula are fairly lengthy, but it is interesting to ask why there is this difference between Eqs. (19a) and (19b). A short answer is that it depends on the $B$ terms of the dipolar Hamiltonian (11). We have seen that for identical nuclei $B$ behaves as a first-order perturbation. If the nuclei are different their Zeeman energies are not the same and the zero order basic spin functions change from (2) to (1), while $B$ becomes a second-order perturbation which no longer affects the second moment.

## 3·5　STRUCTURAL STUDIES BY THE METHOD OF MOMENTS

Second moment studies are of particular value in distinguishing between alternative molecular structures, particularly when the different structures each contain characteristic groupings of magnetic nuclei. An early example was the study of so-called infusible white precipitate, prepared by the addition of mercuric chloride to aqueous ammonia. Three alternative chemical formulae had been suggested, namely, (a) $NHg_2Cl$, $NH_4Cl$, (b) $NH_2.HgCl$, and (c) $xHgO$, $(1-x)HgCl_2.2NH_3$. These three possible structures each have distinctive groupings of the protons: structure (a) contains a tetrahedral arrangement of protons, (b) contains a pair, while (c) possesses a triangle of protons. The expected line shapes for these arrangements are quite distinctive and are shown in Fig. 3.3. The experimental n.m.r. line shape corresponds closely with that expected for protons interacting strongly in pairs and the second moment [18.6 (gauss)$^2$] is close to the calculated value of 20 (gauss)$^2$. Structures (a) and (c) are expected to yield second moments close to 50 and 38 (gauss)$^2$, respectively.

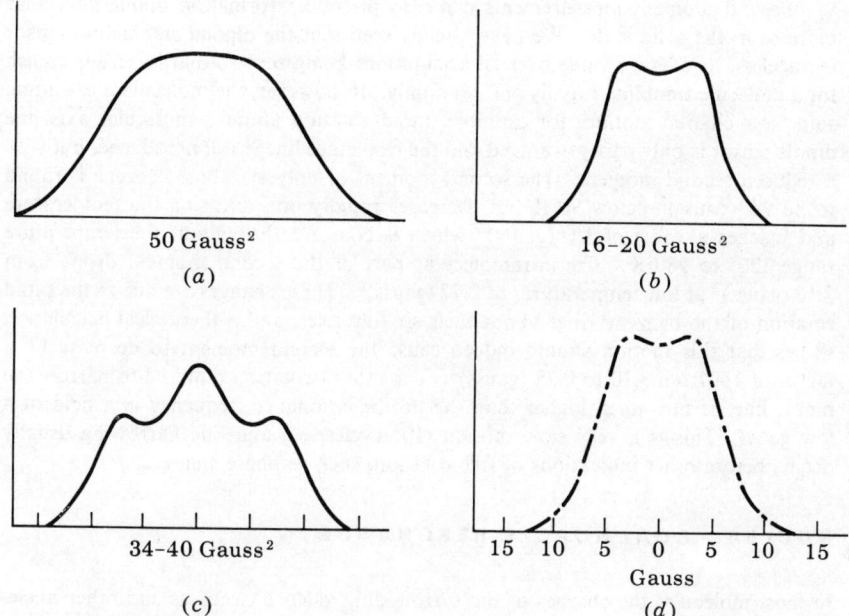

FIG. 3.3. Proton resonance line shapes expected for various alternative structures of "Infusible white precipitate." (a) Tetrahedral groups of protons. (b) Pairs. (c) Triangles. (d) Observed spectrum.

A similar application is to the monohydrates of nitric, sulphuric, perchloric, and oxalic acids. The point of interest is whether these compounds crystallize as oxonium salts (containing $H_3O^+$) or normal hydrates ($H_2O$). Clearly one might hope to distinguish between a structure containing triangles of nuclei, and one containing pairs. The n.m.r. spectra of polycrystalline materials at 90°K do indeed show quite clearly that oxalic acid crystallizes as a true hydrate, but the other acids do not.

One of the difficulties of interpreting second moment measurements is to separate the contributions to the line width due to pairs of nuclei inside the same molecule from the part due to interaction with the surrounding molecules. The intramolecular part is of principal interest because it depends upon the chemical structure, whereas the intermolecular part depends upon the crystal structure. It is therefore often necessary to have some knowledge of the crystal structure in order to estimate the intermolecular contribution. However there are other ways to separate the two terms. For example, the second moments of benzene and 1:3:5-trideuterobenzene have been measured at temperatures below 90°K, where the molecules do not rotate. The magnetic moment of the deuteron is very small and dipolar interactions which involve deuterium can actually be neglected. Deuteration reduces the overall value of the second moment, but it reduces the intramolecular and intermolecular contributions by different amounts, so that their relative contributions can be calculated. The distance between adjacent protons in the benzene molecule was deduced from the n.m.r. spectrum, giving $r = 2.495 \pm 0.018$ Å, as compared with the value of $2.473 \pm 0.025$ Å derived from x-ray crystal data.

Second moment measurements can also provide information about molecular motions in the solid state. We have already seen that the dipolar interaction tensor is traceless, its average value over all orientations being zero so that its effects vanish for a molecule tumbling rapidly and randomly. If, however, the molecule is executing some less chaotic motion, for example, rapid rotation about a molecular axis, the dipole tensor is only partly averaged and the resonance line is still broadened, but with a reduced second moment. The second moment of polycrystalline benzene is. found to be 9.7 (gauss)$^2$ below 90°K but decreases rapidly on increasing the temperature and reaches a value of 1.6 (gauss)$^2$ which is constant throughout the temperature range 120° to 280°K. The intramolecular part of the second moment drops from 3.10 (gauss)$^2$ at low temperatures to 0.77 (gauss)$^2$. These changes are due to the rapid rotation of the benzene rings about their six-fold axes, and a theoretical calculation shows that this motion should indeed cause the second moment to decrease by a factor of 1/4 from 3.10 to 0.78 (gauss)$^2$. The rate of rotation required to narrow the n.m.r. lines is not much higher than the proton resonance frequency in a field of a few gauss. This is a very slow motion ($10^4$ cycles/sec) and line narrowing usually occurs before other indications of free rotation, such as phase changes.

## 3·6  NUCLEAR QUADRUPOLE RESONANCE

In most molecules the charges of the surrounding valence electrons and other nuclei produce a large nonuniform electric field gradient at each nucleus. The electric quadrupole moments of nuclei with spins greater than 1/2 will interact with the field gradient to give a series of quantized energy levels, even if there is no external magnetic field. The field gradient, caused by charges outside a small sphere drawn round the nucleus, is described by a traceless tensor with components $\partial^2 V/\partial x^2$, $\partial^2 V/\partial x\partial y$, and so on, where $V$ is the electrostatic potential. In many molecules the tensor has axial symmetry about some direction $z$; then the splitting of the spin energy levels depends only on $\partial^2 V/\partial z^2$ and is given by the formula

$$E = \frac{1}{4} eQ \frac{\partial^2 V}{\partial z^2} \left( \frac{3m^2 - I(I+1)}{3I^2 - I(I+1)} \right) \tag{21}$$

A small oscillating magnetic field can now produce resonant transitions between these levels, and give a "pure quadrupole resonance," as shown in Fig. 3.4. Nuclei of spins 1 or 3/2 (e.g., $N^{14}$ and $Cl^{35}$) give only a single resonance line; but a spin of 5/2, such as $I^{127}$ gives two, with one frequency exactly double the other. The field gradient $\partial^2 V/\partial z^2$ is usually denoted by the symbol $eq$, where $q$ is measured in $cm^{-3}$, and a quadrupole resonance experiment yields the value of the quantity $e^2qQ/h$ rather than $q$ itself. If a precise value of the quadrupole moment $Q$ is known, as it is for deuterium and chlorine nuclei, one may deduce the field gradient in the molecule directly and learn something about the nature of the chemical bonding. However, nuclear quadrupole moments are hard to measure and one often has to be content with a value of the "quadrupole coupling constant" $e^2qQ/h$. This quantity ranges from 0.224992 Mc/s for the deuteron in $D_2$ to 2420 Mc/s for iodine in ICN, and may therefore be either negligible or enormous compared with the nuclear magnetic Zeeman energy $g_N\beta_N H$.

An alternative to pure quadrupole resonance is to perform the resonance experiment in a magnetic field. In the so-called high-field experiment $H$ is chosen to make $g_N\beta_N H$ much larger than $e^2qQ$, so that the nuclear spin is quantized along the field

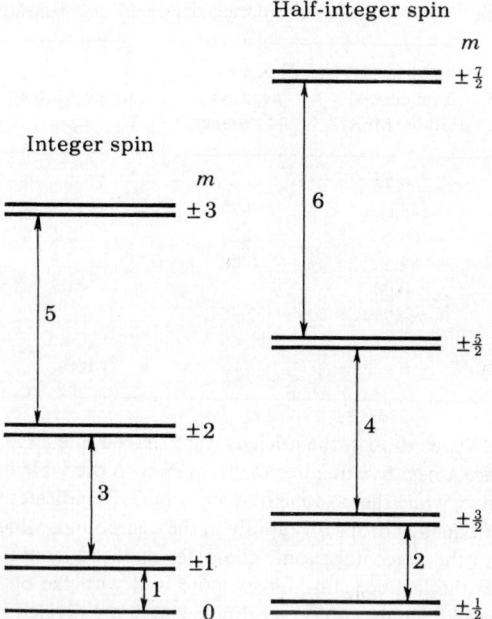

FIG. 3.4. Energy levels and resonance transitions of a nuclear quadrupole in an axial field gradient. The energy is measured in units of $\frac{3}{4}e^2qQ/[3I^2 - I(I + 1)]$.

direction, making some angle $\theta$ with the molecular symmetry axis. Under these particular conditions the effective field gradient resolved along the direction of $H$ is $\frac{1}{2}eq(3\cos^2\theta - 1)$ and the first-order energy levels are

$$E = -g_N\beta_N Hm + \frac{3}{8}e^2qQ\left(\frac{3m^2 - I(I + 1)}{3I^2 - I(I + 1)}\right)(3\cos^2\theta - 1) \qquad (22)$$

The allowed transitions are those in which the nuclear quantum number $m$ changes by $\pm 1$, and the resonance spectrum in a fixed field $H$ has $2I$ lines at the frequencies

$$hv = g_N\beta_N H - \frac{3}{8}e^2qQ\frac{(2m - 1)}{[3I^2 - I(I + 1)]}(3\cos^2\theta - 1) \qquad (23)$$

which move as the sample is rotated, just as we have described in Section 3.3. In the second-order approximation the lines are not equidistant, and weak $\Delta m = \pm 2$ transitions are possible. In the alternative weak field experiment the nuclear spins are quantized along the molecular axis and the magnetic field breaks the degeneracy of the $\pm m$ energy levels to give a small splitting of the quadrupole resonance lines.

One of the most interesting applications of nuclear quadrupole resonance (n.q.r.) has been the study of chemical bonding in chlorides. The chlorine atom has a vacant $p$ orbital in its valence shell and the unbalanced electronic charges produce a large field gradient at the $Cl^{35}$ or $Cl^{37}$ nucleus. On the other hand, a free $Cl^-$ ion is spherical and $q$ vanishes. Thus the $Cl^{35}$ resonance frequency in any molecule is directly proportional to the number of unbalanced $p$ electrons in the chlorine atom, which depends in turn on the amount of sp hybridization and the ionic character of the bond. Table 3.1 illustrates this trend in a wide range of molecules. (The quadrupole moment

TABLE 3.1.  $Cl^{35}$ Quadrupole Resonance Study of the Bonding in Chlorides

| Compound | Quadrupole Frequency $e^2qQ/h$ (Mc/s) | Cl-X Electro-negativity Difference | Dipole Moment $\mu/er$ | Calculated Ionic Character |
|---|---|---|---|---|
| Cl (atom) | −109.74 | — | 0 | 0 |
| FCl | −146.0 | 0.75 | 11.3 | 25.9 |
| BrCl | −103.6 | 0.2 | 5.6 | 5.6 |
| ICl | − 82.5 | 0.50 | 5.8 | 11.5 |
| KCl | ± 0.04 | 2.35 | 81.8 | 100 |
| RbCl | + 0.774 | 2.35 | — | 99.2 |
| CsCl | ± 3 | 2.45 | 74.5 | 96.8 |
| $Cl^-$ (ion) | 0 | — | 100 | 100 |

$Q$ of $Cl^{35}$ is negative, as though the nucleus were shaped like a rotating planet rather than an egg. Hence a negative quadrupole frequency in the table means that the field gradient $q$ is positive, while the positive frequency in CsCl indicates that $q$ has changed sign.) The n.q.r. frequency drops off rapidly in the compounds which are known to be strongly ionic, and the percentage ionic character deduced from the resonance data, which is shown in the last column, agrees quite well with the other "experimental" values ($\mu/er$) deduced from the molecular dipole length $\mu/e$ divided by the bond length.

The ionic character and the n.q.r. frequency also correlate well with the electronegativity difference between the two bonded atoms.

## PROBLEMS

1.  One of two identical nuclei with spin 1/2 is fixed at the origin, while the other moves rapidly round the circumference of a circle in the $xy$ plane, centered at the origin, with radius $r$. Calculate the average value of each component of the dipolar coupling tensor, Eq. (7).

2.  Use the basis functions (1) and the operators (11) to write out the energy matrix of $\mathcal{H}_D$. Verify that it is Hermitian. Now recalculate the matrix using the functions (2), and check that $(A + B)$ is diagonal.

3.  Calculate the transition probabilities for two identical dipole coupled spins of 1/2 in an $H_1$ field along the $x$-axis.

4.  Suppose the nuclei $I_1$ and $I_2$ are not the same. Describe what changes are necessary in Problems 2 and 3.

5.  Verify by direct calculation that the moment formula (19a) gives the right answer for the special case of gypsum. What would be the second moment of a polycrystalline sample of gypsum in which $D_2O$ replaces $H_2O$? (Assume the molecular dimensions do not change.)

6.  Use Eq. (20) to find the intramolecular part of the proton resonance second moment of (a) benzene; (b) 1-3-5 deuterobenzene with $D$ interactions included; (c) 1-3-5 deuterobenzene with $D$ neglected. Take bond lengths C—C = 1.40 Å, C—H = 1.09 Å.

7.  Use the result of Problem 1 to find the second moment of a single benzene molecule which rotates in the ring plane, (a) when $H$ is perpendicular to the ring plane; (b) when $H$ is oblique at an angle $\theta$ off the perpendicular.

8.  Use Eqs. (19a) and (19b) to calculate the ratio of the *intramolecular* contribution to the second moment in 1-3-5 deuterobenzene to that in benzene. What is the ratio for the *intermolecular* contributions?

**9.** Calculate the quadrupole coupling constants $eq$ and $e^2qQ/h$ for a bare deuteron placed 1 Å from a proton. Calculate $q$ in the $D_2$ molecule from the frequency of 0.225 Mc/s and the known value of $Q$. Compare the two results.

**10.** The $N^{14}$ quadrupole constant $e^2qQ/h$ for nitrobenzene in solution has been found to be 2.6 Mc/s. Compare this with the frequency of $N^{14}$ n.m.r. at 10,000 gauss, and with the e.s.r. frequency of the nitrobenzene negative ion radical. Hence assess the relative importance of quadrupole effects in n.m.r. and e.s.r. spectra.

## SUGGESTIONS FOR FURTHER READING

Slichter: Chapter 3. Proof of Van Vleck's moment formula.

Abragam: Chapters 4 and 7. The full theory of second moments and of nuclear quadrupole effects.

Deeley and Richards: *J. Chem. Soc.*, 3697 (1954). Infusible white precipitate.

Pake: *J. Chem. Phys.*, **16**: 327 (1948). The n.m.r. spectrum of a single crystal of gypsum.

Andrew and Eades: *Proc. Roy. Soc. (London)*, **A218**: 537 (1953). Discusses n.m.r. and molecular motion in solid benzene.

Pound: *Phys. Rev.*, **79**: 675 (1950). Classic early paper on quadrupole effects in n.m.r.

Dailey and Townes: *J. Chem. Phys.*, **23**: 118 (1955). Electric field gradients and chemical bonding in chlorine compounds.

Das and Hahn: "Nuclear Quadrupole Resonance Spectroscopy," *Solid State Physics*, Supplement 1, Seitz and Turnbull, eds. (New York: Academic Press Inc., 1958). An excellent review article.

CHAPTER 4

# THE ANALYSIS OF N.M.R. SPECTRA IN LIQUIDS

## 4·1 INTRODUCTION

The nuclear resonance lines obtained from liquids are exceedingly narrow, and consequently very small magnetic interactions can be detected. These interactions are characteristic of the chemical environment of the nuclei and their study has led to the development of n.m.r. as an indispensable tool in chemical structure analysis. We will not attempt to review this enormous field; instead we will concentrate on the fundamental interactions which are involved and describe the methods used to analyze the spectra, principally proton resonance spectra.

There are two types of interaction which are important: the nuclear Zeeman interaction and nuclear spin-spin coupling. As we saw in Chapter 2, the surrounding electrons in a molecule produce shielding effects which change the Zeeman term from $\mathscr{H} = -g_N\beta_N H I_z$ to $-g_N\beta_N(1 - \sigma)H I_z$. This is because the actual field $H_{\text{nucl.}}$ at the nucleus is the sum of two magnetic fields, $H$ the applied field, and $H' = -\sigma H$, the local field at the nucleus arising from the induced electronic currents. $H'$ varies according to the chemical environment of the proton, and if measurements are made with a fixed frequency, the applied field necessary to satisfy the resonance condition also varies from one type of proton to another.

The strongest coupling between two nuclear moments is the dipole-dipole coupling discussed in Chapter 3. However we noted that the dipolar interaction tensor is traceless and hence vanishes for a rotating molecule in the liquid phase. Fortunately there are other mechanisms which couple the nuclear spins and although the magnitudes of the couplings are extremely small, they do not vanish for molecules in solution. The most important mechanism leading to correlation of nuclear spins involves polarization of the intervening electron spins. We will see in the next chapter how this occurs.

Before embarking on a detailed discussion of the two main interactions, let us consider the familiar example of acetaldehyde, $CH_3CHO$, which illustrates the most important features. First we note the presence of two chemically distinct types of protons, so that the proton resonance spectrum (Fig. 4.1) shows two groups of lines with different relative intensities. The stronger group arises from the three equivalent methyl protons, and the weaker group from the aldehyde proton. Under conditions of normal resolution these two groups reveal the presence of further structure. The

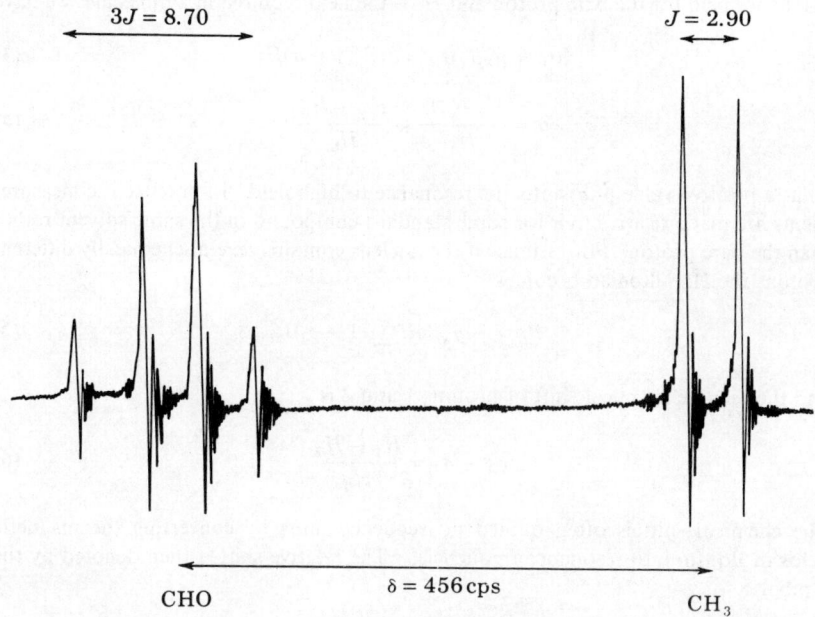

$3J = 8.70$                  $J = 2.90$

$\delta = 456 \, cps$

CHO                                  $CH_3$

FIG. 4.1. The n.m.r. spectrum of acetaldehyde $CH_3CHO$. The magnetic field increases from left to right.

methyl line is split into two components of equal intensity, corresponding to the two allowed orientations of the aldehyde proton spin. The aldehyde proton line shows a more complex splitting into four lines.

The separation between the methyl and aldehydic peaks is called the chemical shift; the fine structure within each peak is called the spin-spin splitting. To account for the splitting it is necessary to postulate an isotropic coupling Hamiltonian of the type

$$\mathscr{H} = J\mathbf{I}_1 \cdot \mathbf{I}_2 \tag{1}$$

between each pair of proton spins in the molecule and then analyze the energy levels and transitions within the spin system by the methods used in Chapter 2. $J$ is called the coupling constant and is measured in cycles per second (cps). As usual, molecules in a liquid rotate rapidly and $J$ is actually the average overall orientations of an anisotropic spin-spin coupling tensor.

## $4 \cdot 2$ THE CHEMICAL SHIFT

In liquids the screening tensor $\sigma$ reduces to an isotropic screening constant $\sigma$ and the nuclear Zeeman energy becomes

$$\mathscr{H}_0 = -g_N \beta_N (1 - \sigma) H I_z \tag{2}$$

In principle the spectrum is obtained by using a fixed frequency $\nu_0$, scanning the magnetic field, and determining the value of $H$ required for each proton. If $H_0$ is the resonance field for the bare proton and $H_1$ is the field required in a molecule, we have

$$h\nu_0 = g_N\beta_N H_0 = g_N\beta_N(1 - \sigma)H_1 \tag{3}$$

or
$$\sigma = \frac{H_1 - H_0}{H_1} \approx \frac{H_1 - H_0}{H_0} \tag{4}$$

Thus a positive value of $\sigma$ shifts the resonance to high field. In practice the measurements are made relative to $\sigma$ for some standard compound in the same solvent rather than the bare proton. For instance if the nucleus contains several chemically different protons the Hamiltonian becomes

$$\mathscr{H}_0 = -g_N\beta_N H \sum_i (1 - \sigma_i)I_{zi} \tag{5}$$

and the relative chemical shift of protons 1 and 2 is

$$\sigma_1 - \sigma_2 = \frac{H_1 - H_2}{H_0} \tag{6}$$

The chemical shift is often quoted in frequency units by converting the magnetic fields in Eq. (6) into resonance frequencies. The relative shift is then denoted by the symbol $\delta$.

$$\delta_1 - \delta_2 = g_N\beta_N(H_1 - H_2)/h = (\sigma_1 - \sigma_2)g_N\beta_N H_0/h$$
$$= (\sigma_1 - \sigma_2)\nu_0 \tag{7}$$

The disadvantage of using frequency units is that the $\delta$ values are field dependent, whereas the shifts $\sigma$ are dimensionless constants measured in parts per million.

Precise measurements of the chemical shift are complicated by various small corrections. The simplest is due to the shielding effects of the bulk diamagnetism of the liquid sample and it depends on the shape of the container. For a long cylindrical sample the local magnetic field $H$ acting on the dissolved molecules is related to the external field $H_{\text{mag}}$ of the magnet by the formula

$$H = H_{\text{mag}}\left(1 - \frac{2\pi}{3}\chi_v\right) \tag{8}$$

where $\chi_v$ is the bulk susceptibility of the liquid, and the true chemical shift differs from the apparent observed one by an amount

$$\sigma - \sigma_{\text{obs}} = \frac{2\pi}{3}\chi_v \tag{9}$$

There are also important corrections due to complex chemical interactions between the molecules and the solvent, so that it is essential to specify the solvent and its concentration when quoting observed values of the chemical shift.

In practice the bulk susceptibility corrections are usually eliminated by using an "internal" standard such as $t$-butyl alcohol or tetramethylsilane in the same solvent. If it is necessary to use an "external" standard then it is usually placed in a small capillary inside the sample tube.

# 4·3 THE SPIN-SPIN COUPLING

The acetaldehyde spectrum is a straightforward example of spin-spin coupling, and we shall now interpret it in more detail. The first stage is to write down a suitable spin Hamiltonian. It is most convenient to work in frequency units, so that the Zeeman energy is

$$\mathcal{H}_0 = -v_0(1 - \sigma_A)(I_{z1} + I_{z2} + I_{z3}) - v_0(1 - \sigma_B)I_{z4} \tag{10}$$

Here $I_1$, $I_2$, $I_3$ are the methyl proton spins, shielding constant $\sigma_A$, and $I_4$ is the aldehyde proton with shielding $\sigma_B$. The spin $I_4$ must couple equally on the average with the three methyl spins, giving a term

$$\mathcal{H}_{AB} = J(I_1 + I_2 + I_3) \cdot I_4 \tag{11}$$

Couplings will also exist between the methyl protons, but these can be ignored, as we shall see in Section 4.4. The important feature which makes a simple analysis possible is that $J$ is much smaller than the relative chemical shift

$$\delta = v_0(\sigma_A - \sigma_B) \tag{12}$$

of the two kinds of proton. The stationary states of the Zeeman Hamiltonian have the four nuclei quantized along the magnetic field, with quantum numbers $m_1$, $m_2$, $m_3$, and $m_4$, while the zero order energy levels depend only on the total resolved spins of the two groups $A$ and $B$. Let us define total spin vectors for the groups

$$\mathbf{F}_A = (\mathbf{I}_1 + \mathbf{I}_2 + \mathbf{I}_3), \qquad \mathbf{F}_B = \mathbf{I}_4 \tag{13}$$

and new quantum numbers for the operators $F_{zA}$ and $F_{zB}$

$$M_A = (m_1 + m_2 + m_3) \tag{14}$$

$$M_B = m_4 \tag{15}$$

The Zeeman energy now becomes

$$\mathcal{H}_0 = -v_0(1 - \sigma_A)F_{zA} - v_0(1 - \sigma_B)F_{zB} \tag{16}$$

The first order spin-spin coupling energy $J\mathbf{F}_A \cdot \mathbf{F}_B$ depends only on the diagonal matrix elements of operators like $J\mathbf{I}_1 \cdot \mathbf{I}_4$, that is on the value of $I_{z1}I_{z4}$ or $m_1 m_4$. So we easily see that the total coupling energy is proportional to $M_A M_B$. To sum up, the first order energy levels of the spin system are

$$E/h = -v_0(1 - \sigma_A)M_A - v_0(1 - \sigma_B)M_B + JM_A M_B \tag{17}$$

There are two types of allowed n.m.r. transition. In the methyl transitions one of the protons 1,2,3 turns over, so that $M_A$ changes by $\pm 1$ and the coupling energy changes by $\pm JM_B$. Hence there are two lines at frequencies

$$v = v_0(1 - \sigma_A) - JM_B \quad (A \text{ resonance})$$

$$= v_0(1 - \sigma_A) \mp \tfrac{1}{2}J \tag{18}$$

The aldehyde proton makes transitions at the frequency

$$v = v_0(1 - \sigma_B) - JM_A, \quad (B \text{ resonance}) \tag{19}$$

but there are now four possible lines corresponding to the different arrangements of the methyl spins.

$$
\left.
\begin{array}{ll}
\alpha\alpha\alpha & M_A = \tfrac{3}{2} \\
\alpha\beta\alpha,\ \alpha\alpha\beta,\ \beta\alpha\alpha & M_A = \tfrac{1}{2} \\
\beta\beta\alpha,\ \alpha\beta\beta,\ \beta\alpha\beta & M_A = -\tfrac{1}{2} \\
\beta\beta\beta & M_A = -\tfrac{3}{2}
\end{array}
\right\}
\tag{20}
$$

At room temperature all orientations of the spins are equally probable, and the two methyl n.m.r. lines are equally intense. However, the aldehyde resonance consists of four equidistant lines a distance $J$ apart with intensities in the ratio $1:3:3:1$ which correspond to the number of ways of arranging the methyl spins in Eq. (20).

Unfortunately n.m.r. spectra become increasingly complex as the number of chemically distinguishable nuclei increases, particularly when the couplings and chemical shifts are of about the same magnitude. In these spectra a full solution of the problem involves finding the exact energy levels of the spin Hamiltonian and calculating transition probabilities between every possible pair of stationary states.

## 4·4 THE ANALYSIS OF COMPLEX SPECTRA

### 4.4.1 Classification of Spectra

In the next section we shall describe the analysis of two relatively simple spectra, conventionally described as $AB$ and $A_2B_2$. An $AB$ system is one in which two protons are coupled together, having a coupling constant $J$ comparable to the chemical shift $\delta$; an example is the proton resonance spectrum of 2-bromo-5-chlorothiophene.

$(AB)$

On the other hand the HF molecule, with an enormous shift between the $H^1$ and $F^{19}$ resonances, is classed as $AX$. In an $A_2B_2$ system two distinct pairs of protons with a small shift are involved, as in $CH_2Cl—CH_2Br$, or any benzene derivative like

$(A_2B_2)$

If the chemical shift between $A$ and $B$ is very large, or if $A$ and $B$ are different nuclei, as in 1,1-difluoroethylene, the system is denoted $A_2X_2$. In our previous discussion of acetaldehyde we could have called it an $A_3X$ spectrum.

### 4.4.2 The Analysis of an $AB$ Spectrum

The spin Hamiltonian for the $AB$ spectrum is clearly

$$
\mathcal{H} = -v_0(1 - \sigma_A)I_{zA} - v_0(1 - \sigma_B)I_{zB} + J\mathbf{I}_A \cdot \mathbf{I}_B
\tag{21}
$$

(We are now working in frequency units.) As usual we choose the set of basis functions

$$\phi_1 = |\alpha\alpha\rangle \qquad \phi_2 = |\alpha\beta\rangle \qquad \phi_3 = |\beta\alpha\rangle \qquad \phi_4 = |\beta\beta\rangle \tag{22}$$

and write down the energy matrix in the same way as for the hydrogen atom in Chapter 2. It helps to split the operator $\mathbf{I}_A \cdot \mathbf{I}_B$ into the two parts

$$\mathbf{I}_A \cdot \mathbf{I}_B = I_{zA}I_{zB} + \tfrac{1}{2}(I_A^+ I_B^- + I_A^- I_B^+) \tag{23}$$

and as before the term $\tfrac{1}{2}(I_A^+ I_B^- + I_A^- I_B^+)$ mixes $\phi_2$ with $\phi_3$. We also see that $\phi_1$ and $\phi_4$ are already eigenfunctions of the spin Hamiltonian with energies

$$E_1/h = v_0(-1 + \tfrac{1}{2}\sigma_A + \tfrac{1}{2}\sigma_B) + \tfrac{1}{4}J, \qquad \psi_1 = |\alpha\alpha\rangle$$

$$E_4/h = v_0(1 - \tfrac{1}{2}\sigma_A - \tfrac{1}{2}\sigma_B) + \tfrac{1}{4}J, \qquad \psi_4 = |\beta\beta\rangle \tag{24}$$

In order to obtain the other two energies it is necessary to solve the $2 \times 2$ secular determinant

$$
\begin{array}{cc}
& \phi_2 \qquad\qquad\qquad\qquad \phi_3 \\
\begin{array}{c} \phi_2 \\ \phi_3 \end{array}
\begin{vmatrix}
[-\tfrac{1}{2}v_0(\sigma_A - \sigma_B) - \tfrac{1}{4}J] - E & \tfrac{1}{2}J \\
\tfrac{1}{2}J & +[\tfrac{1}{2}v_0(\sigma_A - \sigma_B) - \tfrac{1}{4}J] - E
\end{vmatrix} = 0
\end{array} \tag{25}
$$

involving $\phi_2$ and $\phi_3$. This is conveniently accomplished by making the substitutions

$$\tfrac{1}{2}v_0(\sigma_A - \sigma_B) = \tfrac{1}{2}\delta = C\cos 2\theta$$

$$\tfrac{1}{2}J = C\sin 2\theta \tag{26}$$

where

$$C = \tfrac{1}{2}\sqrt{J^2 + \delta^2} \tag{27}$$

The determinant then becomes

$$
\begin{vmatrix}
-C\cos 2\theta - \tfrac{1}{4}J - E & C\sin 2\theta \\
C\sin 2\theta & +C\cos 2\theta - \tfrac{1}{4}J - E
\end{vmatrix} = 0 \tag{28}
$$

the solutions being

$$E_2/h = -\tfrac{1}{4}J - C, \qquad \psi_2 = \cos\theta|\alpha\beta\rangle - \sin\theta|\beta\alpha\rangle$$

$$E_3/h = -\tfrac{1}{4}J + C, \qquad \psi_3 = \sin\theta|\alpha\beta\rangle + \cos\theta|\beta\alpha\rangle \tag{29}$$

We now determine the transition frequencies relative to the center of the pattern, i.e., relative to $v_0(1 - \tfrac{1}{2}\sigma_A - \tfrac{1}{2}\sigma_B)$, and the corresponding relative intensities. The results are:

| Transition | Frequency (fixed field) | Intensity |
|---|---|---|
| $(4 \rightarrow 2)$ | $\tfrac{1}{2}J + C$ | $1 - \sin 2\theta$ |
| $(3 \rightarrow 1)$ | $-\tfrac{1}{2}J + C$ | $1 + \sin 2\theta$ |
| $(4 \rightarrow 3)$ | $\tfrac{1}{2}J - C$ | $1 + \sin 2\theta$ |
| $(2 \rightarrow 1)$ | $-\tfrac{1}{2}J - C$ | $1 - \sin 2\theta$ |

$$(30)$$

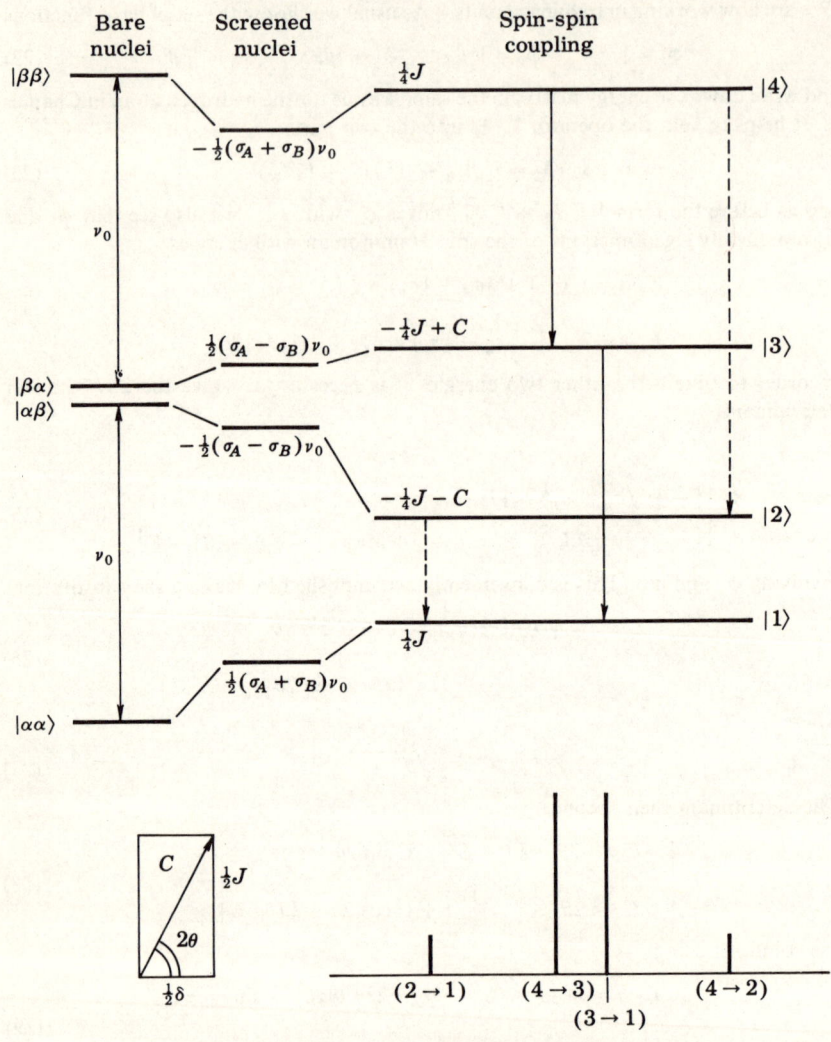

FIG. 4.2. Energy levels and n.m.r. transitions of the $AB$ spectrum.

These transitions are illustrated in Fig. 4.2. The intensity calculation is done in the usual way. For instance, the $(3 \rightarrow 1)$ transition probability depends on the factor

$$|\langle \psi_1 | I_{xA} + I_{xB} | \psi_3 \rangle|^2 = |\langle \alpha\alpha | I_{xA} + I_{xB} | \beta\alpha \cos \theta + \alpha\beta \sin \theta \rangle|^2$$

$$= \tfrac{1}{4}(\cos \theta + \sin \theta)^2$$

$$= \tfrac{1}{4}(1 + \sin 2\theta) \tag{31}$$

We are now in a position to calculate the appearance of the spectrum for different relative values of $J$ and $\delta$.

1. When $J = 0$ the pair of transitions $(3 \rightarrow 1)$ and $(4 \rightarrow 2)$ are degenerate giving one line. Similarly $(2 \rightarrow 1)$ and $(4 \rightarrow 3)$ are also degenerate. All four transitions are equally intense, since $\sin 2\theta = 0$, but only two lines are seen (Fig. 4.3a).

2. If $J$ is so small compared with $\delta$ that only first order perturbation effects are important we have an $AX$ spectrum. The $A$ and $B$ lines are each split into doublets with a splitting equal to $J$.

3. When the coupling is fairly small compared with $\delta$ one sees a pair of close doublets, as shown in Fig. 4.3b. The outer lines of each doublet are weaker than the inner ones.

4. When $\sigma_A$ and $\sigma_B$ are almost equal $\sin 2\theta$ is close to 1 and the transitions $(3 \rightarrow 1)$, $(4 \rightarrow 3)$ almost disappear. The two allowed transitions have nearly the same frequency and intensity, giving a single close doublet (Fig. 4.3c) at $\nu_0$, while the forbidden lines appear as weak satellites at $\nu_0 \pm J$. Finally when $\sigma_A = \sigma_B$ the two protons are equivalent and the spectrum reduces to a single line with no observable splitting!

This last result is very important since it illustrates a general rule that spin couplings within a group of magnetically equivalent nuclei often produce no observable splitting of the n.m.r. lines.

We see from this analysis that the appearance of an n.m.r. spectrum is markedly dependent on the relative values of the spin-spin couplings and chemical shifts, and that detailed calculations become necessary when these quantities are of comparable magnitude.

One final point is that the sign of $J$ may be either positive or negative. The spectrum has precisely the same appearance in either case, but lines at corresponding positions represent different transitions. If $J$ is negative $\sin 2\theta < 0$ and $(4 \rightarrow 2)$ and $(2 \rightarrow 1)$ change places with $(3 \rightarrow 1)$ and $(4 \rightarrow 3)$.

### 4.4.3 The Analysis of an $A_2B_2$ Spectrum

The analysis of an $A_2B_2$ system can be extremely complicated and we shall not deal with the spectrum completely. However we will pursue the matter far enough to see what the problems are, and to illustrate further rules which are of general importance in analyzing complex spectra.

First we notice from Fig. 4.4 that two symmetrical pairs of nuclei may have four different spin-spin coupling constants, so that the spin Hamiltonian will take the form

$$\mathscr{H} = -\nu_A(I_{z1} + I_{z2}) - \nu_B(I_{z3} + I_{z4}) + J_A \mathbf{I}_1 \cdot \mathbf{I}_2 + J_B \mathbf{I}_3 \cdot \mathbf{I}_4 + J(\mathbf{I}_1 \cdot \mathbf{I}_4 + \mathbf{I}_2 \cdot \mathbf{I}_3)$$

$$+ J'(\mathbf{I}_2 \cdot \mathbf{I}_4 + \mathbf{I}_1 \cdot \mathbf{I}_3) \tag{32}$$

Here $\nu_A$ and $\nu_B$ are resonance frequencies $\nu_0(1 - \sigma_A)$ and $\nu_0(1 - \sigma_B)$ for the two kinds of nuclei.

At this point it is worth noting that there are several different systems of nomenclature for the types of n.m.r. spectra. $A_2B_2$ or $A_2X_2$ spectra may be of two kinds. If $J \neq J'$ then the $A$ and $B$ nuclei are said not to be completely equivalent, whereas if $J = J'$ they are magnetically equivalent. The two molecules

$$
\begin{array}{cc}
\underset{H}{\overset{F}{\diagdown}} C = C \underset{H}{\overset{F}{\diagup}} \qquad & \underset{H}{\overset{H}{\diagdown}} C = C = C \underset{F}{\overset{F}{\diagup}}
\end{array}
$$

illustrate the difference. In the first $J \neq J'$ because there are different cis and trans couplings, while the second has a common H-F coupling constant. Some scientists

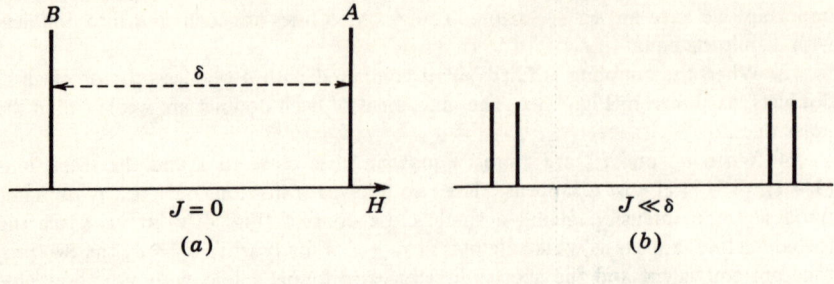

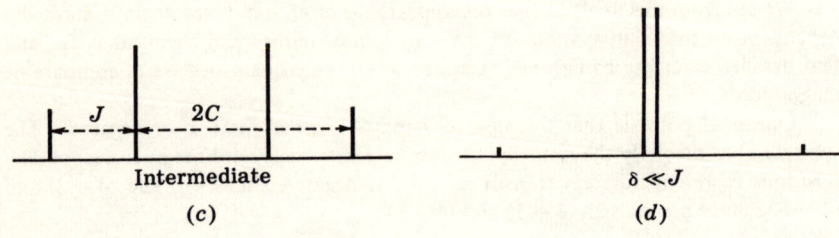

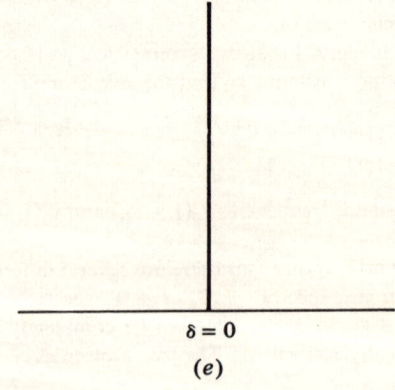

FIG. 4.3. *AB* spectrum with various ratios of the chemical shift $\delta = \nu_0(\sigma_A - \sigma_B)$ to the spin-spin coupling constant *J*.

use the notation $AA'XX'$ and $A_2X_2$ to distinguish these cases. Others call them $A_2X_2$ and $A_2^*X_2^*$. We now return to the analysis of our $A_2B_2$ or $AA'BB'$ spectrum.

The choice of basis spin functions is extremely important because, provided they are chosen intelligently, the Hamiltonian matrix is relatively simple. First we note that the spin system has an effective symmetry plane bisecting the lines between nuclei 1 and 2 and between 3 and 4. Reflection interchanges the nuclei of each pair. The basis functions are chosen to be either symmetric or antisymmetric under this operation. Thus we have

$$
\begin{array}{cc}
\textit{A nuclei} & \textit{B nuclei} \\
\end{array}
$$

$$
\left.\begin{array}{c}
\alpha_1\alpha_2 \\[4pt]
\dfrac{1}{\sqrt{2}}(\alpha_1\beta_2 + \beta_1\alpha_2) \\[4pt]
\beta_1\beta_2
\end{array}\right\} \text{ symmetric } \left\{\begin{array}{c}
\alpha_3\alpha_4 \\[4pt]
\dfrac{1}{\sqrt{2}}(\alpha_3\beta_4 + \beta_3\alpha_4) \\[4pt]
\beta_3\beta_4
\end{array}\right. \tag{33}
$$

$$
\dfrac{1}{\sqrt{2}}(\alpha_1\beta_2 - \beta_1\alpha_2) \quad \text{antisymmetric} \quad \dfrac{1}{\sqrt{2}}(\alpha_3\beta_4 - \beta_3\alpha_4)
$$

By multiplying any $A$ function and any $B$ function together we construct sixteen basis functions. The complete function $A \times B$ is symmetric (denoted by the symbol $s$) if both $A$ and $B$ are symmetric or antisymmetric; if one part is symmetric and the other antisymmetric, then the complete basis function is antisymmetric (symbol $a$).

A further division of the states is to separate them according to their total value of $I_z$, which is denoted by the symbol $F_z$, as in Eq. (14). The states, with their $F_z$ values as subscripts, are as follows:

$$s_2 = \alpha\alpha\alpha\alpha$$

$$\left.\begin{array}{l} 1s_1 = \dfrac{1}{\sqrt{2}}(\alpha\beta + \beta\alpha)\alpha\alpha \\[8pt] 2s_1 = \dfrac{1}{\sqrt{2}}\alpha\alpha(\alpha\beta + \beta\alpha) \end{array}\right\} \qquad \left.\begin{array}{l} 1a_1 = \dfrac{1}{\sqrt{2}}(\alpha\beta - \beta\alpha)\alpha\alpha \\[8pt] 2a_1 = \dfrac{1}{\sqrt{2}}\alpha\alpha(\alpha\beta - \beta\alpha) \end{array}\right\}$$

$$\left.\begin{array}{l} 1s_0 = \beta\beta\alpha\alpha \\[8pt] 2s_0 = \alpha\alpha\beta\beta \\[8pt] 3s_0 = \dfrac{1}{2}(\alpha\beta - \beta\alpha)(\alpha\beta - \beta\alpha) \\[8pt] 4s_0 = \dfrac{1}{2}(\alpha\beta + \beta\alpha)(\alpha\beta + \beta\alpha) \end{array}\right\} \qquad \left.\begin{array}{l} 1a_0 = \dfrac{1}{2}(\alpha\beta + \beta\alpha)(\alpha\beta - \beta\alpha) \\[8pt] 2a_0 = \dfrac{1}{2}(\alpha\beta - \beta\alpha)(\alpha\beta + \beta\alpha) \end{array}\right\} \tag{34}$$

$$\left.\begin{array}{l} 1s_{-1} = \dfrac{1}{\sqrt{2}}(\alpha\beta + \beta\alpha)\beta\beta \\[8pt] 2s_{-1} = \dfrac{1}{\sqrt{2}}\beta\beta(\alpha\beta + \beta\alpha) \end{array}\right\} \qquad \left.\begin{array}{l} 1a_{-1} = \dfrac{1}{\sqrt{2}}(\alpha\beta - \beta\alpha)\beta\beta \\[8pt] 2a_{-1} = \dfrac{1}{\sqrt{2}}\beta\beta(\alpha\beta - \beta\alpha) \end{array}\right\}$$

$$s_{-2} = \beta\beta\beta\beta$$

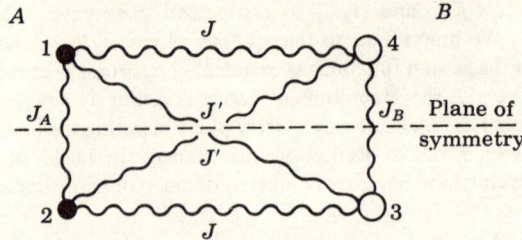

FIG. 4.4. Spin-spin coupling constants for a symmetrical $A_2B_2$ spectrum.

The only nonzero matrix elements of the Hamiltonian (32) are those between functions with the same symmetry and the same $F_z$ value. So the full $16 \times 16$ matrix reduces to one $4 \times 4$, five $2 \times 2$ and two $1 \times 1$ matrices which can be diagonalized separately. The possible transitions are then limited by the rules:

1. Only transitions between states of the same symmetry are allowed.
2. The selection rule $\Delta F_z = \pm 1$ holds.
3. Only one of the group spin variables $F_{zA}$ or $F_{zB}$ is allowed to change in each transition.

    One finds that there are altogether 24 allowed transitions and consequently 24 absorption lines. Twelve of these are centered about the frequency $\nu_A$ and are described as $\Delta F_{zA} = \pm 1$ transitions. The remaining twelve form an identical pattern centered about the frequency $\nu_B$ and satisfying the selection rule $\Delta F_{zB} = \pm 1$. It is not possible to determine the signs of the coupling constants, or to distinguish $J_A$ from $J_B$, and $J$ from $J'$, from the n.m.r. spectrum alone. There are also 4 very weak forbidden lines which disappear as the chemical shift $\delta$ becomes large.

    The analysis of an $A_2B_2$ spectrum is quite difficult, and each case must be examined separately. Sometimes simplifications due to equivalence of coupling constants are possible; in other cases one or more of the coupling constants may be too small to observe. For example, the proton resonance spectrum of naphthalene at 40 Mc/s can be analyzed as two equivalent $A_2B_2$ systems, the coupling between protons in different rings being negligible. The results obtained are indicated below; $J_A$ is too small to detect.

$$J = 8.6 \quad A$$
$$B$$
$$J_B = 6.0 \quad \quad \quad \quad J' = 1.4 \quad \quad \quad \quad \delta = 14.3 \text{ cps}$$
$$B$$
$$A$$

    When $\delta$ is large the analysis of an $A_2X_2$ spectrum such as that of 1,1-difluoroethylene is considerably simpler, since many of the spin functions in (34) are approxi-

$$J_{\text{cis}} = 0.7$$
$$J_A = 4.8 \quad \text{H} \quad \quad \text{F}$$
$$\text{C}\!=\!\!=\!\!\text{C} \quad \quad J_B = 36.4$$
$$\text{H} \quad \quad \text{F}$$
$$J_{\text{trans}} = 33.9$$

mate eigenfunctions of the spin Hamiltonian. The only states which are appreciably mixed are the degenerate pairs $3s_0$ with $4s_0$ and $1a_0$ with $2a_0$. The $H^1$ and $F^{19}$ spectra are mirror images of one another, each consisting of four strong lines and eight weak lines. Analysis of this spectrum shows that $J_{cis}$ and $J_{trans}$ have the same sign; it has also been shown that $J_{HH}$ and $J_{FF}$ have opposite signs.

### 4.4.4 Splitting from Magnetically Equivalent Nuclei

One puzzling fact which we have not yet explained is that the couplings between the three $CH_3$ protons in acetaldehyde $CH_3CHO$ do not appear at all in the n.m.r. spectrum, and the $AB$ coupling apparently disappears when two $AB$ nuclei have the same chemical shift. To explain this let us look at the n.m.r. energy levels of the methyl group by itself. The spin Hamiltonian is

$$\mathcal{H} = -\nu_0(1 - \sigma_A)F_{zA} + J_A(\mathbf{I}_1 \cdot \mathbf{I}_2 + \mathbf{I}_2 \cdot \mathbf{I}_3 + \mathbf{I}_3 \cdot \mathbf{I}_1) \tag{35}$$

[see Eqs. (13) and (16)], where $J_A$ is the methyl-methyl internal coupling constant. The important point now is that the energy depends only on the total spin $\mathbf{F}_A$ of the methyl group and on its $z$ component $F_{zA}$. This is because the scalar products in (35) can be written

$$\mathbf{I}_1 \cdot \mathbf{I}_2 + \mathbf{I}_2 \cdot \mathbf{I}_3 + \mathbf{I}_3 \cdot \mathbf{I}_1 = \tfrac{1}{2}(\mathbf{I}_1 + \mathbf{I}_2 + \mathbf{I}_3)^2 - \tfrac{1}{2}(\mathbf{I}_1^2 + \mathbf{I}_2^2 + \mathbf{I}_3^2)$$

$$= \tfrac{1}{2}\mathbf{F}_A^2 - \tfrac{9}{8} \tag{36}$$

The spin Hamiltonian therefore becomes

$$\mathcal{H} = -\nu_0(1 - \sigma_A)F_{zA} + \tfrac{1}{2}J_A[\mathbf{F}_A^2 - \tfrac{9}{4}] \tag{37}$$

while the spin states are eigenfunctions of the two commuting operators $\mathbf{F}_A^2$ and $F_{zA}$ (see Appendixes A and C) with the eigenvalues $F_A(F_A + 1)$ and $M_A$. Transitions are only allowed between states with the same values of $\mathbf{F}_A^2$, so the spin-spin coupling energy does not change and the nuclei only give a single n.m.r. line at $\nu_A = \nu_0(1 - \sigma_A)$.

For three spins of $1/2$ we have already seen in Section 4.3 that there are 8 possible spin functions. They may be separated into a quartet and two doublets as shown below:

$$\left.\begin{array}{c} \alpha\alpha\alpha \\[4pt] \dfrac{1}{\sqrt{3}}(\alpha\alpha\beta + \alpha\beta\alpha + \beta\alpha\alpha) \\[4pt] \dfrac{1}{\sqrt{3}}(\beta\beta\alpha + \beta\alpha\beta + \alpha\beta\beta) \\[4pt] \beta\beta\beta \end{array}\right\} \quad \text{Quartet } \left(F = \dfrac{3}{2}\right)$$

$$\left.\begin{array}{c} \dfrac{1}{\sqrt{2}}(\alpha\beta\alpha - \alpha\alpha\beta) \\[4pt] \dfrac{1}{\sqrt{2}}(\beta\alpha\beta - \beta\beta\alpha) \end{array}\right\} \quad \text{Doublet I } \left(F = \dfrac{1}{2}\right)$$

$$\left.\begin{array}{c} \dfrac{1}{\sqrt{6}}(2\beta\alpha\alpha - \alpha\beta\alpha - \alpha\alpha\beta) \\[4pt] \dfrac{1}{\sqrt{6}}(2\alpha\beta\beta - \beta\alpha\beta - \beta\beta\alpha) \end{array}\right\} \quad \text{Doublet II } \left(F = \dfrac{1}{2}\right).$$

$$(38)$$

In acetaldehyde the quantum states of the four spins again have definite values of $F_A^2$ which do not change in a transition, and so the $J_A$ coupling is still invisible.

Our $A_2X_2$ or $A_2B_2$ spectra do however show effects from $J_A$ and $J_B$ provided that $J$ and $J'$ are not equal. But if $J = J'$ the effects disappear. These curious observations are special cases of two useful rules:

1. If a molecule contains $n$ nuclei with the same chemical shift, *and no others*, the spin-spin couplings between them produce no splitting of the n.m.r. line, even if the coupling constants are not all equal. (Examples: benzene, $Si(CH_3)_4$.)

2. If a molecule has two groups or more of nuclei, say $A_mB_n$ and *all the AB coupling constants are equal*, then the $AA$ and $BB$ couplings have no effect on the n.m.r. spectrum.

## 4·5 PRACTICAL CONSIDERATIONS

Enough has been said, perhaps, to indicate the sort of complications which arise in the analysis of high-resolution n.m.r. spectra. Analyses for many characteristic types of spectrum have now been tabulated so that in many cases the interpretation is routine. As we have seen the principal difficulties arise when the chemical shift and spin coupling constants are of comparable magnitude; the present trend in spectrometer design is to employ higher frequencies (100 Mc/s is now quite common), and among other advantages is the fact that the chemical shift becomes increasingly large while the spin coupling constants are unaffected. Many spectra which are difficult to analyze with a spectrometer frequency of 40 Mc/s become fairly simple at 60 or 100 Mc/s. On the other hand, one does often lose information about the relative signs of coupling constants.

High-speed computers are invaluable for setting up the spin Hamiltonian matrices and calculating the transitions in complex spectra. With their aid, the analysis even of exceedingly complicated spectra becomes relatively easy.

In accurate work it is important to reduce solvent effects to a bare minimum. This can be done in several ways; using solvents which have no chemical interaction with the sample molecules; working at the lowest possible concentrations; measuring shifts relative to an internal standard in the same solvent to avoid bulk susceptibility corrections.

### PROBLEMS

1. Using the parameters $|\delta| = 4.7$ cps and $|J| = 3.9$ cps for the proton resonance spectrum of 2-bromo-5-chlorothiophene at 30.5 Mc/s, calculate how the spectrum should look at 60 Mc/s.

2. The n.m.r. proton resonance of a cylindrical sample of pure liquid acetone at 60 Mc/s shows a shift of 38.4 cps to high field relative to a dilute solution of acetone in carbon tetrachloride. How much of this is due to bulk susceptibility effects? ($\chi_v = -0.460 \times 10^{-6}$ for acetone; $-0.684 \times 10^{-6}$ for carbon tetrachloride.)

3. Work through the theory of the $AB$ spectrum when $J$ is negative and identify the new transitions. Check that the appearance of the spectrum is independent of the sign of $J$.

4. Use symmetry and $F_z$ values to classify the 8 basic spin states of a symmetrical $AB_2$ system with coupling constants $J_B$ and $J$. Set up the two $2 \times 2$ energy matrices and find the energy levels of the spin system.

**5.** In the $AB_2$ spectrum find out which transitions are allowed. (Do not attempt to calculate the intensities!) Show that the frequencies do not depend on $J_B$ at all.

**6.** The $F^{19}$ resonance spectrum of $ClF_3$ at 10 Mc/s is an $AB_2$ type with $J = 378$ cps and $\delta = 1114$ cps. Calculate the frequencies of the allowed transitions and compare with Muetterties and Phillips, *J. Am. Chem. Soc.*, **79**: 323 (1957).

**7.** What would the $A_2X_2$ spectrum of $H_2C = CF_2$ look like if $J$ and $J'$ were both equal to 17 cps?

**8.** Set up the $8 \times 8$ matrix of the operator $J(I_1 \cdot I_2 + I_2 \cdot I_3 + I_3 \cdot I_1)$. Verify that the spin functions in Eq. (38) are correct eigenfunctions, and find the eigenvalues.

## SUGGESTIONS FOR FURTHER READING

Pople, Schneider, and Bernstein: Chapter 6. Fuller discussion of the material in this chapter.

Abragam: Chapter 11. Details of the $AB$ and $A_2B_2$ spectra.

Roberts: *An Introduction to the Analysis of Spin-Spin Splitting in High Resolution Nuclear Magnetic Resonance Spectra* (New York: W. A. Benjamin, Inc., 1961). A clear and simple step-by-step description for organic chemists, with many examples.

McConnell, McLean, and Reilly: *J. Chem. Phys.*, **23**: 1152 (1955). Expounds principles of spectral analysis. Applied to $CH_2 = CF_2$.

Pople, Schneider, and Bernstein: *Can. J. Chem.*, **35**: 1060 (1957). Complete analysis of the n.m.r. spectrum of naphthalene.

Anderson: *Phys. Rev.*, **102**: 151 (1956). Discusses n.m.r. spectra of hydrocarbons. Analyzed by perturbation theory.

Bothner-By and Glick: *J. Chem. Phys.*, **26**: 1647 (1957). Solvent effects on the chemical shift.

Martin and Dailey: *J. Chem Phys.*, **37**: 2594 (1962). Discusses n.m.r. spectra of substituted benzenes.

CHAPTER 5

# INTERPRETATION
# OF CHEMICAL SHIFTS
# AND SPIN-SPIN COUPLINGS

## 5·1 INTRODUCTION

The complete analysis of high-resolution n.m.r. spectra in solution is only a prelude to the far more important and interesting task of interpreting the observed chemical shifts $\sigma$ and coupling constants $J$. We shall now look at the origin of the shifts and couplings in more detail, and see how some typical features are explained in terms of the molecular electronic structure. The theoretical principles involved are fairly simple, but their application to individual molecules raises many difficult questions which are still unanswered.

We begin with chemical shifts. Fig. 5.1 illustrates proton shifts for a variety of organic compounds. Proton shifts are usually measured on the so-called "$\tau$ scale" relative to the protons of tetramethylsilane, $Si(CH_3)_4$, as an "internal" standard in the same solvent. This compound gives a single sharp line whose position is taken to be $+10.00$ ppm. High $\tau$ values correspond to resonances at high field, implying that the nucleus is effectively screened by the surrounding electrons.

## 5·2 ORIGINS OF THE CHEMICAL SHIFT

### 5.2.1 Molecular Electronic Currents

We saw in Chapter 2 that a uniform magnetic field causes the whole electron cloud of a helium atom to rotate, or rather to precess, about the direction of $H$, causing electronic currents to flow. These give a secondary magnetic field at the nucleus and lead to the chemical shift.

If the nucleus is part of the molecule, the electrons are, in general, no longer completely free to rotate about the direction of $H$. This leads to two new effects; first that the secondary field $H'$ is not necessarily parallel to $H$ and the shielding must be represented by an anisotropic tensor $\sigma$; second, the theoretical expression for the screening involves an additional term which often has the opposite sign to the Lamb term.

54

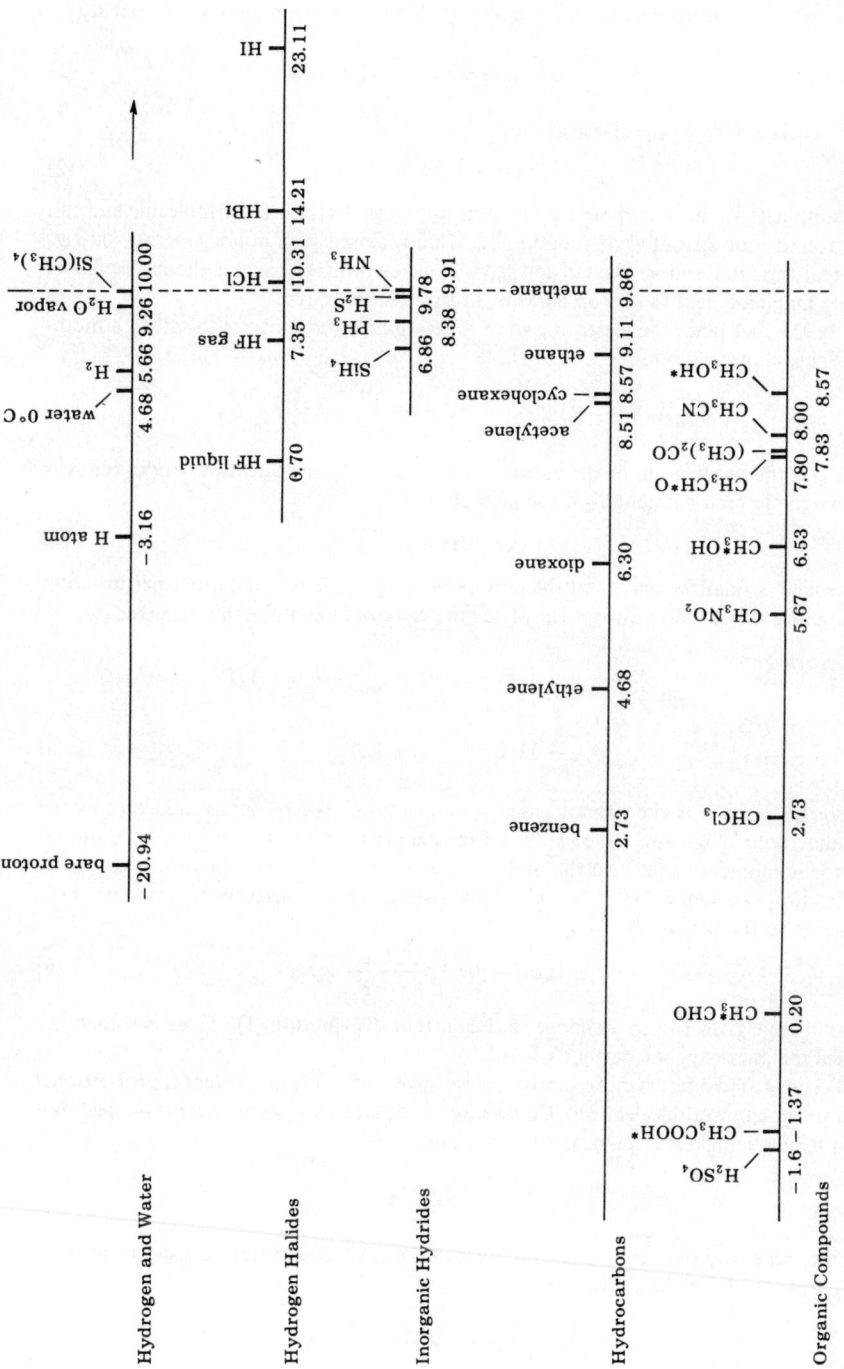

Fig. 5.1. Proton chemical shifts of typical organic compounds, measured on the τ-scale relative to Si(CH₃)₄ = 10.00. Positive shifts are to high field, and all values are given in parts per million. All the compounds are in the liquid state unless otherwise stated.

In quantum mechanics an electron in a state $\psi$ has a probability distribution $\rho(r) = \psi^*\psi$, and *in the absence of a magnetic field* there is an electric current density

$$\mathbf{j} = \frac{e\hbar i}{2mc}(\psi^*\nabla\psi - \psi\nabla\psi^*) \qquad (H = 0) \tag{1}$$

which corresponds to the classical expression

$$\mathbf{j} = -(e/c)\rho\mathbf{v} \tag{2}$$

For simplicity we now suppose that there is only one electron in the molecule and call the ground state spatial wave function $\psi_0$. In any closed shell nondegenerate state $\psi_0$ is a real function and so the current $\mathbf{j}_0(\mathbf{r})$ vanishes in every part of the molecule. A steady magnetic field $\mathbf{H}$ sets up currents in two distinct ways.

In the first place the electronic wave function changes, and has to satisfy a modified Schrodinger equation

$$\left(\frac{-\hbar^2}{2m}\nabla^2 + V\right)\psi + \left(\frac{e}{mc}\mathbf{A}\cdot\mathbf{p} + \frac{e^2}{2m^2c^2}\mathbf{A}^2\right)\psi = E\psi \tag{3}$$

Here $\mathbf{p}$ is the momentum of the electron, represented by the operator $-i\hbar\nabla$, and $\mathbf{A}$ is the magnetic vector potential of the field $\mathbf{H}$

$$\mathbf{A} = \tfrac{1}{2}(\mathbf{H} \times \mathbf{r}) \tag{4}$$

If the field is small we may treat the term $(e/mc)\mathbf{A}\cdot\mathbf{p}$ in (3) as a first order perturbation and ignore the $\mathbf{A}^2$ term altogether. Thus the electronic Hamiltonian changes by

$$\frac{e}{mc}\mathbf{A}\cdot\mathbf{p} = \frac{e}{2mc}[(\mathbf{H} \times \mathbf{r})\cdot\mathbf{p}] = \frac{e}{2mc}[\mathbf{H}\cdot(\mathbf{r} \times \mathbf{p})]$$

$$= \frac{e\hbar}{2mc}\mathbf{H}\cdot\mathbf{L} \tag{5}$$

where $\mathbf{L}\hbar = \mathbf{r} \times \mathbf{p}$ is the orbital angular momentum. We recognize $(e\hbar/2mc)$ as the familiar Bohr magneton $\beta$, so (5) is just the energy of interaction between the orbital magnetic moment of the electron and the external field $H$. Since the field is along the $z$ direction, we replace $\mathbf{H}\cdot\mathbf{L}$ by $HL_z$, and write down the perturbed electronic wave function in the new form

$$\psi = \psi_0 - \beta H \sum_n \frac{\langle n|L_z|0\rangle}{E_n - E_0}\psi_n \tag{6}$$

This change gives rise to a change in the current distribution (1). (This is sometimes called the paramagnetic current.)

The second effect is more subtle. As we have seen in (2) the current is proportional to the velocity $\mathbf{v}$ of the electron. Classically $\mathbf{v}$ is equal to $\mathbf{p}/m$ when there is no field, but in a magnetic field the classical relation changes to

$$\mathbf{v} = \frac{1}{m}\left(\mathbf{p} + \frac{e}{c}\mathbf{A}\right) \tag{7}$$

In the same way the formula for the quantum mechanical current density changes in the presence of a field, becoming

$$\mathbf{j} = \frac{e\hbar i}{2mc}(\psi^*\nabla\psi - \psi\nabla\psi^*) - \frac{e^2}{mc^2}\mathbf{A}\psi^*\psi \tag{8}$$

Thus even if the wave function of the electrons does not change at all, there is still an induced current $(e^2/2mc^2)\psi_0^*\psi_0[\mathbf{H} \times \mathbf{r}]$ which corresponds to a free precession of the entire electron cloud about the field direction (this is usually called the diamagnetic current).

### 5.2.2 Ramsey's Formula

Next we consider the magnetic field $\mathbf{H}'$ produced by the induced currents. We shall simplify the calculation by taking a classical point of view, introducing quantum mechanics at the last stage of the argument. The field produced by an electron at $\mathbf{r}$ with velocity $\mathbf{v}$ is

$$\mathbf{H}' = -\left(\frac{e}{c}\right) \frac{\mathbf{r} \times \mathbf{v}}{r^3} = -\left(\frac{e}{mc}\right) \frac{\mathbf{r} \times [\mathbf{p} + (e/c)\mathbf{A}]}{r^3} \tag{9}$$

by using (7). If we now substitute $\mathbf{L}h$ for $\mathbf{r} \times \mathbf{p}$ and $\frac{1}{2}(\mathbf{H} \times \mathbf{r})$ for $\mathbf{A}$ this becomes

$$\mathbf{H}' = -\left(\frac{eh}{2mc}\right) \frac{2\mathbf{L}}{r^3} - \left(\frac{e^2}{2mc^2}\right) \frac{\mathbf{r} \times (\mathbf{H} \times \mathbf{r})}{r^3} \tag{10}$$

The classical expression (10) has to be replaced by a quantum mechanical average over the perturbed electronic state $\psi$, and for simplicity we only work out the $z$ component of $\mathbf{H}'$, which has the value

$$H_z' = -\sigma_{zz}H = -\frac{eh}{2mc}\left\langle \psi \left| \frac{2L_z}{r^3} \right| \psi \right\rangle - \left(\frac{e^2}{2mc^2}\right)H\left\langle \psi \left| \frac{x^2 + y^2}{r^3} \right| \psi \right\rangle \tag{11}$$

All that remains is to substitute the correct wave function from (6) and evaluate (11) up to first order in $H$, with the result:

$$\sigma_{zz} = \frac{e^2}{2mc^2}\left\langle 0 \left| \frac{x^2 + y^2}{r^3} \right| 0 \right\rangle$$

$$- \left(\frac{eh}{2mc}\right)^2 \sum_n \left\{ \frac{\langle 0|L_z|n\rangle\left\langle n \left| \frac{2L_z}{r^3} \right| 0 \right\rangle}{E_n - E_0} + \frac{\left\langle 0 \left| \frac{2L_z}{r^3} \right| n \right\rangle\langle n|L_z|0\rangle}{E_n - E_0} \right\} \tag{12}$$

Equation (12) is called Ramsey's shielding formula and applies to any closed shell molecule. The quantities $L_z$ and $(x^2 + y^2)/r^3$ must be summed over all the electrons. The other components $\sigma_{xx}$ and $\sigma_{yy}$ of the tensor are given by similar equations, and one may then work out the average isotropic screening constant $\sigma$. The two parts of $\sigma$ in (12) are usually called the "diamagnetic" and "paramagnetic" terms

$$\sigma = \sigma_d + \sigma_p \tag{13}$$

The diamagnetic part, $\sigma_d$, is fairly easy to estimate theoretically, since it depends only on the electron distribution in the electronic ground state. The paramagnetic contribution depends also on the excited states. It vanishes for electrons in $s$ orbitals, which have zero angular momentum; but it may become very large when there is an asymmetric distribution of $p$ and $d$ electrons close to the nucleus and these electrons have low-lying excited states. Fluorine chemical shifts are largely dominated by the paramagnetic term. (It is important to realize that this paramagnetic chemical shift is not in any way connected with electron spin.)

## 5·3 PROTON CHEMICAL SHIFTS

The diamagnetic and paramagnetic parts of the shielding are both small for hydrogen nuclei because only a small number of electrons can occupy the $1s$ orbital surrounding the proton and chemical shifts all lie in a small range of about 30 ppm. This fact complicates the interpretation because various small effects which are unimportant for other nuclei become relatively large.

The shielding of any particular proton may be regarded as arising from four separate contributions. The basic starting point in this discussion is that one can divide up the electronic currents into separate parts on the different atoms. The main terms are as follows.

1. The diamagnetic screening from the circulation of electrons on the same atom as the nucleus in question. This is calculated from the first term of Eq. (12), the integration being limited to the local atomic electron density. An increase in the electron density in the hydrogen $1s$ orbital causes an increased shielding. The ring protons of substituted benzenes are a good example of this effect (Fig. 5.2). Here the shifts are strongly correlated with the electron-withdrawing power of the substituent, the changes being largest at the ortho and para positions. The most likely explanation is that the build up of charge on the ring carbon atoms polarizes the C—H bonds and increases the electron density at the proton. There is also some correlation with the electronegativity of the adjacent atom in other compounds, though not consistently. We might also note the small shielding of the acid protons in sulphuric and acetic acids.

2. Corrections to (1) due to paramagnetic currents on the same atom. This term is generally small for hydrogen; however one idealized example is the screening of a hydrogen atom in a strong electric field $E$. The field distorts the atom, and reduces $\sigma$ by an amount

$$\Delta\sigma = -\left(\frac{e^2}{3mc^2}\right)\frac{881}{24}\frac{E^2 a_0^4}{e^2} \tag{14}$$

which is proportional to $E^2$. Here $a_0$ is the Bohr radius. Typically a unit charge $e$ at a distance of $3a_0$ (1.587 Å) would reduce $\sigma$ by 2.6 ppm. Strong electric fields do occur in the hydrogen bond, and are believed to account for part of the characteristic "hydrogen bond shift" between the protons in water vapour ($\tau = 9.26$) and water at its freezing point ($\tau = 4.68$). However the electron density effect, represented by structures like $O^-\cdots H^+\cdots O$ is probably also important.

3. The combined effect of the diamagnetic and paramagnetic currents on other atoms. This effect is largest if the electrons on a near neighbor atom have a *large* and *anisotropic* magnetic susceptibility. It is particularly important in linear closed shell molecules like acetylene. In particular the paramagnetic term is much more anisotropic than the diamagnetic one, so that the neighbor contribution to the nuclear screening is largest if the atom has low-lying excited states. Such a situation holds for the hydrogen halides HCl, HBr, HI.

We shall now do a simple calculation of this effect (Fig. 5.3). We suppose that the nucleus is at a distance $R$ from an axially symmetric anisotropic molecular grouping whose magnetic susceptibility is $\chi_\parallel$ along the $z$ axis and $\chi_\perp$ along the $x$ and $y$ axes. A field $H$ along the $z$ axis induces a magnetic dipole $H\chi_\parallel$ in the group, and if the nucleus lies at an angle $\theta$ off the axis in the $xz$ plane, the dipole magnetic field has a component

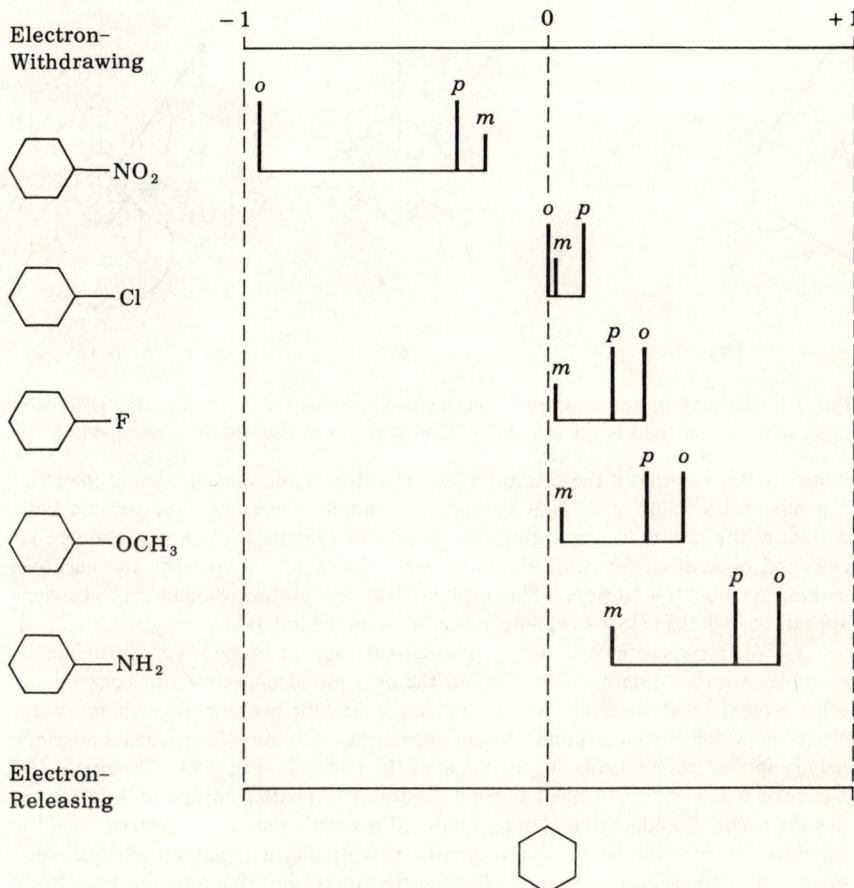

FIG. 5.2. Proton shifts in substituted benzenes, compared with normal benzene.

$H\chi_\parallel (3\cos^2\theta - 1)/R^3$ in the $z$ direction. This means a contribution of $\chi_\parallel (1 - 3\cos^2\theta)/R^3$ to $\sigma_{zz}$. Taking the field along the three axes in turn we find

$$\sigma_{xx} = \frac{\chi_\perp (1 - 3\sin^2\theta)}{R^3}$$

$$\sigma_{yy} = \frac{\chi_\perp}{R^3} \qquad (15)$$

$$\sigma_{zz} = \frac{\chi_\parallel (1 - 3\cos^2\theta)}{R^3}$$

The averaged correction to the screening for a molecule in solution is therefore

$$\Delta\sigma = \frac{\chi_\parallel - \chi_\perp}{3R^3} (1 - 3\cos^2\theta) \qquad (16)$$

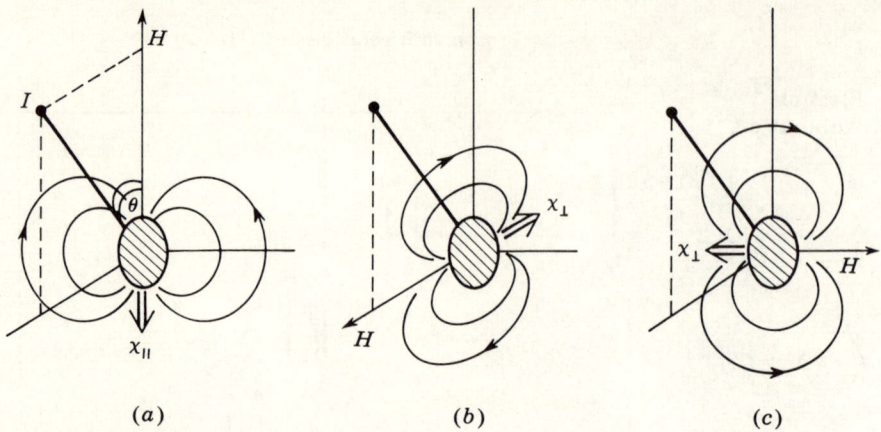

$(a)$                     $(b)$                     $(c)$

FIG. 5.3. Shielding by the anisotropic magnetic susceptibility of a neighboring group for three different directions of the external field ($\chi_\parallel$ and $\chi_\perp$ are assumed to be negative).

Notice that it vanishes if the susceptibility $\chi$ is isotropic and that the sign of the effect depends on the value of $\theta$. In acetylene, for example, there is a large paramagnetic current in the carbon atoms of the triple bond when the field is across the molecular axis, and none at all for fields along the axis. Hence $\chi_\parallel - \chi_\perp$ is large and negative, making $\Delta\sigma$ about $+10$ ppm. This explains why the proton resonance of acetylene appears at high field close to ethane, rather than on the low side of ethylene (Fig. 5.1).

4. Electronic currents flowing round closed rings of atoms. The most famous examples are the "ring currents" due to the delocalized $\pi$-electrons in benzene and other aromatic hydrocarbons. On a very simple view the benzene ring contains 6 free electrons, which circulate round with an angular velocity of $eH/2mc$ when a magnetic field is applied perpendicular to the plane of the molecule (Fig. 5.4). The ring is like a circular wire of radius $R$ (the CC bond length 1.40 Å) with a current of $3e^2H/2\pi mc^2$ flowing round it, and so it acquires an induced magnetic moment of $3e^2HR^2/2mc^2$ in the direction opposite to $H$. The magnetic susceptibility is highly anisotropic, since there is no free electron current if $H$ lies in the ring plane. We now use Eq. (16) to calculate the screening constant. The result is

$$\Delta\sigma = \frac{e^2}{2mc^2}\frac{R^2}{(R+d)^3} = -1.75 \times 10^{-6} \tag{17}$$

for the ring protons, where $d$ is the C—H bond length (1.09 Å). The ring currents probably account for most of the shift of $-1.95$ ppm in going from ethylene to benzene. Protons occupying a position above or below the center of the ring would be expected to experience a local field opposing the applied field, and have larger screening constants. This prediction is borne out by studies of 1,4-polymethylene benzenes like

$$CH_2{-}(CH_2)_6{-}CH_2$$

where some of the methylene protons do indeed show a shift to high field.

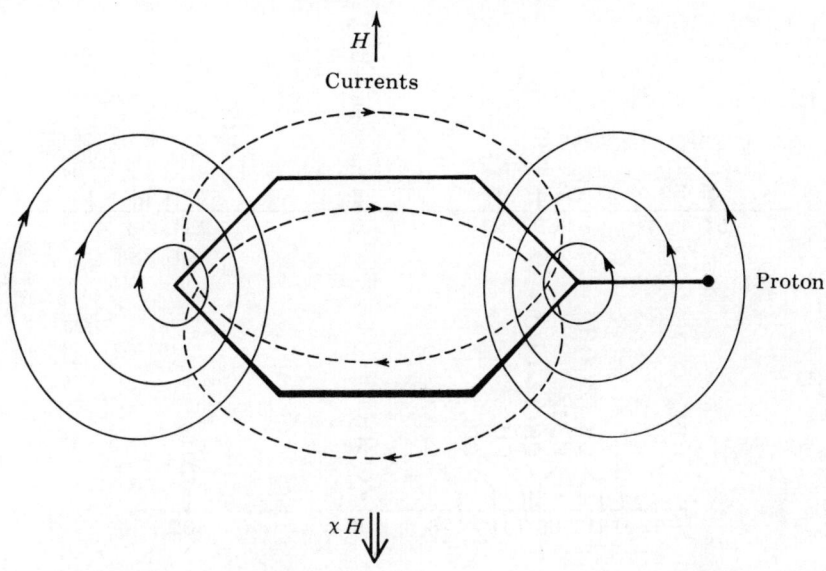

FIG. 5.4. Free electron ring currents in benzene.

## 5·4 SHIFTS FROM OTHER NUCLEI

### 5.4.1 Fluorine $F^{19}$

Typical fluorine shifts are shown in Fig. 5.5. They cover a large range of about 1,000 ppm and it is notable that the nuclei of $F_2$ and $UF_6$ have a negative screening relative to the bare nucleus, due to a very large paramagnetic term. At the other extreme, fluoride ion, $F^-$, has a purely diamagnetic shift, which amounts to $+338.1$ ppm. In general, fluorine chemical shielding is dominated by paramagnetic shifts, which vary over a wide range of values.

The theory of the paramagnetic shifts is complicated, but we shall now illustrate the principles involved by doing a very simple model calculation for the $F_2$ molecule. It is convenient to take the origin of coordinates at one nucleus, $A$, and choose the F—F bond as the $x$ axis. Each fluorine atom has filled $2s$, $2p_y$ and $2p_z$ orbitals, but the two $2p_x$ orbitals, which we denote by $x_a$ and $-x_b$, form bonding and antibonding molecular orbitals

$$\psi_g = \frac{1}{\sqrt{2}}(x_a + x_b)$$

$$\psi_u = \frac{1}{\sqrt{2}}(x_a - x_b)$$

(18)

$\psi_g$ contains two electrons in the ground state, while $\psi_u$ is empty. When the magnetic field is along the bond there is no paramagnetic contribution to $\sigma_{xx}$, because the molecular ground state is an eigenstate of $L_x$, with $L_x = 0$. However according to Eqs. (6) and (12) a field across the axis couples the ground state with excited states for which

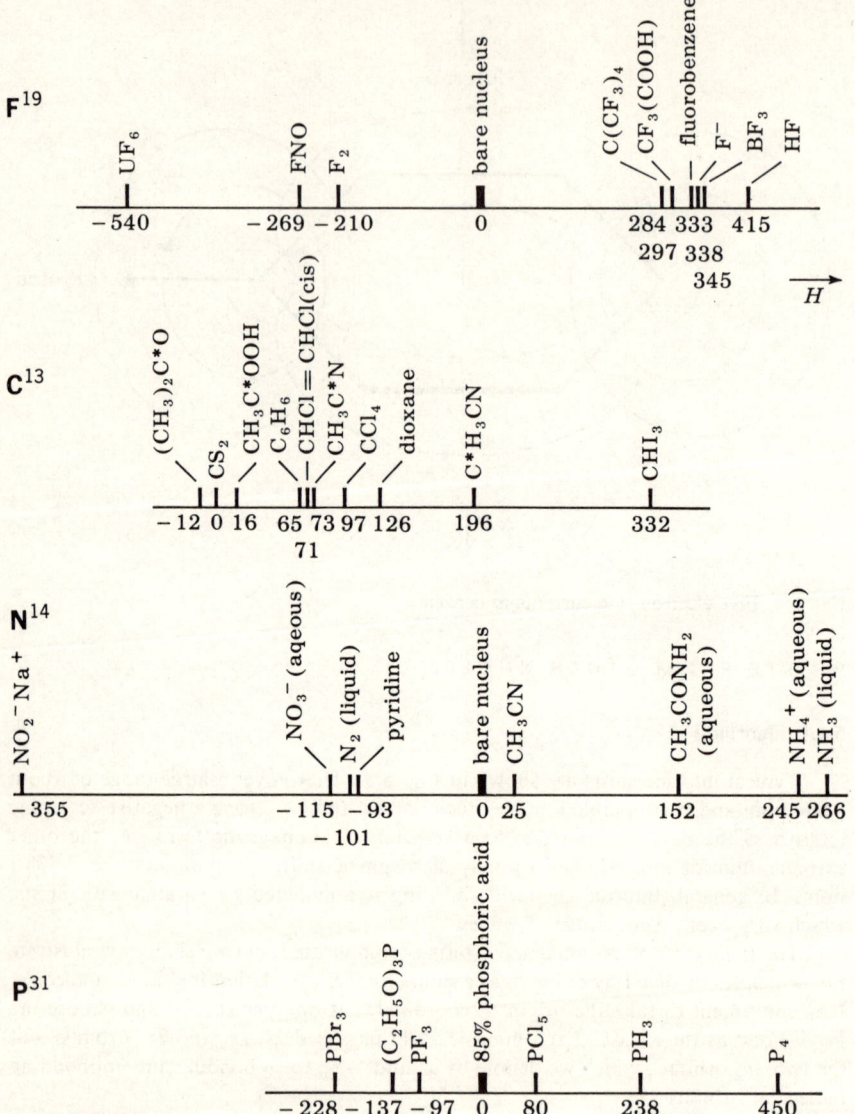

FIG. 5.5. Chemical shifts of $F^{19}$, $C^{13}$, $N^{14}$, and $P^{31}$ relative to the bare nucleus.

there is a large matrix element $\langle 0|L_z|n\rangle$. The most important of these for nucleus $A$ is the promotion of a $2p_y$ electron into the antibonding orbital $\psi_u$, and the matrix element for the process is

$$\langle y_a|L_z|\psi_u\rangle = \frac{1}{\sqrt{2}}\{\langle y_a|L_z|x_a\rangle - \langle y_a|L_z|x_b\rangle\} \tag{19}$$

$L_z$ being the angular momentum about nucleus $A$. We may neglect the second term in (19) which is considerably smaller than the first, and use the fact that the complex orbitals $(p_x \pm ip_y)$ are proper eigenstates of $L_z$ with eigenvalues of $\pm 1$.

$$L_z|x_a + iy_a\rangle = |x_a + iy_a\rangle$$
$$L_z|x_a - iy_a\rangle = -|x_a - iy_a\rangle$$
(20)

Thus $L_z|x_a\rangle$ is equal to $i|y_a\rangle$ and the matrix element is $i/\sqrt{2}$. According to Ramsey's formula (12) the paramagnetic shift now reduces to

$$(\sigma_{zz})_p = -\left(\frac{e\hbar}{2mc}\right)^2 \left\langle \frac{1}{r^3} \right\rangle \frac{2}{\Delta E}$$
(21)

where $\langle 1/r^3 \rangle$ is the average of $1/r^3$ for a fluorine $2p$ electron and $\Delta E$ is the excitation energy. Equation (21) must finally be multiplied by a factor of two because the excited electron may have either $\alpha$ or $\beta$ spin. $\sigma_{yy}$ has the same value as $\sigma_{zz}$, and the average isotropic shift is approximately

$$\sigma_p = -\frac{2}{3} \frac{e^2\hbar^2}{m^2c^2} \left\langle \frac{1}{r^3} \right\rangle \frac{1}{\Delta E}$$
(22)

A very rough numerical estimate is that $\sigma_p = -2000$ ppm, for the $F_2$ molecule. This is much larger than the experimental estimate that $\sigma_p = -800$, but part of the discrepancy is due to the partial ionic character of the bond, which will reduce $\sigma_p$ considerably from its calculated value.

Large paramagnetic shifts also occur in the $Co^{59}$ resonances of many $d^6$ cobalt complexes, and they are found to vary linearly with the inverse excitation energy $1/\Delta E$ of the $d$ electrons.

### 5.4.2 Carbon $C^{13}$

The n.m.r. spectra from $C^{13}$ in its natural abundance of $1.1\%$ are by no means easy to detect, but even so carbon chemical shifts have been studied in considerable detail. The shifts cover a range of more than 350 ppm and are dominated by paramagnetic effects. One particularly interesting group of compounds is the substituted benzenes. Here there are characteristic $C^{13}$ shifts at the ortho, meta, and para positions which are closely correlated with the electron-withdrawing power of the substituent. The reason for this is that the paramagnetic shift is proportional to the $\pi$-electron charge densities in the carbon $2p_z$ orbitals and in the neighboring bonds. Another example of the same effect is the series of cyclic aromatic rings $C_7H_7^+$, $C_6H_6$, $C_5H_5^-$, and $C_8H_8^=$ whose $C^{13}$ shifts vary linearly with the $\pi$-electron charge, having the values $-27.6$, $0$, $+25.7$, and $+42.5$ relative to benzene.

### 5.4.3 Nitrogen $N^{14}$ and $N^{15}$

The quadrupole moment of $N^{14}$ couples the nuclear spin to the molecular rotational motion, and most $N^{14}$ spectra are strongly broadened, especially in amides and nitriles, so that no spin-spin splittings are visible. The broadening also afflicts other nuclei, like the protons in $NH_3$ or the $C^{13}$ in $CH_3C^*N$ which have spin-spin couplings with nitrogen. Unfortunately the $N^{15}$ isotope, with spin 1/2, is too rare to observe in natural abundance, but recently a few specially enriched compounds have been studied.

⌐

**5.4.4 Other Nuclei**

The most important remaining nuclei are $Si^{29}$ and $P^{31}$, both with spin 1/2. Their chemical shifts involve no new principles but have a different range of applications. Silicon shifts are fairly small and often show effects from the bonding of electrons in $d$ orbitals. Phosphorus shifts and spin couplings have proved extremely useful in determining structures for organic phosphorus compounds as well as for the many kinds of diphosphate and polyphosphate ions.

**5.4.5 Solvent Effects**

Although one would usually like to know the chemical shifts of an isolated molecule, this ideal is hardly ever attainable in practice. Measurements of n.m.r. spectra in gases are difficult and require conditions which are unsuitable for most chemical studies, while in liquids the solvent affects the chemical shift in many different ways, which are difficult to allow for. Apart from the bulk susceptibility correction, which we have already discussed, there are specific physical and chemical interactions between neighboring molecules. These will depend on the concentration of solvent. For example the proton chemical shift of pure benzene is 0.70 ppm higher than for benzene at infinite dilution in carbon tetrachloride, even after correcting for the susceptibility. In this case ring currents in other benzene molecules produce local fields which shift the proton resonance to high field. Hydrogen bonding, molecular association, electric fields from polar molecules, and Van der Waals forces all alter the chemical shifts and so it is best to avoid as many of these effects as possible by choosing a very inert solvent. As a further precaution the observed shifts should be extrapolated to an infinitely dilute solution. This, of course, does not remove the solvent effects, but it does make them reproducible.

# 5·5  THE ORIGIN OF NUCLEAR SPIN-SPIN COUPLINGS

We now address ourselves to the problem of how the spins of two nuclei are coupled via the intervening electrons. The coupling is a two-stage process, in which the first nucleus perturbs the valence electrons and the electrons in turn produce a small magnetic field at the other nucleus. The spin of the first nucleus, $A$, interacts with the electrons in three ways.

1. Electronic currents are set up by the action of the magnetic field of the nuclear dipole on the *orbital* magnetic moments of the electrons. The corresponding term in the Hamiltonian is

$$g_N \beta_N \beta \frac{2\mathbf{L} \cdot \mathbf{I}}{r^3} \qquad (23)$$

as may be seen from Eq. (10). The induced currents then produce a field at nucleus $B$.

2. The dipolar coupling [Chapter 2, Eq. (8)] between the electron *spin* and the nucleus polarizes the electron spins in the parts of the molecule close to nucleus $A$. The magnetic field of the electron spins acts directly on the second nucleus.

3. Most important of all, the contact coupling $a\mathbf{I} \cdot \mathbf{S}$ tends to align the spins of the electrons at $A$ antiparallel to the nuclear spin. As the directions of the electron spins in different parts of the molecule are coupled via the chemical bonds, the electron

spin polarization at $A$ is accompanied by a slight excess of $\alpha$ or $\beta$ electrons near the other nucleus $B$, which orients its spin antiparallel or parallel to that of $A$ (Fig. 5.6).

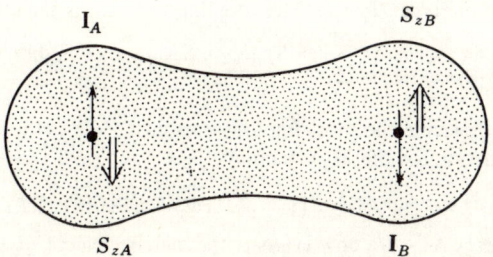

FIG. 5.6. Contact spin-spin coupling in the hydrogen molecule.

Let us now consider the contact effect in the hydrogen molecule and see how the proton spins are coupled through the two electrons in the $\sigma$ bond. For convenience we suppose that the spin of nucleus $A$ is definitely quantized along the $z$ axis, so that the first-order contact interaction energy with the two electrons can be written

$$\mathscr{H}_A = \frac{8\pi}{3} g\beta g_N \beta_N I_{zA} S_{zA} \tag{24a}$$

where

$$S_{zA} = S_{z1}\delta(\mathbf{r}_1 - \mathbf{r}_A) + S_{z2}\delta(\mathbf{r}_2 - \mathbf{r}_A) \tag{24b}$$

is the "spin angular momentum density" of the two electrons at the first nucleus. The $\delta$ functions impose the condition that the electrons must be at $\mathbf{r}_A$. Normally the two electron spins of the bond are paired, and if we denote the two hydrogen $1s$ orbitals around nuclei $A$ and $B$ by $\phi_a$ and $\phi_b$, respectively, the bonding and antibonding molecular orbitals are

$$\psi_g = \frac{1}{\sqrt{2}}(\phi_a + \phi_b)$$
$$\psi_u = \frac{1}{\sqrt{2}}(\phi_a - \phi_b) \tag{25}$$

The complete singlet electronic wave function for the ground state is

$$\Psi_0 = \frac{1}{\sqrt{2}}\psi_g(\mathbf{r}_1)\psi_g(\mathbf{r}_2)[\alpha_1\beta_2 - \beta_1\alpha_2] \tag{26}$$

Owing to the perturbation $\mathscr{H}_A$ the singlet wave function gets mixed with excited states, and takes the form

$$\Psi = \Psi_0 - \frac{8\pi}{3} g\beta g_N \beta_N I_{zA} \sum_n \frac{\langle n|S_{zA}|0\rangle}{E_n - E_0} \Psi_n \tag{27}$$

Thus the electronic wave function now depends on the direction of the nuclear spin $I_{zA}$. If we now go to nucleus $B$ and evaluate the local electron spin polarization

$$S_{zB} = S_{z1}\delta(\mathbf{r}_1 - \mathbf{r}_B) + S_{z2}\delta(\mathbf{r}_2 - \mathbf{r}_B) \tag{28}$$

in the state $\Psi$ we find that it differs from zero by a small amount proportional to $I_{zA}$. Each excited state gives a first order contribution proportional to $\langle 0|S_{zB}|n \rangle$, and the interaction of this electron spin polarization with the second nuclear spin gives an energy of the type $J I_{zA} I_{zB}$. The coupling constant is calculated to be

$$J = -\left(\frac{8\pi}{3} g\beta g_N \beta_N\right)^2 \sum_n \frac{\langle 0|S_{zA}|n\rangle\langle n|S_{zB}|0\rangle + \langle 0|S_{zB}|n\rangle\langle n|S_{zA}|0\rangle}{E_n - E_0} \tag{29}$$

In the hydrogen molecule the most important excited state is the lowest triplet $\Psi_1$, with the wave function for the $M_S = 0$ component

$$\Psi_1 = \tfrac{1}{2}[\psi_g(\mathbf{r}_1)\psi_u(\mathbf{r}_2) - \psi_u(\mathbf{r}_1)\psi_g(\mathbf{r}_2)][\alpha_1\beta_2 + \alpha_2\beta_1] \tag{30}$$

and excitation energy $\Delta E$. We now evaluate the matrix element $\langle 1|S_{zA}|0\rangle$. Operating on the spin part of $\Psi_0$ with $S_{z1}$ and $S_{z2}$ gives

$$S_{z1}[\alpha_1\beta_2 - \beta_1\alpha_2] = \tfrac{1}{2}[\alpha_1\beta_2 + \beta_1\alpha_2]$$
$$S_{z2}[\alpha_1\beta_2 - \beta_1\alpha_2] = -\tfrac{1}{2}[\alpha_1\beta_2 + \beta_1\alpha_2] \tag{31}$$

so that after integrating over the spin coordinates of both electrons we find

$$\langle 1|S_{zA}|0\rangle = \frac{1}{2\sqrt{2}} \langle \psi_g(1)\psi_u(2) - \psi_u(1)\psi_g(2)|\delta(\mathbf{r}_1 - \mathbf{r}_A) - \delta(\mathbf{r}_2 - \mathbf{r}_A)|\psi_g(1)\psi_g(2)\rangle$$
$$= -\frac{1}{\sqrt{2}} \psi_u(\mathbf{r}_A)\psi_g(\mathbf{r}_A) = -\frac{1}{2\sqrt{2}} \{\phi_a^2(\mathbf{r}_A) - \phi_b^2(\mathbf{r}_A)\} \tag{32}$$

The dominant term in (32) is $\phi_a^2(\mathbf{r}_A)$, and after a similar calculation of $\langle 0|S_{zB}|1\rangle$ the coupling constant is predicted to be

$$J = \left(\frac{8\pi}{3} g\beta g_N \beta_N\right)^2 \frac{\phi_a^2(\mathbf{r}_A)\phi_b^2(\mathbf{r}_B)}{4\Delta E} \approx 200 \text{ cps} \tag{33}$$

This agrees well with the experimental value of 280 cps, which cannot be measured directly, but may be deduced from the coupling of 43.5 cps in HD.

The orbital and dipolar parts of the coupling are much smaller for $H_2$, being, respectively, about 3 and 20 cps.

## 5·6   PROTON SPIN-SPIN COUPLINGS

The coupling constant of hydrogen is exceptionally large, since it is the only example of two directly bonded protons. In other molecules the coupling has to be transmitted successively through several bonds and in organic compounds the value of $J$ often lies between 1 and 10 cps. The coupling occurs predominantly through the $\sigma$ electrons and tends to decrease rapidly as the number of bonds between the nuclei increases. For example, analysis of the n.m.r. spectrum of liquid pyridine and deuterated pyridines gave the results below (in cps):

|  | ortho | meta | para |
|---|---|---|---|
|  | $J_{23} = 5.5$ | $J_{24} = 1.9$ | $J_{25} = 0.9$ |
|  | $J_{34} = 7.5$ | $J_{26} = 0.4$ |  |
|  |  | $J_{35} = 1.6$ |  |

However there are some remarkable exceptions to this rule. In the substituted ethylenes couplings between trans protons are much larger than either the cis or geminal couplings. The values for $C^{13}$—$C^{12}$ ethylene are typical:

$$
\begin{array}{ccc}
& gem & cis & trans \\
& J_{12} = +2.5 & J_{23} = +11.6 & J_{13} = +19.1
\end{array}
$$

As we have seen in Chapter 4, the absolute sign of the spin-spin coupling cannot be determined from the n.m.r. spectrum; but it is possible to find relative signs, and the geminal couplings are negative in several of the substituted ethylenes.

A rather different kind of long-range coupling carried by $\pi$-electrons occurs between protons in chloroallene

$$J_{12} = 6.1 \text{ cps}$$

but this is exceptional. The longest coupling yet observed runs through eight consecutive bonds in $CH_3$—$(C{\equiv}C)_3$—H, with $J = 0.65$!

The ways in which the perturbations produced by a nuclear spin are transmitted to another nucleus through the electron cloud of a molecule vary widely from one compound to another, and the sign of the coupling constant need not necessarily be positive. According to the conventional model of an electron-pair bond the electrons at opposite ends of a bond are coupled together, with antiparallel spins. Thus electron spins which are two bonds apart prefer to be parallel; the spin directions in a long chain of bonds should be alternately up and down. This picture leads to the conclusion that the positive or negative sign of the coupling constant between two protons should simply depend on whether they are an odd or even number of bonds apart. In fact nature is less simple and the couplings behave in a rather subtle fashion. The most successful interpretation is in terms of molecular orbital theory. All the electrons in a molecule are supposed to move independently in a set of orbitals $\psi_1 \cdots \psi_n$, with two electrons of opposite spin in each orbital. The molecule can then be excited to a higher state by promoting an electron from a filled orbital $\psi_i$ to an empty one $\psi_j$. By an extension of the reasoning we used in Section 5.5 one can show that the coupling constant $J$ (in energy units) is given by the formula

$$J = -\left(\frac{8\pi}{3} g\beta g_N \beta_N\right)^2 \sum_{ij} \frac{\psi_i \psi_j(\mathbf{r}_A)\psi_j \psi_i(\mathbf{r}_B)}{\varepsilon_j - \varepsilon_i} \tag{34}$$

Here $\varepsilon_i$ and $\varepsilon_j$ are the energies of the orbitals. $J$ is the sum of contributions from all the excited states, which will in general be of either sign. The contributions are weighted with an energy factor $1/(\varepsilon_j - \varepsilon_i)$, so that the low-lying states are usually the most important. With the aid of Eq. (34) it is possible to account for many otherwise inexplicable trends in the observed couplings.

Nuclear magnetic resonance is a powerful tool for elucidating fine details in the structure of organic molecules, because the proton couplings in saturated molecules are quite sensitive to changes of conformation, axial-axial (trans) couplings being much larger than equatorial-equatorial (gauche) ones. Dioxane,

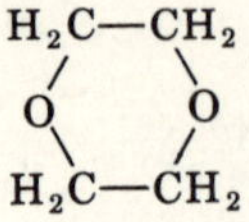

$$H_2C—CH_2$$
$$O \qquad O$$
$$H_2C—CH_2$$

which is stable in the "chair" form is found to have trans and gauche couplings of 9.4 and 2.7 cps and similar results are found for the substituted ethanes. Here the results are complicated by rotation about the C—C bond. Thus liquid $Cl_2HC—CHCl_2$ may exist in one trans or two gauche configurations.

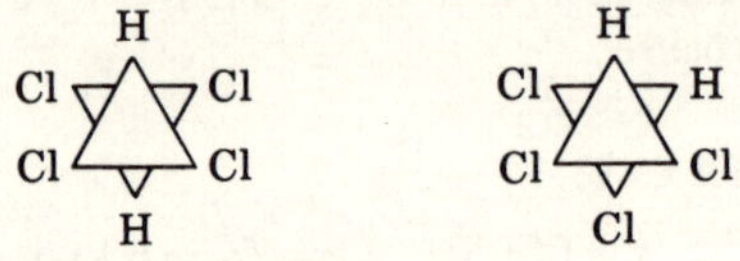

trans: $J = 15.3$      gauche: $J = 3.4$

The gauche form has a slightly lower energy, and the observed proton-proton coupling in the n.m.r. spectrum is a weighted statistical average of the two constants, varying with temperature.

The geminal coupling between two H—C—H protons also varies strongly with angle, and shows characteristic changes depending on the electronegativity of the neighboring atoms.

Spin-spin couplings involving other nuclei, such as $C^{13}$—H, $F^{19}$—$F^{19}$, are much larger, and may be several hundred cycles. They may also be used in the same way, for studying molecular conformations. $C^{13}$—H couplings are not intrinsically different. They are only large if the atoms are directly bonded, and are usually measured by looking at the proton rather than the $C^{13}$ resonance.

## 5·7 MOLECULAR STRUCTURE STUDIES BY N.M.R.

One of the earliest and most interesting chemical applications of n.m.r. was to help decide the structures of the many hydrides of boron. A notable example is diborane, $B_2H_6$. At one time two alternative structures were considered: the bridge structure (a) with two "three-center" bonds and the ethane structure (b)

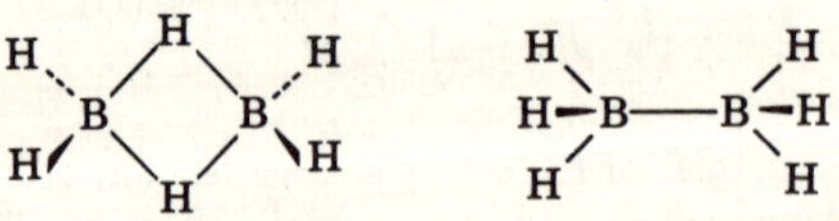

(a)            (b)

Both the protons and the two boron isotopes $B^{11}$ (spin 3/2) and $B^{10}$ (spin 3) give resonance signals. The proton spectrum from structure (b) would show a single chemical shift with a symmetrical spin-spin splitting from the B—H couplings, but the observed spectrum shows two kinds of hydrogen and definitely rules out this possibility. It shows that the bridge structure is correct. Moreover, the two bridge hydrogens are more shielded than the four terminal ones, since the three-center bond gives a larger electron density in the middle of the bridge. The boron resonances show a large $BH_2$ coupling of 128 cps and a much smaller splitting from the bridge H atoms, but no $B^{10}$—$B^{11}$ splitting. This again confirms the bridge structure. N.M.R. has also given important clues to the structures of other hydrides, including the remarkable polyhedral $B_{10}H_{10}^=$ ion. We should add that the interpretation of these spectra is far from simple, and often requires special double resonance techniques (see Chapter 13), as well as a sound understanding of the origin of chemical shifts and spin couplings.

Another application is to the study of spatial conformations of flexible molecules (e.g., boat and chair forms of cyclohexane). Frequently the interchange between different forms takes place so fast that the n.m.r. spectrum reflects only the time-averaged shifts and couplings, rather than two distinct superposed sets of lines. However, analysis of the spectrum at different temperatures reveals systematic changes in the spectrum which reflect a change in the statistical weights of the different forms. A phenyl substituted cyclobutane ring, for example, may exist in the two bent forms

(X)                                    (Y)

with the phenyl group occupying either the axial or equatorial position, (X) or (Y). If two fluorine atoms are now attached opposite the substituent they will have slightly different chemical shifts $\delta_1$ and $\delta_2$ and yield an $AB$ n.m.r. spectrum. The relative shift $\delta = (\delta_1 - \delta_2)$ is different for the two forms, and if the fractions of molecules of each type are $x$ and $y$ the observed shift is

$$\delta = \delta_X - y(\delta_X - \delta_Y) \tag{35}$$

Careful studies over a wide temperature range show that at 85°C the equatorial form is favored by 79% of the molecules, and is more stable by about 950 kcal/mole. The chemical shifts of the two forms (at 60 mc/s) are, respectively, 40 and 1175 cps. The H—F coupling constants also vary in a characteristic way as the bonds twist.

## PROBLEMS

**1.** A boron atom with the electron configuration $(1s)^2(2s)^2 2p$ is placed in an asymmetric environment which causes the $p_x$, $p_y$, and $p_z$ atomic orbitals to have different energies, which are respectively $-\Delta E$, $+\Delta E$, and 0. Initially the outer electron is in the $2p_x$ orbital. A uniform field $H$ is now applied along the $z$ axis. Use Eq. (6) to show that the $p_x$ orbital changes to $p_x + i\lambda p_y + i\mu p_z$ and calculate the values of $\lambda$ and $\mu$.

**2.** Continuing Problem 1, calculate the paramagnetic and diamagnetic current densities for the $p$ electron. Show that they produce antishielding and shielding secondary magnetic fields at the nucleus.

3. Calculate the change of the chemical shift of the H atom in a uniform electric field of 10,000 volts/cm.

4. Estimate the ring current chemical shifts in naphthalene, assuming that each ring contains six free electrons, and there is no interference between the currents in the two rings.

5. Theoretical arguments suggest that the cobalt 59 chemical shift in octahedral $Co^{3+}$ complexes should obey a formula of the type

$$\sigma = D - B/\Delta E$$

where $D$ is a constant, $\Delta E$ is the excitation energy of the lowest electronic excited state, and $B$ is proportional to the average of $1/r^3$ for a cobalt $d$ electron.

$$B = 8(e\hbar/2mc)^2 \langle 1/r^3 \rangle$$

The n.m.r. frequency $\nu$ in a fixed field of 4370.9 gauss and the long wavelength optical electronic absorption peak were measured for several complexes with the following results:

| Compound | Resonance Frequency (M/cs) | Absorption Wavelength $\lambda$ (in Ångstroms) |
|---|---|---|
| Potassium hexacyanocobaltate (III) | 4.4171 | 3110 |
| Tris(ethylenediamine) cobalt (III) chloride | 4.4488 | 4700 |
| Hexammine–cobalt (III) chloride | 4.4534 | 4750 |
| Cobalt (III) trisacetylacetonate | 4.4731 | 5970 |
| Tricarbonate cobalt (III) nitrate | 4.4795 | 6450 |

Analyze them graphically and estimate the size of the cobalt $d$ orbital.

6. The spin-spin coupling between the protons in $CHCl_2$—$CHCl_2$ varies with temperature in the following way:

| Temperature (°K) | $J$ (cps) |
|---|---|
| 238 | 2.67 |
| 264 | 2.90 |
| 300 | 3.06 |
| 361 | 3.40 |
| 414 | 3.67 |

On the assumption that $J$ is a thermal average of the gauche and trans coupling constants $J_g$ and $J_t$ prove the theoretical relation

$$2(J - J_g) = (J_t - J)e^{-\Delta E/kT}$$

where $\Delta E$ is the energy difference between the two isomers. Given that $J_t = 16.35$ cps and $J_g = 2.01$ cps plot a suitable graph and deduce the value of $\Delta E$.

## SUGGESTIONS FOR FURTHER READING

Pople, Schneider, and Bernstein: Chapters 7, 8, and 11. A comprehensive account, complete up till 1959.

Slichter: Sections 4.1–4.7. Theory of chemical shifts.

Nachod and Phillips: *Determination of Organic Structures by Physical Methods*, Vol. 2 (New York: Academic Press Inc., 1962). Chapters 6 and 7. Proton and $F^{19}$ magnetic resonance spectra of organic molecules. $C^{13}$ resonance spectra.

Richards: *Advances in Spectroscopy*, **2**: 121 (1961). A good brief survey of the whole n.m.r. field.

McConnell: *J. Chem. Phys.*, **27**: 226 (1957). Theory of chemical shifts.

Pople: *Proc. Roy. Soc. (London)*, **A239**: 541, 550, (1957). Paramagnetic chemical shifts in acetylene and other molecules.

Baker, Anderson, and Ramsey: *Phys. Rev.*, **133A**: 1533 (1964). Chemical shifts related to results of molecular beam experiments.

Spiesecke and Schneider: *J. Chem. Phys.*, **35**: 731 (1961). $C^{13}$ and proton shifts in substituted benzenes.

Ramsey: *Phys. Rev.*, **91**: 303 (1953). Nuclear spin-spin couplings. General theory, with application to HD and $H_2$.

Pople and Santry: *Mol. Phys.*, **8**: 1 (1964). Molecular orbital theory of spin-spin couplings.

Bothner-By and Pople: *J. Chem. Phys.*, **42**: 1339 (1965). Theory of proton spin-spin couplings in $CH_2$ groups.

Lynden-Bell and Sheppard: *Proc. Roy. Soc.* (London), **A269**: 385 (1962). Proton and $C^{13}$ couplings in ethylene.

Sheppard and Turner: *Proc. Roy. Soc.* (London), **A252**: 506 (1959). Hindered rotation and couplings in substituted ethanes.

Lipscomb: *Boron Hydrides*, (New York: W. A. Benjamin, Inc., 1963). Chapter 4. Structural studies by n.m.r.

Shoolery: *Disc. Faraday Soc.*, **19**: 215 (1955). Boron hydrides.

# CHAPTER 6

# E.S.R. SPECTRA
# OF ORGANIC RADICALS
# IN SOLUTION

## 6·1 THE SPIN HAMILTONIAN: HYPERFINE SPLITTINGS

The analysis of liquid phase e.s.r. spectra is, in principle, quite straightforward. One is interested in molecules possessing one unpaired electron which interacts with magnetic nuclei, particularly protons. In Chapter 2 we saw that the electron-nuclear hyperfine interaction is, in general, represented by the term

$$\mathbf{S \cdot T \cdot I} \tag{1}$$

where $\mathbf{T}$ is the hyperfine tensor. This tensor may be decomposed into the sum of two terms

$$a\mathbf{S \cdot I} + \mathbf{S \cdot T' \cdot I} \tag{2}$$

where $a$ is the isotropic contact part, and $\mathbf{T'}$ is the magnetic dipolar tensor (Section 2.2.3). In the solid state it gives rise to anisotropic splitting which we shall investigate in Chapter 7; in the liquid phase, however, the average value of $\mathbf{T'}$ is zero and the observed hyperfine splitting gives the isotropic coupling constant $a$ directly.

Let us first consider a free radical with just two protons, the isotropic couplings being $a_1$ and $a_2$. The spin Hamiltonian for this system is

$$\mathcal{H} = g\beta H S_z - g_N \beta_N H (I_{1z} + I_{2z}) + a_1 \mathbf{S \cdot I_1} + a_2 \mathbf{S \cdot I_2} \tag{3}$$

We can omit the terms describing the nuclear Zeeman interactions because, as we saw in Chapter 2, they do not affect the positions of the absorption lines in a liquid phase spectrum. Most free radical spectra can be satisfactorily interpreted by first-order perturbation theory, so we shall adopt the simpler Hamiltonian

$$\mathcal{H} = g\beta H S_z + S_z(a_1 I_{1z} + a_2 I_{2z}) \tag{4}$$

If our basis spin functions are chosen by multiplying $\alpha_e$ or $\beta_e$ for the electron with each of the four nuclear states $\alpha_1\alpha_2$, $\alpha_1\beta_2$, $\beta_1\alpha_2$, and $\beta_1\beta_2$, we see that they are actually the correct eigenfunctions of $\mathcal{H}$, and their energies are proportional to the three quantum numbers $m_S, m_1, m_2$:

$$E = m_S[g\beta H + a_1 m_1 + a_2 m_2] \tag{5}$$

These energy levels are shown in Fig. 6.1. The allowed e.s.r. transitions are those in which $m_S$ alters by $\pm 1$ but the nuclear spins stay unchanged; and so the corresponding

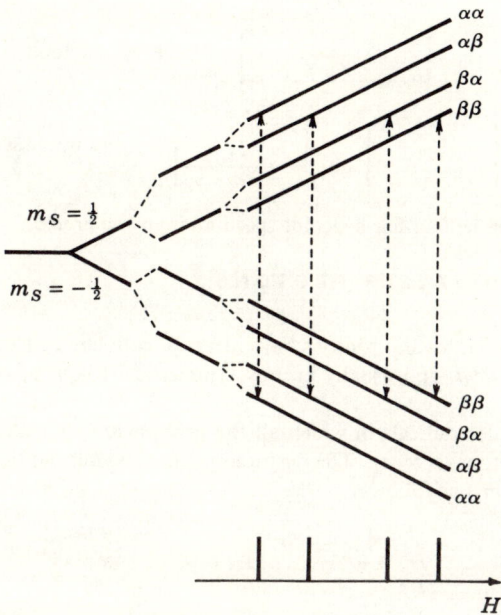

FIG. 6.1. Hyperfine energy levels and transitions for two protons ($a_1$ and $a_2$ are both assumed to be positive).

resonance frequency is

$$h\nu = g\beta H + a_1 m_1 + a_2 m_2 \tag{6}$$

Usually one works at a fixed frequency $\nu$ and varies $H$ through a small range on either side of the field $H^*$ where the isolated electron spin would be in resonance

$$h\nu = g\beta H^* \tag{7}$$

If we now convert the coupling constants $a_1$ and $a_2$ into gauss (1 gauss = 2.80 Mc/s) and substitute (7) into (6) resonance will be observed at the positions

$$H = H^* - a_1 m_1 - a_2 m_2 \tag{8}$$

$$= H^* - \sum_i a_i m_i \tag{9}$$

There are four transitions, and since each nuclear spin state is equally probable, four equally intense absorption lines are obtained. Each line may be identified by the values of the two quantum numbers $m_1$ and $m_2$.

A simple way of describing the two-proton spectrum is to say that proton 1 splits the original single line due to the electron spin into a doublet, each line being split into a further two lines because of the interaction with proton 2, as shown in Fig. 6.2.

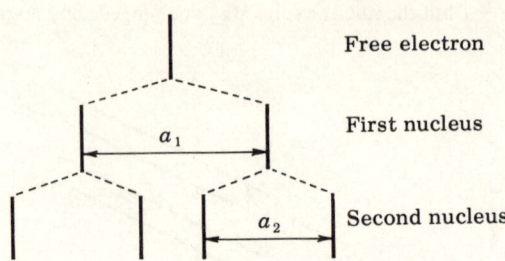

FIG. 6.2. Origin of the four-line hyperfine spectrum from two protons.

## 6·2 SETS OF EQUIVALENT PROTONS

It often happens that several protons have identical coupling constants. Such protons are termed *equivalent*, and usually occupy symmetrically equivalent positions in the molecule.

Let us consider radicals in which all the protons are equivalent and commence with just two protons, $a_1 = a_2$. The positions of the lines depend on the total resolved spin $M = (m_1 + m_2)$ as follows:

| Field | $(m_1 + m_2)$ | Spin states |
|---|---|---|
| $H^* - a$ | 1 | $\alpha\alpha$ |
| $H^*$ | 0 | $\alpha\beta, \beta\alpha$ |
| $H^* + a$ | $-1$ | $\beta\beta$ |

There are now only three lines, and their relative intensities are $1:2:1$ because of the degeneracy of the $M = 0$ level.

We have already discussed (Section 4.4.4) the nuclear spin states which arise from three equivalent protons, forming a quartet state and two doublets, with two different values of the total spin $\mathbf{F}^2$. However, the hyperfine energy depends only on the value of $F_z$ and not on $\mathbf{F}^2$, so we can equally well write the wave functions as simple products (Chapter 4, Eq. 20). We see that the $M = \pm 3/2$ states are single, but the $\pm 1/2$ components are each triply degenerate. The resulting hyperfine pattern is shown in Fig. 6.3. There are eight allowed transitions which give rise to four absorption lines with relative intensities $1:3:3:1$

It will now be apparent that interaction of the odd electron with $n$ equivalent protons results in $n + 1$ lines whose relative intensities are proportional to the coefficients of the binomial expansion of $(1 + x)^n$. The results for $n = 1$ to 8 are as follows:

| electron | | | | | | 1 | | | | | | |
|---|---|---|---|---|---|---|---|---|---|---|---|---|
| 1 proton | | | | | 1 | | 1 | | | | | |
| 2 | | | | | | 1 | 2 | 1 | | | | |
| 3 | | | | | 1 | 3 | | 3 | 1 | | | |
| 4 | | | | 1 | 4 | | 6 | | 4 | 1 | | |
| 5 | | | 1 | 5 | | 10 | | 10 | | 5 | 1 | |
| 6 | | 1 | 6 | | 15 | | 20 | | 15 | | 6 | 1 |
| 7 | 1 | 7 | | 21 | | 35 | | 35 | | 21 | 7 | 1 |
| 8 | 1 | 8 | 28 | | 56 | | 70 | | 56 | 28 | 8 | 1 |

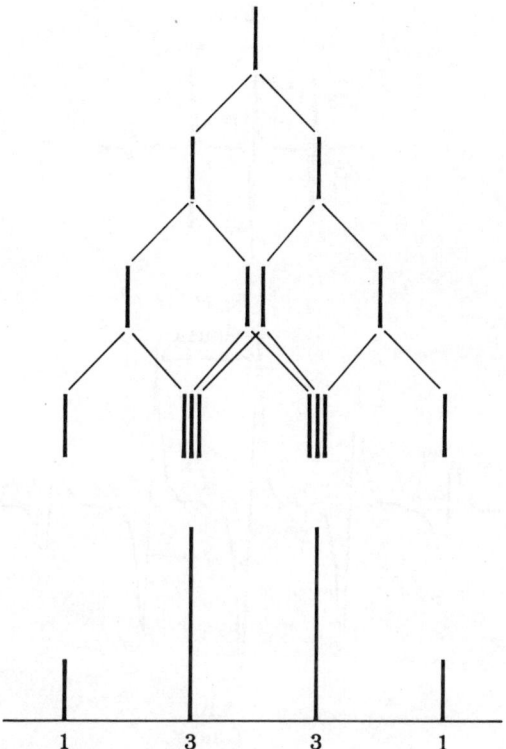

FIG. 6.3. Hyperfine pattern for three equivalent nuclei.

Figure 6.4 shows examples of spectra from radicals in which all protons are equivalent, namely, the p-benzosemiquinone anion with four protons, the benzene anion with six, and the cyclooctatetraene anion with eight.

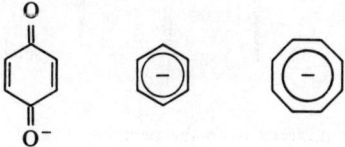

The separation between any two adjacent lines gives the value of the hyperfine coupling constant.

In more complex radicals there may be several distinct sets $A, B, C \cdots$ of equivalent protons and the spin Hamiltonian may be written

$$\mathscr{H} = g\beta HS_z + S_z(a_A F_{zA} + a_B F_{zB} + \cdots) \tag{10}$$

where $F_{zA}$ is the total $I_z$ value for the set $A$, and so on. Let us take the naphthalene

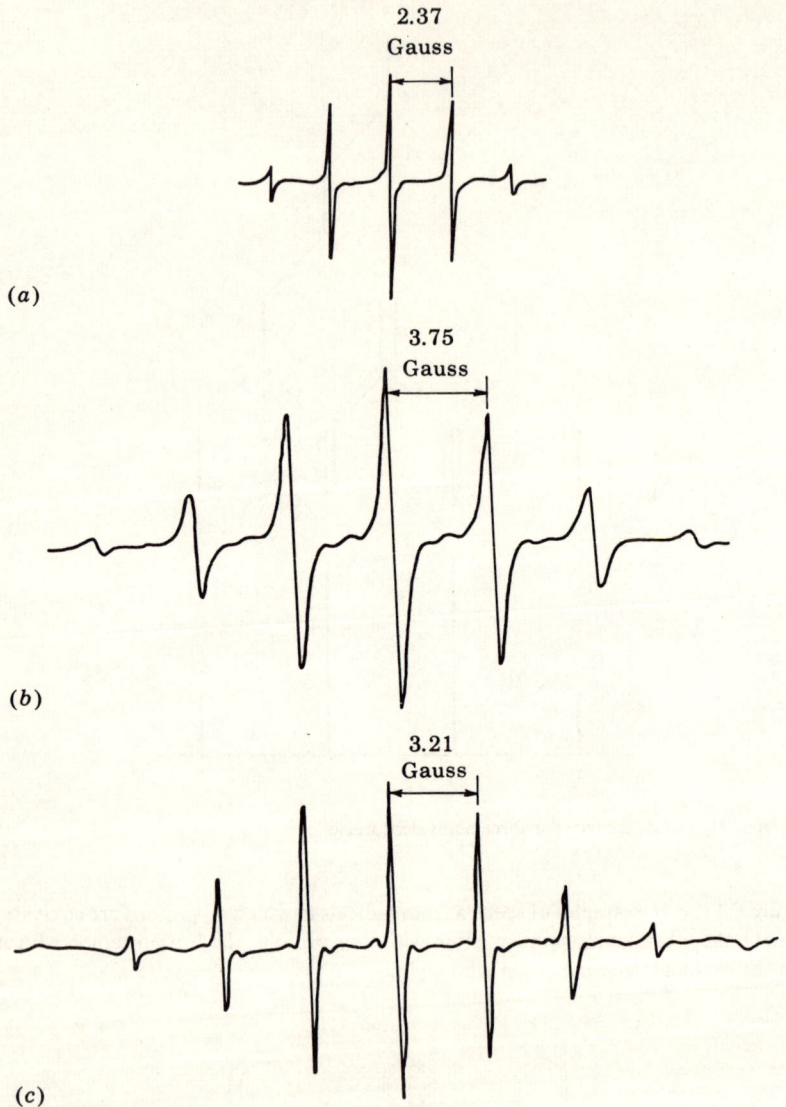

FIG. 6.4. Proton hyperfine patterns from the negative ions of (a) *p*-benzosemiquinone, (b) benzene, and (c) cyclooctatetraene.

negative ion, where the odd electron interacts with four *A*-type protons and four *B*-type protons.

Interaction with the four $A$ protons alone would give rise to a pattern of five lines with relative intensities $1:4:6:4:1$. However, each of these lines is split into a further quintet due to interaction with the $B$ protons. Hence the complete spectrum contains 25 lines and its analysis yields the two coupling constants $a_A$ and $a_B$. The analysis would be an almost trivial task if one coupling constant were very much greater than the other, for the spectrum would then consist of five separated quintets with relative intensities,

$$\begin{array}{ccccc}
1, & 4, & 6, & 4, & 1 \\
4, & 16, & 24, & 16, & 4 \\
6, & 24, & 36, & 24, & 6 \\
4, & 16, & 24, & 16, & 4 \\
1, & 4, & 6, & 4, & 1
\end{array}$$

However, the splittings are, in fact, of more comparable magnitude so that the quintets overlap. Nevertheless the analysis is straightforward and is illustrated in Fig. 6.5. The

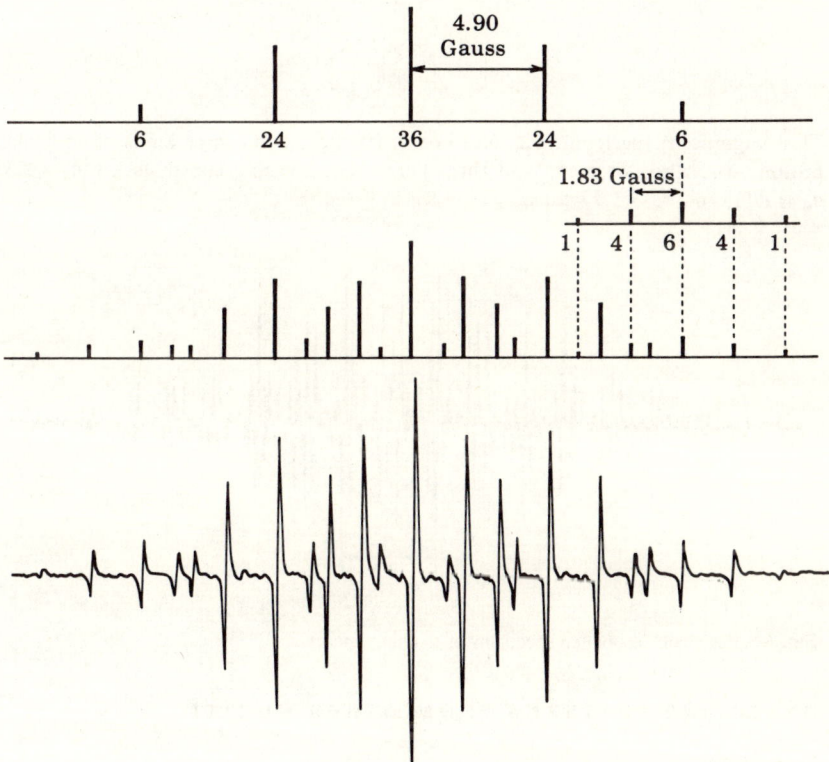

FIG. 6.5. Analysis of the hyperfine pattern for naphthalene negative ion.

two splitting constants are $a_A = 4.90$, $a_B = 1.83$ gauss. These splittings may, if desired, be converted into frequency units (Mc/s) simply by multiplication by 2.80. Note that this is *only* true for $g = 2$.

The analysis of liquid phase free radical spectra thus reduces to the problem of relating regularities in the hyperfine pattern to regularities in the spatial disposition of the nuclei. Spectra which are completely resolved are easy to analyze; however, the number of lines increases rapidly with increasing number of protons so that different components become superimposed, leading to distorted line shapes. Line broadening effects may also add to the confusion. Thus although the analysis is, in principle, very straightforward, it is not difficult to make mistakes. The hyperfine structure is independent of the signs of the coupling constants, and in solution signs can only be determined indirectly from subtle second-order effects on the line widths and hyperfine splittings.

Finally, to illustrate what can be deduced from the analysis of a very complex spectrum we show the hyperfine pattern of triphenylmethyl

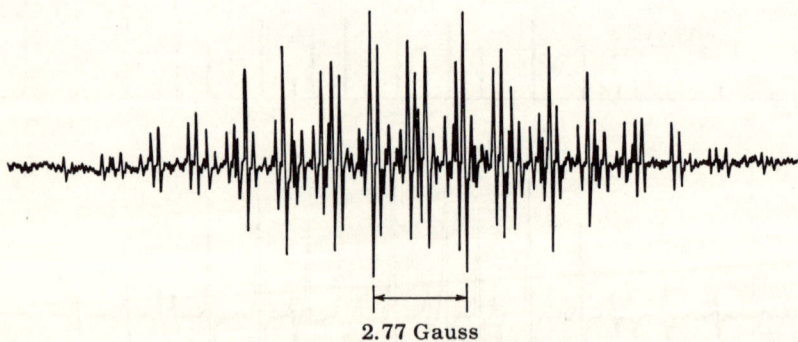

This magnificent spectrum (Fig. 6.6) boasts 196 lines from three kinds of equivalent proton: six ortho, six meta, and three para! The splitting constants are $a_o = 2.53$, $a_m = 1.11$, and $a_p = 2.77$ gauss.

2.77 Gauss

FIG. 6.6. Electron resonance spectrum of triphenylmethyl.

## 6·3 HYPERFINE PATTERNS FROM OTHER NUCLEI

The energy level diagrams and selection rules which govern the analysis of proton hyperfine patterns are valid for any nucleus of spin 1/2. However, if the nucleus is present as one of several isotopes some care must be exercised in calculating relative intensities. Thus structure from $C^{13}$ nuclei ($I = 1/2$) is frequently observed but since the natural abundance is only 1.1%, the hyperfine components are very weak in comparison with proton lines.

The methyl radical $CH_3$ illustrates this point. Of the radicals 98.9 % have the $C^{12}$ nucleus, spin zero, and give a 1:3:3:1 proton hyperfine spectrum with a splitting constant of 23.04 gauss. The remaining 1.1 % have an additional $C^{13}$ coupling of 41 gauss which splits each of the four lines into doublets. The two types of spectra are superposed and the $C^{13}$ splitting manifests itself by the presence of weak satellite lines (Fig. 6.7).

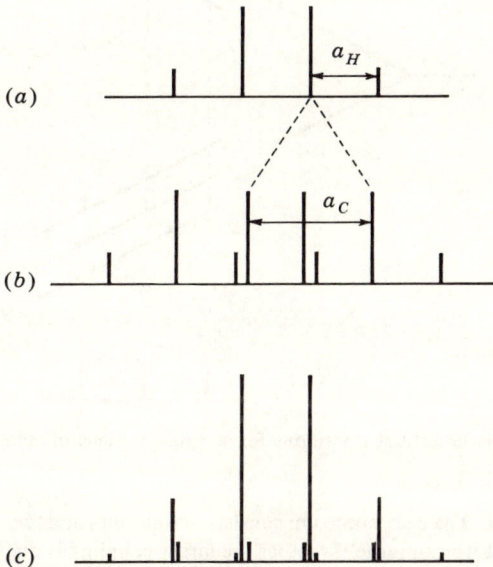

FIG. 6.7. Carbon 13 splittings in the methyl radical. (a) Protons in $C^{12}H_3$; (b) spectrum of $C^{13}H_3$; (c) natural mixture of $C^{12}$ and $C^{13}$. Splittings are $a_H = -23$ gauss, $a_C = +41$ gauss.

Nuclei with spin $I = 1$ are frequently encountered, common examples being $N^{14}$ and $D^2$ (deuterium). A nucleus with $I = 1$ has three allowed spin orientations with components in the z direction corresponding to $m_I$ values of $+1$, 0, and $-1$. Hence interaction of an unpaired electron with one $N^{14}$ nucleus gives rise to three equally intense, equally spaced lines, as will be deduced from the energy level diagram shown in Fig. 6.8. Interaction with a second $N^{14}$ nucleus splits each of these three lines into a further 1:1:1 triplet, so that if the two splittings are unequal, nine equally intense lines will be obtained. If the nitrogen atoms are equivalent the three triplets overlap:

$$
\begin{array}{ccccc}
1 & 1 & 1 & & \\
 & 1 & 1 & 1 & \\
 & & 1 & 1 & 1 \\
\hline
1 & : & 2 & : 3 : 2 : 1
\end{array}
$$

and only five lines are observed with the intensities indicated.

Similar rules apply to the analysis of deuterium structure. Although the natural abundance of deuterium is too low for it to yield observable hyperfine structure, deuterium substitution is frequently used to resolve ambiguities in the assignment of proton splittings. Consider, for example, a radical possessing two inequivalent

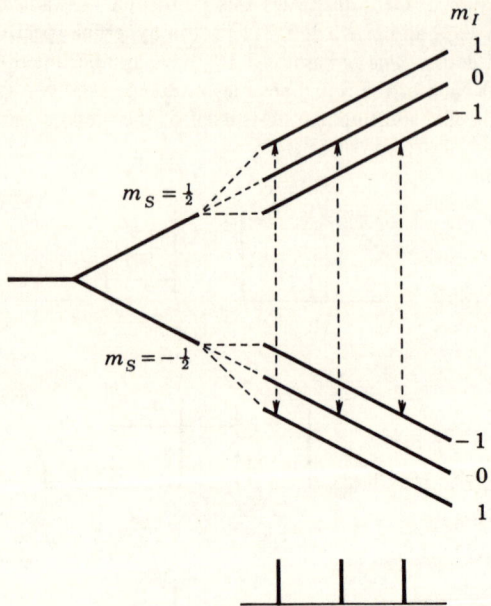

FIG. 6.8. Hyperfine levels and transitions for a single nucleus of spin $I = 1$ with positive coupling constant.

protons, $A$ and $B$. The e.s.r. spectrum consists of four lines and two coupling constants. are obtained, but we cannot tell whether the larger coupling is due to nucleus $A$ or $B$. This problem can be resolved by replacing proton $A$ with deuterium. The deuteron, as we have seen, has a spin $I = 1$ so that a triplet splitting replaces one of the doublet splittings. Moreover, the hyperfine splitting is reduced by a factor $g_D/g_H = 0.1531$. Hence replacement of proton $A$ or proton $B$ by deuterium will give completely different patterns.

The naphthalene negative ion spectrum, discussed in Section 6.2 presents just such an ambiguity since there are four protons of each type. Deuteration experiments have confirmed that the larger splitting is indeed due to protons in positions 1, 4, 5 and 8.

## $6 \cdot 4$ MECHANISM OF THE HYPERFINE COUPLING

### 6.4.1 The Unpaired Spin Density

The isotropic hyperfine interaction in the hydrogen atom is represented by a term in the Hamiltonian proportional to $|\psi(0)|^2$, the probability of finding the electron at the nucleus, and the splitting constant for the hydrogen $1s$ orbital was found to be 506.8 gauss. The unpaired electron in a molecular radical such as vinyl is often distributed over many atoms, and gives proportionately smaller splittings. Thus in vinyl the odd electron occupies a $\sigma$ orbital extending over the entire system of bonds, and the probability that it is on a particular hydrogen atom is quite small (a $\sigma$ orbital is one which is symmetrical on reflection through the molecular plane). The splitting of 68.5 gauss at the *trans* position indicates that the unpaired spin spends a fraction of roughly $68.5/506.8 = 0.135$ of its time in the *trans* hydrogen $1s$ orbital. This fraction is often called the "unpaired spin density" in the $1s$ orbital, and is denoted by the symbol $\rho_H$. There is a direct proportionality between the spin density in the $1s$ (or other $s$) orbitals and the hyperfine splittings.

We shall soon see that this view of the situation is dangerously oversimplified. In fact a molecule contains many electrons, all of whose spins are coupled together, and it is not in general possible to say that there is just one "unpaired electron" in a certain orbital, while the other electrons are perfectly paired together.

The correct contact Hamiltonian for a molecule is represented by the operator

$$\mathscr{H}_c = \frac{8\pi}{3}\, g\beta g_N\beta_N \sum_k \delta(\mathbf{r}_k - \mathbf{r}_N)\mathbf{S}_k\cdot\mathbf{I} \tag{11}$$

where the sum runs over all the electrons, and the value of the splitting constant is found by averaging over the full many-electron wave functions. The result can be written

$$a = \frac{4\pi}{3}\, g\beta g_N\beta_N\rho(\mathbf{r}_N) \tag{12}$$

where $\rho(\mathbf{r}_N)$, the unpaired electron density at the nucleus is defined as

$$\rho(\mathbf{r}_N) = \int \Psi^* \sum_k 2S_{zk}\delta(\mathbf{r}_k - \mathbf{r}_N)\Psi\, d\tau \tag{13}$$

and the spin component of the state $\Psi$ is taken to be $S_z = 1/2$. $\rho(\mathbf{r}_N)$ is simply the difference between the average number of electrons at the nucleus which have spin $\alpha$ and the number with spin $\beta$; the operator $2S_{zk}$ gives a factor $\pm 1$ in (13) depending on the spin, while the $\delta$ function ensures that the electron is at the nucleus. If the nucleus is in a position where an excess of electrons have $\beta$ spin the value of $\rho(\mathbf{r}_N)$ will be negative, and one may speak of a "negative spin density."

It is important to distinguish between the two related quantities which we have just introduced. The unpaired electron density or spin density $\rho(\mathbf{r}_N)$ *at the nucleus* or at some other point in space is a probability density, measured in electrons/ (angstrom)$^3$. The spin density *in an orbital*, $\rho_H$, is a number, representing the fractional population of unpaired electrons on an atom. For instance, if the electrons on a hydrogen atom are mainly in the $1s$ orbital, the two quantities are connected by the relation

$$\rho(\mathbf{r}_H) = \rho_H|\psi_{1s}(r_H)|^2 \tag{14}$$

### 6.4.2 Indirect Coupling Through a C—H Bond

In aromatic radicals, like the ions whose spectra appear in Fig. 6.4, the odd electron occupies a molecular $\pi$ orbital delocalized over the carbon atom framework of the molecule. This orbital, formed by overlap of the carbon $2p_z$ atomic orbitals, has a node in the plane of the molecule, where the ring hydrogen atoms are. How then do the hydrogen $1s$ orbitals acquire the unpaired spin density necessary to account for the observed hyperfine splittings?

The answer to this question is that the exchange forces between the electrons couple together the spins of the $\sigma$ electrons in the C—H bond and the $\pi$ electrons in the ring. The coupling may be explained in terms of either valence bond or molecular orbital theory, and it is simplest to consider an isolated $>$C—H fragment with one $\pi$ electron on it, occupying the $2p_z$ carbon orbital, perpendicular to the plane of the three trigonal bonds. So far as the spins of the two electrons forming the C—H $\sigma$ bond are concerned, one could draw the two structures indicated below.

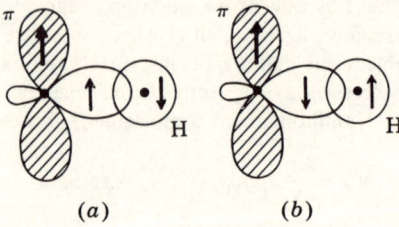

(a)                (b)

In the approximation of perfect pairing one would consider structures (a) and (b) to be equally important. If, however, the interaction between the $\sigma$ and $\pi$ systems is taken into account, structure (a) is slightly preferred because of the favorable exchange inter-action between the $\pi$ electron and the carbon $\sigma$ electron, whose spins are parallel. The spins of the electrons in the C—H $\sigma$ bond are therefore polarized slightly. If the odd electron has spin $\alpha$, there is slight excess $\alpha$ spin in the carbon $\sigma$ orbital and corre-sponding excess $\beta$ spin in the hydrogen $1s$ orbital, which gives rise to the isotropic proton splitting. Note that $\alpha$ spin in the carbon $p_\pi$ orbital induces $\beta$ spin at the proton; for this reason the spin density at the proton and the hyperfine coupling con-stant are both negative.

An alternative description of the spin polarization is provided by molecular orbital theory. The normal doublet ground state wave function for the C—H fragment would be written as the Slater determinant,

$$\Psi_0 = \|\sigma_b(1)\sigma_b(2)\pi(3)\|\alpha\beta\alpha \tag{15}$$

where two electrons occupy the C—H bonding orbital $\sigma_b$ and one electron occupies the $2p_z$ orbital, $\pi$. Equation (15) is an abbreviation for the determinant

$$\Psi_0 = \frac{1}{\sqrt{6}}\begin{vmatrix} \sigma_b\alpha(1) & \sigma_b\alpha(2) & \sigma_b\alpha(3) \\ \sigma_b\beta(1) & \sigma_b\beta(2) & \sigma_b\beta(3) \\ \pi\alpha(1) & \pi\alpha(2) & \pi\alpha(3) \end{vmatrix} \tag{16}$$

There are also two excited states with total spin $S = 1/2$ and $m_S = 1/2$ where one $\sigma$ bonding electron is promoted into the $\sigma$ antibonding orbital $\sigma_a$. The wave functions are constructed by combining several Slater determinants with different spin arrange-ments.

$$\Psi_1 = \|\sigma_b(1)\sigma_a(2)\pi(3)\|\frac{1}{\sqrt{2}}(\alpha\beta\alpha - \beta\alpha\alpha)$$

$$\Psi_2 = \|\sigma_b(1)\sigma_a(2)\pi(3)\|\frac{1}{\sqrt{6}}(2\alpha\alpha\beta - \alpha\beta\alpha - \beta\alpha\alpha) \tag{17}$$

Notice that these are the same spin functions as we wrote down for three nuclei in Chapter 4, and they are eigenfunctions of the operator $S^2$. The exchange couplings mix the ground configuration $\Psi_0$ with a small amount of the excited state $\Psi_2$ and produce a small negative spin polarization at the hydrogen nucleus. Introduction of $\Psi_1$ is in fact unnecessary, because it does not affect the spin distribution.

Theoretical calculations estimate that a single $\pi$ electron induces a negative spin density of about $-0.05$ in the hydrogen $1s$ orbital, corresponding to a hyperfine splitting of 20–25 gauss ($a$ is negative). This agrees splendidly with observed $\alpha$-proton splittings of $-23.04$ and $-22.38$ gauss in the methyl and ethyl radicals.

### 6.4.3 McConnell's Relation

In aromatic radicals the odd electron is delocalized, so that the average unpaired spin population of a particular carbon $2p_z$ orbital is considerably smaller than unity. The extent to which the C—H $\sigma$ electrons are polarized is directly proportional to the net unpaired electron population, or "$\pi$-electron spin density" $\rho_\pi$ on the carbon atom. This is summed up in the simple equation

$$a_H = Q\rho_\pi \tag{18}$$

which is called McConnell's relation. $Q$ is the proportionality constant and has a value of about $-22.5$ gauss. The cyclic polyene radicals $C_nH_n$ illustrate this equation nicely. In each radical the odd electron is distributed uniformly round the ring, and the $\pi$-electron spin density is determined by symmetry. For example, in the benzene negative ion, the spin density is $1/6$ and the hyperfine splitting is $-3.75$ gauss, so that $Q = a_H/\rho_\pi = 6 \times -3.75 = -22.5$ gauss.

As Table 6.1 shows, McConnell's relation is not generally as accurate as this result implies. Nevertheless, as we shall soon see, it is good enough to provide a means of interpreting the spectra of virtually all aromatic free radicals.

TABLE 6.1. Hyperfine Splittings and Spin Distributions
in the Cyclic Polyene Radicals $C_nH_n$

| Radical | Spin Density $\rho_\pi$ | Hyperfine Splitting $a_H$ | Effective Value of $Q$ |
|---|---|---|---|
| $CH_3$ | 1 | 23.04 | $-23.04$ |
| $C_5H_5$ | 1/5 | $-5.98$ | $-29.9$ |
| $C_6H_6^-$ | 1/6 | $-3.75$ | $-22.5$ |
| $C_7H_7$ | 1/7 | $-3.91$ | $-27.4$ |
| $C_8H_8^-$ | 1/8 | $-3.21$ | $-25.7$ |

### 6.4.4 Hyperconjugation

All of the simple alkyl radicals have now been studied in the liquid phase, being formed by continuous electron irradiation of the liquid hydrocarbons at low temperatures. The proton couplings for the ethyl, iso-propyl, and t-butyl radicals are as follows.

$$\cdot CH_2\!\!-\!\!CH_3 \qquad\qquad 22.11\ \overset{\displaystyle CH_3}{\underset{\displaystyle CH_3}{\overset{|}{\underset{|}{\dot{C}H}}}} \qquad CH_3\!\!-\!\!\overset{\displaystyle CH_3}{\underset{\displaystyle CH_3}{\overset{|}{\underset{|}{\dot{C}}}}}$$

$$22.38 \quad 26.87 \qquad\qquad\qquad 24.68 \qquad\qquad 22.72$$

The most notable feature of these results is that the methyl, or $\beta$ proton, coupling is actually larger than that for the $\alpha$ protons.

The mechanism of the $\beta$-proton coupling almost certainly involves hyperconjugation rather than exchange interactions and can be formulated in terms of molecular orbital theory as follows (Fig. 6.9). The three hydrogen $1s$ orbitals $a$, $b$, $c$ can be combined to form three new orthogonal group orbitals which are

$$\psi_1 = \frac{1}{\sqrt{3}}(a + b + c)$$

$$\psi_2 = \frac{1}{\sqrt{2}}(b - c) \tag{19}$$

$$\psi_3 = \frac{1}{\sqrt{6}}(2a - b - c)$$

$\psi_1$ is roughly symmetric about the axis of the C—C bond, but $\psi_2$ and $\psi_3$ have nodal planes which pass through the axis and resemble a pair of $p_y$ and $p_z$ orbitals. If we regard the carbon atom of the methyl group as forming two $sp_x$ hybrid orbitals along the C—C direction, then one hybrid forms the C—C bond, while the other joins with $\psi_1$ to form a single C—H$_3$ bond. The remaining carbon $p_y$ and $p_z$ orbitals can, respectively, overlap $\psi_2$ and $\psi_3$, and the $p_z$ orbital can also overlap with the $p_z$ orbital on the $\alpha$ carbon atom. Thus, in the ethyl radical, the odd electron can be thought of as occupying a molecular $\pi$ orbital compounded of $p_z$ orbitals on both carbon atoms, and the "H$_3$" group orbital, $\psi_3$. This is, of course, a rather extreme view of the electronic structure of the ethyl radical but it does indicate a mechanism whereby the odd electron may penetrate directly into the hydrogen $1s$ orbitals. It also implies that the $\beta$-proton coupling constant should be *positive*; the positive sign has indeed been confirmed by n.m.r. experiments.

It must be remembered that the methyl group is normally free to rotate, and the isotropic splitting observed in the liquid phase is, in fact, an *average* splitting. When we come to investigate single crystal spectra, we shall encounter cases where the methyl group is not rotating freely and the e.s.r. spectrum reveals the presence of unequal interactions with the three protons. It is found experimentally that the proton splitting is approximately equal to

$$a_H = B_0 + B_2 \cos^2 \theta \tag{20}$$

where $\theta$ is the angle of twist between the $\alpha$-carbon $2p_z$ orbital and the plane containing the $\beta$-proton C—H bond (see Fig. 6.9b). This $\cos^2 \theta$ dependence of the splitting is expected if the coupling is caused by hyperconjugation; if the methyl group rotates rapidly then an average $B_0 + B_2 \langle \cos^2 \theta \rangle_{av} = B_0 + \frac{1}{2}B_2$ is observed for each proton.

Many $\pi$-electron radicals contain methylene groups where the hydrogen atoms project up and down out of the plane and have a strong overlap with the odd $\pi$ orbital.

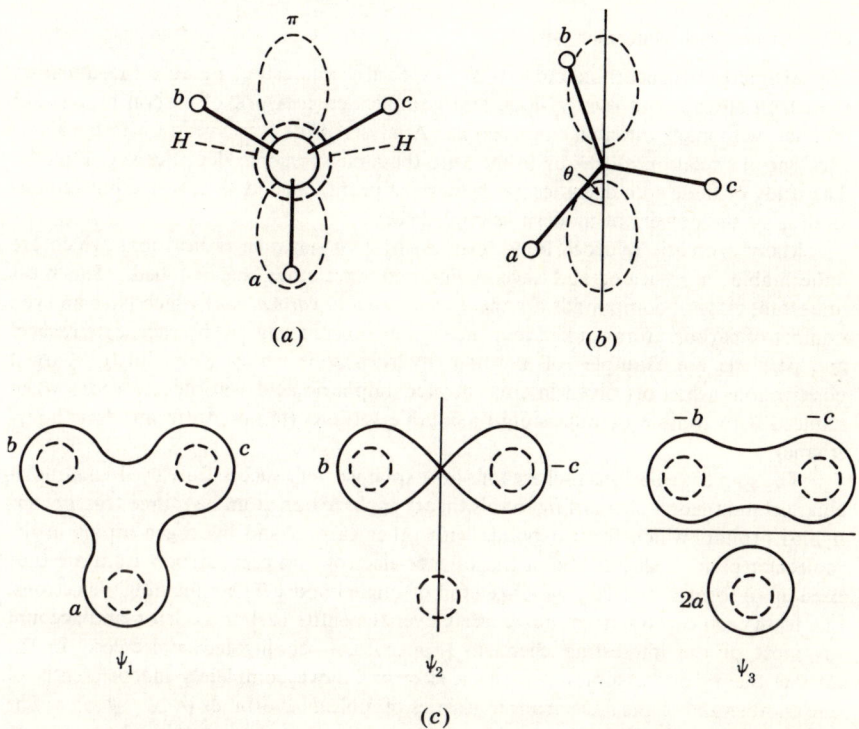

FIG. 6.9. Hyperconjugation in the ethyl radical. (a) End-on-view of the $\pi$ orbital and the methyl group. (b) Rotation about the C–C bond. (c) Group orbitals for the hydrogen atoms.

These molecules also have large hyperfine splittings due to hyperconjugation. For instance, $H_2CN$

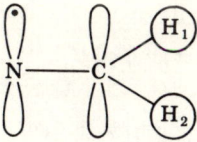

has $a_H = 87.4$ gauss; while the conjugated cyclohexadienyl radical

$$47.71 \quad H \overset{8.99}{\underset{H}{\diagup}} \overset{2.65}{\underset{}{\diagdown}} 13.04$$

still has a big methylene proton splitting even though the $\pi$ electron is spread out over a large ring.

## 6·5   SPIN DISTRIBUTIONS IN ALTERNANT HYDROCARBON IONS

### 6.5.1 Delocalized Molecular Orbitals

Aromatic hydrocarbon radicals are especially interesting because the unpaired $\pi$ electron spreads out over a large and complicated network of carbon atoms and interacts with many different ring protons. Analysis of the e.s.r. spectra with the aid of McConnell's relation allows us to measure the $\pi$-electron spin densities very directly. The study of these spin densities has been most profitable and gives a striking demonstration of the powers of modern valence theory.

Many aromatic hydrocarbons form positive or negative radical ions which are quite stable in solution, and several hundred have now been studied. The most important class of compounds are the *even alternant hydrocarbons* which have an even number of carbon atoms linked to form even-membered rings. Anthracene, tetracene, and perylene are examples of alternant hydrocarbons which form singly charged positive ions when dissolved in concentrated sulphuric acid, and negative ions when reduced with sodium or potassium in suitable solvents (tetrahydrofuran, dimethoxyethane).

The spin distributions in these ions are explained very successfully by Hückel molecular orbital theory. According to this theory each carbon atom has three trigonal $sp^2$ hybrid orbitals which form $\sigma$ bonds with other carbon and hydrogen atoms in the molecular plane. All but one of the valence electrons on each carbon atom are thus used up to form a relatively inert skeleton of single bonds. The remaining $\pi$ electrons, one from each carbon atom, move freely over the entire carbon skeleton and account for most of the interesting chemical properties of conjugated molecules. In the Hückel theory one assumes that the $\pi$ electrons move completely independently of one another and of the $\sigma$ electrons in a series of molecular orbitals $\psi_1, \psi_2, \psi_3 \ldots$. The orbitals are taken to be linear combinations of the carbon $2p_z$ orbitals, which are now denoted $\phi_1, \phi_2 \ldots$. Thus

$$\psi = C_1\phi_1 + C_2\phi_2 + \cdots C_n\phi_n \tag{21}$$

or more generally, distinguishing different molecular orbitals and different atoms by indices $i$ and $r$, we write

$$\psi_i = \sum_r C_{ri}\phi_r \tag{22}$$

The energies of the molecular orbitals are expressed in terms of two parameters (not to be confused with $\alpha$ and $\beta$ spins!)

$$\alpha_r = \int \phi_r^* \mathscr{H} \phi_r \, d\tau$$

$$\beta_{rs} = \int \phi_r^* \mathscr{H} \phi_s \, d\tau \tag{23}$$

which are called the Coulomb integral of the $r$th atom and the resonance integral of the bond $rs$. They are just the matrix elements between the $\pi$ orbitals of the effective $\pi$-electron Hamiltonian. The correct energies and orbitals are found by the procedure described in Appendix A, that is by solving the simultaneous linear equations

$$\begin{bmatrix} (\alpha_1 - E) & \beta_{12} & \beta_{13} \\ \beta_{21} & (\alpha_2 - E) & \beta_{23} \\ \beta_{31} & \beta_{32} & (\alpha_3 - E) \cdots \\ \cdots\cdots\cdots\cdots\cdots\cdots\cdots\cdots\cdots \end{bmatrix} \begin{bmatrix} C_1 \\ C_2 \\ C_3 \\ \vdots \end{bmatrix} = 0 \tag{24}$$

A typical example is the benzene molecule. Here all six carbon atoms have the same Coulomb integral $\alpha$ and the bonds all have the same resonance integral $\beta$. (One always ignores resonance integrals between non-bonded atoms.) The orbital energies are found to be $\alpha + 2\beta$, $\alpha + \beta$, $\alpha - \beta$, and $\alpha - 2\beta$, while the forms of the orbitals are shown in Fig. 6.10. In benzene itself the three bonding orbitals are filled with the

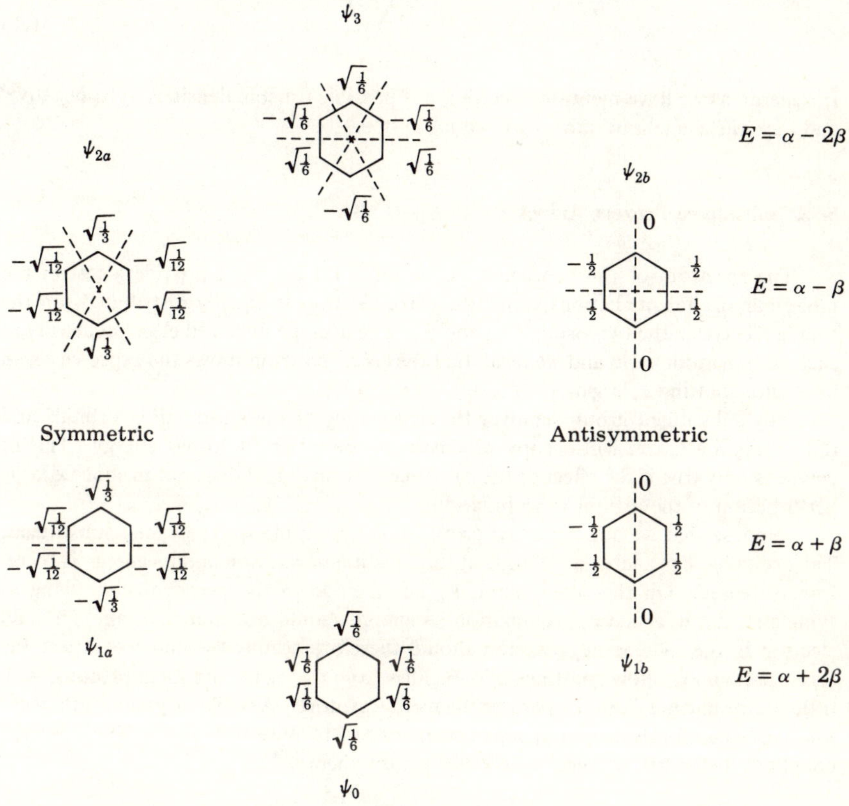

FIG. 6.10. Hückel molecular orbitals of benzene.

necessary six electrons, so that in the benzene negative ion the extra electron can go into either of the two degenerate antibonding orbitals. The probability that the unpaired electron is on the $r$th atom is $|C_{ri}|^2$, and in this case the probability distribution is the same as the spin distribution. Thus the theoretical spin densities in the two orbitals are

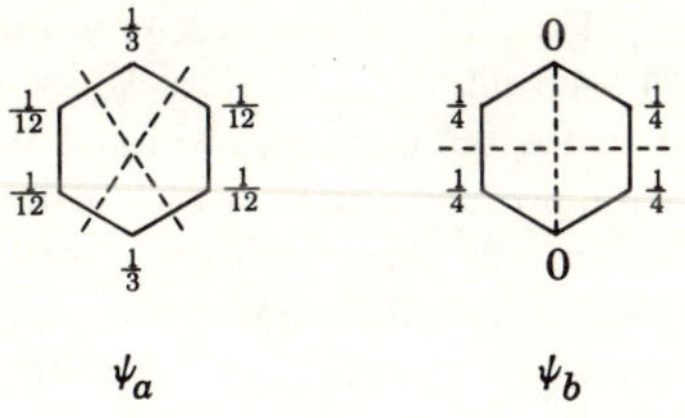

Notice that the normalization of the wave function requires that the probabilities add up to one

$$\sum_r |C_{ri}|^2 = 1 \tag{25}$$

and so the spin densities on the carbon atoms must also be normalized

$$\rho_r = |C_{ri}|^2, \qquad \sum_r \rho_r = 1 \tag{26}$$

In general, as we have mentioned before, it is possible for spin densities to be negative, and then their algebraic sum is still equal to one.

### 6.5.2 Substituted Benzene Anions

The spectrum of the benzene negative ion itself does not provide a test of the molecular orbital predictions, since the extra electron is equally distributed, on the average, between the two orbitals $\psi_a$ and $\psi_b$. The average unpaired electron density at each carbon atom is $1/6$ and we recall that the e.s.r. spectrum shows the expected seven lines with splitting 3.75 gauss.

Any substituent group removes the degeneracy of the antibonding orbitals and the unpaired electron will occupy whichever of them has the lowest energy. This of course is only true if the effect of the substituent is small and does not produce a large perturbation of the orbital wave functions.

Suppose that we add a methyl group, which is an electron-repelling substituent. The orbital $\psi_a$ has a large amplitude at the substituent position and its energy is therefore increased. On the other hand, $\psi_b$ has a node in the perpendicular plane of symmetry and in a first approximation its energy should not change at all. The odd electron in the toluene negative ion should therefore occupy $\psi_b$, and we expect the e.s.r. spectrum to show splittings of 5–6 gauss from the ortho and meta protons, with little or no splitting from the para or the methyl protons. A similar argument applies to $p$-xylene where both methyl groups act in concert to raise orbital $\psi_a$. The observed couplings, listed below, agree very well with the theory.

A rather different situation arises for $m$-xylene where the methyl groups lie on either side of the symmetry plane. The odd electron density at the substituent positions will now be either $1/12$ or $1/4$, depending on whether $\psi_a$ or $\psi_b$ is occupied. Thus the energies of both orbitals increase, but clearly $\psi_a$ will now be lower. One expects large splittings (about 7 gauss) from protons 1 and 4, with smaller ones (about 2 gauss) from protons 3 and 5. Again the results agree with these predictions. Similar reasoning suggests that in the $o$-xylene anion the unpaired electron will occupy an orbital resembling $\psi_a$ which has a lower spin density on the methyl groups.

$$H_3C \underset{7.72}{\overset{6.85}{\diagdown}} CH_3 \quad 2.26 \qquad 1.46$$

$$1.81 \overset{6.93}{\diagdown} CH_3 \quad 2.00$$
$$CH_3$$

The remarkable sensitivity of the spin distribution to exceedingly small perturbations is even better shown by the marked change in the benzene anion spectrum when one ring hydrogen is replaced by deuterium. The hyperfine splittings

$$3.41 \; H \diagup \overset{H \quad H}{\underset{H \quad H}{\diagup}} D \; 0.55$$
$$3.92 \quad 3.92$$

show clearly that the deuterium removes the degeneracy of the two orbitals, although both of them still make some contribution to the spin distribution.

### 6.5.3 The Pairing of Electronic States

One of the most interesting properties of alternant hydrocarbons is the pairing of electronic states. The Hückel molecular orbitals always occur in pairs; if a certain bonding $\pi$ orbital $\psi_i$ exists with energy $\alpha + \varepsilon_i\beta$ ($\varepsilon_i$ being a positive number) then there is a "mirror image" antibonding orbital $\psi'_i$ with energy $\alpha - \varepsilon_i\beta$ (a glance at Fig. 6.10 shows that this is certainly true for benzene). The pairing relation depends on the fact that all the rings of carbon atoms are even-membered. Hence it is possible to divide the carbon atoms in, say, anthracene or the benzyl radical

into two sets, the "starred" and "unstarred" atoms, such that the two kinds of atom alternate round every ring. The atoms of a hydrocarbon radical like azulene negative ion, with odd-membered rings, cannot be labeled in this way. The orbital coefficients of $\psi_i$ and $\psi'_i$ are equal on all the starred atoms, with $C_{ri} = C'_{ri}$; while they are opposite in magnitude on the unstarred atoms, where $C_{ri} = -C'_{ri}$.

Examination of the e.s.r. spectra of the radical ions provides an excellent test of these theoretical predictions. Since the highest bonding orbital and lowest antibonding orbital are paired in the sense we have just described, and these are the orbitals occupied by the unpaired electron in the two ions, it follows that the spin distributions should be identical.

$$|C_{ri}|^2 = |C'_{ri}|^2 \tag{27}$$

or

$$\rho_r^{\oplus} = \rho_r^{\ominus}. \tag{28}$$

Figure 6.11 illustrates the paired odd orbitals of the singly-charged positive and

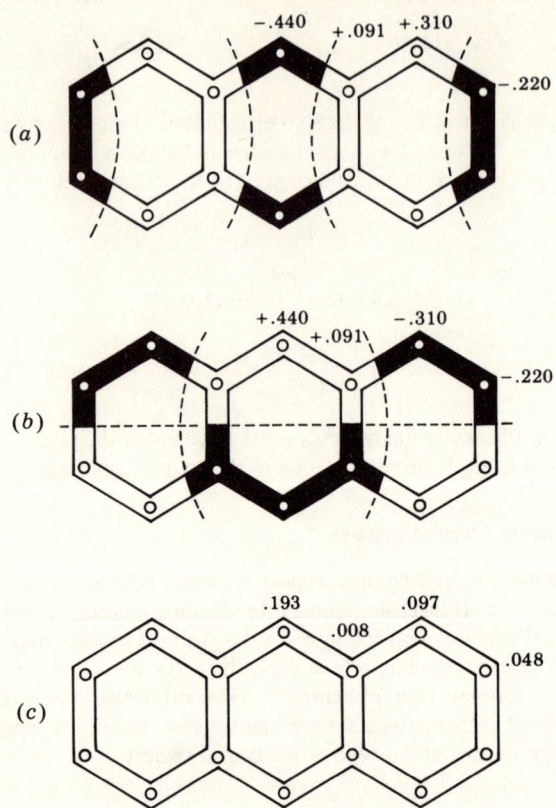

FIG. 6.11. The pairing of the lowest antibonding and highest bonding orbitals of anthracene. (a) Antibonding orbital; (b) bonding orbital; (c) spin distribution.

negative ions of anthracene, with their common spin distribution. The observed proton hyperfine splittings listed below are indeed remarkably similar, although they do tend to be larger in the positive ion.

Hyperfine Splittings in the Anthracene Ions (gauss)

|  | $a_1$ | $a_2$ | $a_9$ |
|---|---|---|---|
| Positive ion | 3.11 | 1.40 | 6.65 |
| Negative ion | 2.74 | 1.57 | 5.56 |

It is also interesting that the simple Hückel molecular orbital theory predicts the ratios of $a_9 : a_1 : a_2$ to be exactly $4 : 2 : 1$ which is very close to the observed ratios. The e.s.r. spectra of the two tetracene and perylene ions are also very similar.

It is highly significant that the pairing theorem does not extend to methyl group splittings in aromatic radicals. As a rule there seems to be a rough proportionality between the observed splitting and the spin density on the adjacent ring carbon atom in many radicals, but marked deviations from a linear relationship are not uncommon. In particular, methyl group splittings are much larger in positive than in negative ions. A striking example is 9, 10-dimethylanthracene where the splitting is over twice as large

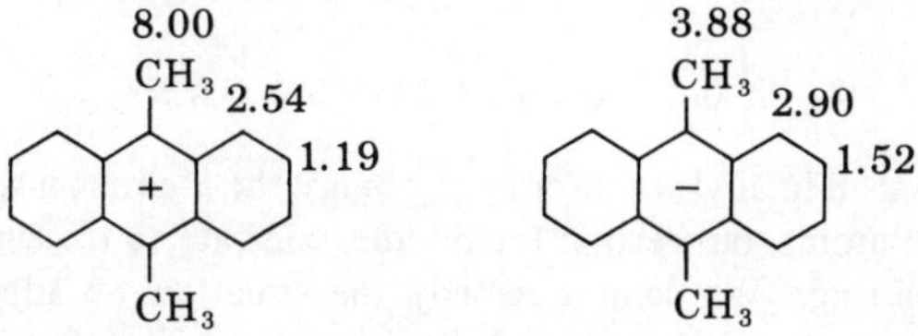

in the positive ion. The most probable explanation of this effect is that the unpaired bonding $\pi$ electron of the positive ion conjugates more readily with the methyl group.

## 6·6  NEGATIVE SPIN DENSITIES IN ODD ALTERNANT RADICALS

Alternant hydrocarbons which possess an odd number of carbon atoms are called "odd-alternant." The pairing theorem relating bonding and antibonding molecular orbitals still holds; in addition such molecules possess a nonbonding orbital of energy $E = \alpha$ which is occupied by one unpaired electron. This orbital is confined to the starred atoms and has nodes at all the unstarred atoms. One of the best known examples is the benzyl radical, for which the distribution of the odd electron is readily calculated as shown below; the experimental proton coupling constants are also shown, and we notice an appreciable splitting from the unstarred meta position

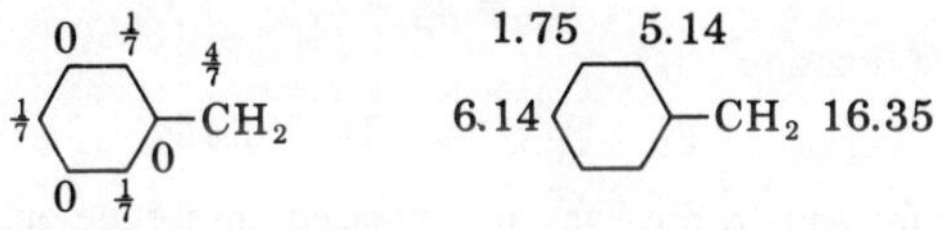

Our earlier spectrum of triphenylmethyl shows the same peculiarity.

In order to understand these observations we shall look at a much simpler molecule, namely the allyl radical. Its e.s.r. spectrum has been obtained in the liquid phase by electron irradiation of cyclopropane, and the coupling constants are as shown below.

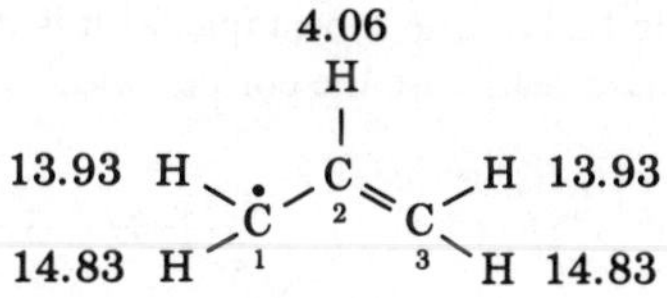

(It is not known which of the $CH_2$ protons has the larger splitting.) The Hückel

molecular orbitals formed from the carbon $2p_z$ atomic orbitals $\phi_1$, $\phi_2$, and $\phi_3$ are calculated to be

$$\psi_1 = \frac{1}{2}(\phi_1 + \sqrt{2}\,\phi_2 + \phi_3) \qquad E = \alpha + \sqrt{2}\,\beta$$

$$\psi_2 = \frac{1}{\sqrt{2}}(\phi_1 - \phi_3) \qquad E = \alpha$$

$$\psi_3 = \frac{1}{2}(\phi_1 - \sqrt{2}\,\phi_2 + \phi_3) \qquad E = \alpha - \sqrt{2}\,\beta$$

The unpaired electron in allyl will occupy $\psi_2$; hence the $\pi$-electron spin density should be 1/2 on the end atoms, but zero in the middle, contrary to the evidence of the 4.06 gauss proton splitting. We cannot remedy the situation by adjusting the orbital $\psi_2$, for it must have a node on atom 2, by symmetry. Therefore it is necessary to consider the other two $\pi$ electrons, and take electron correlation effects into account. This may be accomplished in several different ways.

One of the easiest to visualize is the use of different molecular orbitals for electrons of different spins. The Coulomb repulsion between two electrons of the same spin is lowered by exchange effects when the two electronic wave functions overlap, and so the effective repulsion between the electrons in the orbitals $\psi_1$ and $\psi_2$ is lower if the spin of the electron in $\psi_1$ is $\alpha$ than if it is $\beta$. As the orbital $\psi_2$ is concentrated on the end carbon atoms the Coulomb and exchange forces tend to attract the bonding $\psi_1$ electron of spin $\alpha$ onto the end atoms while at the same time forcing the $\psi_1$ $\beta$ spin towards the center. These arguments suggest that instead of the single determinant wave function

$$\Psi_0 = \|\psi_1(1)\psi_1(2)\psi_2(3)\|\alpha\beta\alpha \qquad (29)$$

it would be better to take two new orbitals

$$\psi_{1a} = \psi_1 + \lambda\psi_3$$
$$\psi_{1b} = \psi_1 - \lambda\psi_3 \qquad (30)$$

and use the wave function

$$\Psi_1 = \|\psi_{1a}(1)\psi_{1b}(2)\psi_2(3)\|\alpha\beta\alpha \qquad (31)$$

The orbital $\psi_{1a}$, for spin $\alpha$, now has an increased amplitude on atoms 1 and 3, while $\psi_{1b}$ has a higher probability that the spin $\beta$ is on atom 2 (see Fig. 6.12). The spin density $\rho(\mathbf{r})$ for the wave function (31) is just the difference between the spin up and spin down probability distributions

$$\rho(\mathbf{r}) = |\psi_2|^2 + |\psi_{1a}|^2 - |\psi_{1b}|^2$$
$$= |\psi_2|^2 + 2\lambda\psi_3\psi_1 \qquad (32)$$

Now, since $\psi_1$ and $\psi_3$ have opposite signs on atom 2, the spin density there is negative. The wave function $\Psi_1$ is in fact a little too simple, as it is not an eigenfunction of the total spin $S^2$, and one should really use the correct linear combination of three determinants

$$\Psi_2 = \|\psi_{1a}(1)\psi_{1b}(2)\psi_2(3)\| \frac{1}{\sqrt{6}} (2\alpha\beta\alpha - \beta\alpha\alpha - \alpha\alpha\beta) \qquad (33)$$

However, this does not alter the main conclusion that the spin density on atom 2 is negative.

Orbitals             Spin distribution

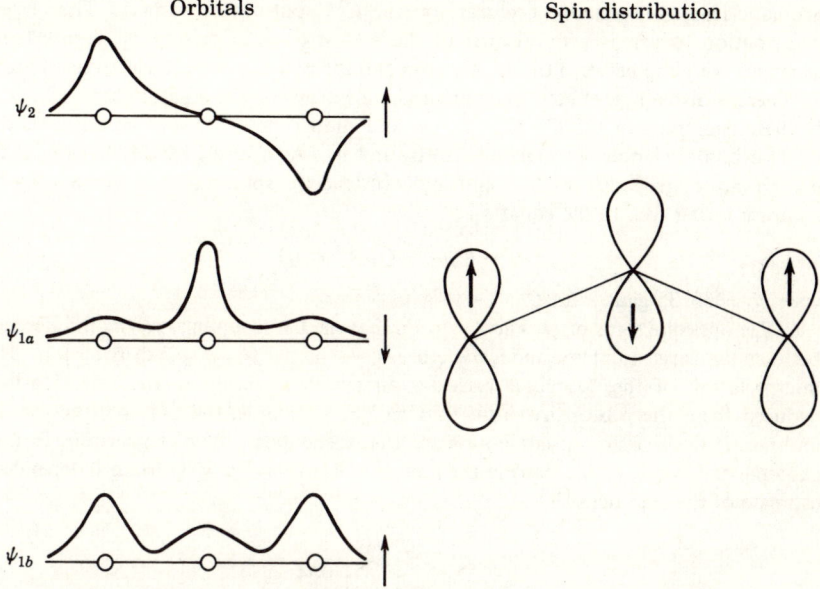

$\psi_2$

$\psi_{1a}$

$\psi_{1b}$

FIG. 6.12. Different orbitals for different spins in the allyl radical.

Similar calculations for the benzyl radical predict negative spin densities at the meta positions. Since the algebraic sum of the spin densities must be one, the presence of a negative spin density at one position means that there must be an extra compensating positive spin density elsewhere in the molecule. Hence the overall spread of the hyperfine pattern (in gauss) tends to be considerably larger than $Q$. In Chapter 13 we shall see how n.m.r. experiments demonstrate that negative $\pi$-electron spin densities also occur in *even* alternant ions, like the pyrene negative ion.

## 6·7 HYPERFINE SPLITTING FROM $C^{13}$ AND $N^{14}$ NUCLEI

### 6.7.1 Carbon 13

The theory of isotropic $C^{13}$ splittings is more complicated than for protons, although it does not involve any radically new principles. We can see this by comparing the methyl radical ($a_C = 41$ gauss) with the benzene negative ion ($a_C = 2.8$). The $C^{13}$ splitting is certainly not directly proportional to the $\pi$-electron spin density at the carbon atom; if it were the benzene splitting ought to be closer to 7 gauss. In fact several different exchange interactions must be considered. Let us look at part of an

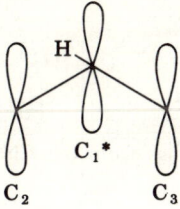

$C_2$      $C_1{}^*$      $C_3$

H

aromatic ring system and concentrate on the $C^{13}$ splitting of atom 1. The largest contribution, which is proportional to the $\pi$-electron spin density $\rho_1$, comes from exchange coupling between the $2p_z$ electron and the paired $1s$ and $2s$ electrons of atom 1. There is also a new effect. The unpaired $\pi$ electrons on the adjacent atoms 2 and 3 polarize the spins in the C—C $\sigma$ bonds and induce negative spin densities in the $C_1$ $2s$ orbital which are proportional to $\rho_2$ and $\rho_3$. As a result McConnell's relation $a = Q\rho$ no longer holds for $C^{13}$ splittings. Instead the splittings in aromatic radicals conform rather well to the equation

$$a_C = Q_1\rho_1 + Q_2(\rho_2 + \rho_3) \tag{34}$$

with $Q_1 = 30\text{--}35$ gauss and $Q_2 = -14$ gauss.

The opposite signs of $Q_1$ and $Q_2$ mean that the $C^{13}$ coupling is a small difference between two large numbers, and is therefore very sensitive to the spin distribution. The interpretation of the couplings provides an excellent check on the spin densities deduced from the proton hyperfine pattern. For example, the $C^{13}$ splittings of the anthracene positive and negative ions are almost identical. Besides providing further experimental proof of the pairing theorem they lead via Eq. (34) to an independent estimate of the spin densities:

### 6.7.2 Nitrogen 14

Virtually all aromatic radicals containing nitrogen have well-resolved $N^{14}$ hyperfine structure if there is any $\pi$-electron spin density on the nitrogen atom. The nuclear quadrupole interaction is always much smaller than the hyperfine energy and so there is no line broadening of the type found in n.m.r. spectra. Typical examples are the Wurster's blue radical ion and the nitrobenzene anion radical which are shown below:

Hückel molecular orbital theory has again been of immense value for interpreting these spectra, and the $N^{14}$ splitting is roughly proportional to the odd electron density in the nitrogen $2p_z$ orbital, with a $Q$ value of about 25 gauss. Finally, an interesting radical which has both $N^{14}$ and $C^{13}$ structure is the tetracyanoethylene negative ion

where $a_N = 1.61$, and the splittings from the CN and C—C carbons are 9.47 and 2.92 gauss.

## 6·8 APPLICATIONS OF E.S.R. TO SOLUTION CHEMISTRY

Our detailed understanding of liquid phase e.s.r. spectra has been accumulated largely through the study of stable free radicals, principally aromatic ions and semiquinones. Clearly the technique has many applications in chemistry and we will complete this review of the subject by listing some of them.

FAST REACTIONS IN SOLUTION. Most of the free radicals involved in chemical reactions are very short-lived and even though it is possible to detect concentrations as small as $10^{-8}$ molar, special techniques have to be devised in many instances. One of the most important recent developments is the introduction of rapid flow techniques, the reactants being mixed just above the center of the resonance cavity and allowed to flow through it at varying speeds. It is not difficult to detect radicals with lifetimes of the order of 100 milliseconds or less. Of the many interesting studies which have been reported we will mention just two. It is well known that attack of benzene by OH radicals results in high yields of diphenyl but the mechanism of the reaction is uncertain. It has now been found that OH addition to benzene occurs, the radical shown below

being detected when aqueous titanous ion and aqueous hydrogen peroxide, both saturated with benzene, are mixed and flowed through the resonance cavity.

The second example concerns semiquinone radicals. The p-benzosemiquinone, formed by one-electron reduction of the quinone or oxidation of the hydroquinone, exists as the free radical *anion* in alkaline solution and is quite stable. It is equally stable in concentrated sulphuric acid where it exists as a doubly protonated *cation*, the e.s.r. spectrum revealing the presence of OH protons. In neutral or near-neutral solutions, however, p-benzosemiquinone exists as an uncharged species and is very short-lived.

Through the use of flow systems it is possible to study the equilibria between the various forms of the radical, and to determine equilibrium constants.

ELECTROCHEMICAL REACTIONS. Aromatic cations and anions can be prepared by chemical oxidation or reduction, but they can also be obtained by electrolysis. Many aromatic molecules can be reduced at the surface of a mercury pool cathode and this method offers several advantages, particularly in the wide range of solvents which are possible. One is, however, restricted to molecules with relatively low reduction potentials. Electrolytic oxidations have been used less extensively, but a good example is the preparation of the *p*-phenylenediamine cation.

$$\text{NH}_2 - \left\langle \overline{\ +\ } \right\rangle - \text{NH}_2$$

Electrolytic techniques have been used mainly as tools in the preparation of specific free radicals but there seems little doubt that e.s.r. studies could also provide information about the details of electrode processes.

PHOTOCHEMICAL REACTIONS. It is possible to study the e.s.r. spectrum of a solution which is subjected to continuous ultraviolet or visible irradiation. Attention so far has been restricted to rather stable free radicals but no doubt techniques for the detection of unstable intermediates could be devised. Among many examples we mention the formation of the nitrobenzene anion by irradiation of alkaline ethanol solutions of nitrobenzene.

SOLVATION EFFECTS. In many cases hyperfine splittings have been found to be solvent dependent, particularly when the free radical contains one or more polar groups. Changes in proton splittings are not usually very large but nitrogen and $C^{13}$ splittings can alter appreciably. The anion of *m*-nitrophenol is a particularly interesting example. Its e.s.r. spectrum in 50% aqueous dimethylformamide at 0°C consists of a superposition of two patterns with different nitrogen splittings. At room temperature, however, a single spectrum is observed. We shall see later that the e.s.r. spectrum of two radical species which are in dynamic equilibrium depends upon the rate of interconversion. When this is very fast, an averaged spectrum is observed. Clearly the *m*-nitrophenol anion is complexed or solvated in two quite distinct ways but interconversion at room temperature is rapid.

ION-PAIR EFFECTS. Aromatic hydrocarbon anions are often prepared by reduction with an alkali metal in tetrahydrofuran or dimethoxyethane. These are solvents of low dielectric constant and one might expect ion-pairing to occur. The existence of ion-pairs is indeed demonstrated very directly by the observation of additional hyperfine structure from the alkali metal nucleus. $Na^{23}$, for example, has a nuclear spin $I$ of 3/2 so that a 1:1:1:1 quartet splitting is observed. The role of the solvent in ion-pairing and the structure of the ion-pairs themselves are subjects of considerable current interest.

KINETIC EFFECTS. We have several times hinted that radicals which are involved in dynamic equilibrium, chemical or otherwise, can give unusual e.s.r. spectra depending upon the rate of the kinetic process. We shall explore some of the details of this subject in Chapter 12. For the moment we just mention that kinetic processes involving electron transfer, proton transfer, rotational isomerism, hydrogen bonding, and so on, have been studied by e.s.r., often with fascinating results.

## PROBLEMS

1. The hyperfine constants of the radical $CH(CH_3)O(C_2H_5)$ formed by reaction of diethyl ether with the hydroxyl radical in solution are 13.8 (C—H proton), 21.9 (CH$_3$ protons) and 1.4 gauss (2 equivalent ethyl protons). Draw a careful reconstruction of the expected hyperfine spectrum and compare it with the one published by Dixon, Norman, and Buley: *J. Chem. Soc.*, 3625 (1964).

**2.** Do a similar reconstruction for the ethyl radical, taking $a = 23.05$ gauss for $CH_2$ and 26.87 for $CH_3$. What would the hyperfine pattern look like under conditions of poor resolution?

**3.** What hyperfine splittings would you expect from the negative ions of 1, 4, 5, 8 deutero-naphthalene and 2, 3, 6, 7 deuteronaphthalene? Make a rough sketch of the hyperfine patterns.

**4.** Construct the hyperfine patterns for radicals having the following nuclei:

(a) Three equivalent spins of 1 (e.g. $CD_3$).

(b) Two spins of 3/2.

(c) Two equivalent protons with splitting constant $a$ and three other protons with splitting constant $2a$.

**5.** Use McConnell's relation $a = Q\rho$ to estimate $\pi$-electron spin densities in the odd alternant hydrocarbon radical perinaphthenyl where the observed proton splittings are $a_1 = 2.2$

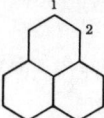

and $a_2 = 7.3$ gauss. Discuss the results and their interpretation in the light of molecular orbital theory.

**6.** The planar cyclooctatetraene negative ion has a pair of doubly degenerate nonbonding orbitals of the type

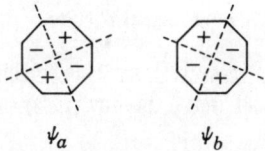

Write down the correct expressions for these orbitals as linear combinations of the carbon $2p_z$ atomic orbitals and find the corresponding spin distribution. What spin distribution do you expect to find in the methyl cyclooctatetraene anion?

**7.** Sketch the atomic orbitals over which the unpaired electron passes in the cyclohexadienyl radical. Can you construct hydrogen group orbitals which hyperconjugate with the $\pi$ electrons? Estimate the $\pi$-electron spin density on the ring carbons next to the $CH_2$ group and deduce a rough $Q$ value for the $CH_2$ protons.

**8.** Use the experimental hyperfine splittings of the allyl radical, and the fact that the spin densities add up to one, to estimate both a value of $Q$ and a $\pi$-electron spin distribution for the allyl radical.

**9.** Use the data of Fig. 6.5 and a $Q$ value of $-24.2$ gauss to deduce $\pi$-electron spin densities in the naphthalene negative ion. Now take Eq. (34) and predict the $C^{13}$ splittings. (For the 9 and 10 positions (34) must be altered to read $a_C = Q_1\rho_1 + Q_2(\rho_2 + \rho_3 + \rho_4)$.) The experimental values are $a_1 = +7.1$, $a_2 = -1.2$, $a_9 = -5.6$.

**10.*** According to the valence bond theory the wave function of the $\pi$ electrons in allyl is a resonance hybrid of the two structures $\dot{C}$—C=C and C=C—$\dot{C}$. It can be written as a sum of three Slater determinants formed from the $2p_z$ atomic orbitals:

$$\Psi = \|\phi_1(1)\phi_2(2)\phi_3(3)\| \frac{1}{\sqrt{6}}(2\alpha\beta\alpha - \alpha\alpha\beta - \beta\alpha\alpha)$$

What are the spin densities in the orbitals? Compare with the result of Problem 8.

**11.*** The Hückel molecular orbitals of butadiene $CH_2{=}CH{-}CH{=}CH_2$ obey the matrix equation

$$\begin{bmatrix} \alpha - E & \beta & 0 & 0 \\ \beta & \alpha - E & \beta & 0 \\ 0 & \beta & \alpha - E & \beta \\ 0 & 0 & \beta & \alpha - E \end{bmatrix} \begin{bmatrix} C_1 \\ C_2 \\ C_3 \\ C_4 \end{bmatrix} = 0$$

By looking for orbitals with coefficients of the form $(a, b, b, a)$ and $(p, q, -q, -p)$, or otherwise, calculate all four orbitals. Find the spin density in the butadiene negative ion, and estimate the probable values of the hyperfine splittings. Experiment gives $a_1 = a_4 = 7.617$, $a_2 = a_3 = 2.791$ gauss for this ion in liquid ammonia. (See Levy and Myers: *J. Chem. Phys.*, **41**: 1062 (1964).)


## SUGGESTIONS FOR FURTHER READING

Coulson: Chapter 9 and Section 13.4. Conjugated molecules. Hyperconjugation.

Carrington: *Quart. Revs.*, **17**: 67 (1963). Review on e.s.r. of aromatic radicals.

Streitwieser: *Molecular Orbital Theory for Organic Chemists* (New York: John Wiley & Sons, Inc., 1961). Chapters 2 and 6. Hückel molecular orbital theory.

Bolton and Fraenkel: *J. Chem. Phys.*, **40**: 3307 (1964). $C^{13}$ splitting in the anthracene negative ion.

Fraenkel: *Pure Appl. Chem.*, **4**: 143 (1962). Review on $C^{13}$ hyperfine structure.

Fessenden and Schuler: *J. Chem. Phys.*, **39**: 2147 (1963). E.S.R. spectra of saturated hydrocarbon radicals.

De Boer and Weissman: *J. Am. Chem. Soc.*, **80**: 4549 (1958). Alternant hydrocarbon ions.

McConnell and Chesnut: *J. Chem. Phys.*, **28**: 107 (1958). Theory of ring proton hyperfine structure.

Gendell, Freed, and Fraenkel: *J. Chem. Phys.*, **41**: 949 (1964). Signs of $C^{13}$ and $N^{14}$ splittings in tetracyanoethylene.

McLachlan: *Mol. Phys.*, **3**: 233 (1960). Theory of negative spin densities.

# CHAPTER 7

# E.S.R. OF TRAPPED ORGANIC RADICALS IN SOLIDS

## 7·1 INTRODUCTION

Many unstable radicals can be trapped in solids, and give highly interesting electron resonance spectra. The radicals may be formed by ultraviolet irradiation, electron bombardment, or exposure to nuclear radiations. Sometimes the radicals are trapped in glasses, frozen rare gas matrices, or in polycrystalline powders, but by far the best results are obtained from single crystals. The radicals are almost always regularly oriented relative to the crystal axes, and this provides the opportunity to rotate the crystal in the magnetic field and measure the anisotropic magnetic interactions. Although the presence of anisotropic tensors makes the analysis and interpretation of spectra more difficult than in solution, the results give valuable information about the structure and orientation of the unstable molecules. We shall confine ourselves in this chapter to radicals of spin 1/2. Most of those studied so far are derived from dicarboxylic acids and amino acids.

## 7·2 THE SPIN HAMILTONIAN

The spin Hamiltonian for a free radical in which one unpaired electron interacts with one proton is

$$\mathscr{H} = \beta \mathbf{H} \cdot \mathbf{g} \cdot \mathbf{S} + \mathbf{S} \cdot \mathbf{T} \cdot \mathbf{I} - g_N \beta_N \mathbf{H} \cdot \mathbf{I} \tag{1}$$

in which the hyperfine term consists of an isotropic part $a\mathbf{S} \cdot \mathbf{I}$ arising from the Fermi contact interaction and an anisotropic part $\mathbf{S} \cdot \mathbf{T}' \cdot \mathbf{I}$ due to the electron-nuclear dipolar interaction. We omit the chemical shielding of the nucleus since its effects are quite unimportant. In this chapter we shall be dealing with organic radicals which show relatively little $g$ tensor anistropy, so we adopt the simpler Hamiltonian,

$$\mathscr{H} = g\beta \mathbf{H} \cdot \mathbf{S} + \mathbf{S} \cdot \mathbf{T} \cdot \mathbf{I} - g_N \beta_N \mathbf{H} \cdot \mathbf{I} \tag{2}$$

in which $g$ is a scalar. The *anisotropic part* of the hyperfine tensor, $\mathbf{T}'$, can always be reduced to diagonal form by choosing as axes the principal axes $X, Y, Z$ (see Appendix

D). We shall call the principal values $t'_1$, $t'_2$, and $t'_3$, so that the dipolar part becomes

$$\mathbf{S \cdot T' \cdot I} = t'_1 S_X I_X + t'_2 S_Y I_Y + t'_3 S_Z I_Z \tag{3}$$

The complete interaction $\mathbf{S \cdot T \cdot I}$ differs only by including the isotropic term $a\mathbf{I \cdot S}$, so $\mathbf{T}$ is also diagonal in the $X, Y, Z$ axis system, and the hyperfine interaction reduces to the form

$$\mathbf{S \cdot T \cdot I} = A S_X I_X + B S_Y I_Y + C S_Z I_Z \tag{4}$$

where

$$A = t'_1 + a$$
$$B = t'_2 + a \tag{5}$$
$$C = t'_3 + a$$

We note that the dipolar tensor has zero trace so that $(t'_1 + t'_2 + t'_3)$ vanishes and $a$ is just the average of the principal values $A$, $B$, $C$. In nearly all single crystal measurements the hyperfine constants are expressed in Mc/s. Division by a factor of 2.80 converts these values to gauss, the unit nearly always used to describe solution splittings. (To add to the confusion, transition metal splittings are usually expressed in $cm^{-1}$. Since no faction seems inclined to give way to the other two and the literature can hardly be rewritten, we might as well be conventional!)

In order to make a reliable measurement of the components of $\mathbf{T}$ it is necessary to study a single crystal in which all the radicals are similarly oriented; the spectra from randomly oriented solids (polycrystalline or glassy) are often difficult to interpret and we shall not pay much attention to them. In this chapter our attention will be concentrated on determining the principal components of $\mathbf{T}$, but first we must study the energy levels and allowed transitions of the spin system.

## 7·3  THE FIRST-ORDER E.S.R. SPECTRUM

Let us begin by defining the direction of the magnetic field to be the $z$ axis. We shall also make two approximations. The first is that the quantization of the electron spin vector along the $z$ axis is not disturbed by the nuclear hyperfine interaction. This is a very good assumption, since the components of $\mathbf{T}$ are all small compared with $g\beta H$ unless the hyperfine interaction is exceptionally large, as with phosphorus. The second, which is often poor, is to ignore the nuclear Zeeman energy.

We may now ignore the spin components $S_x, S_y$ and look at the energy levels of the simpler Hamiltonian

$$\mathscr{H} = g\beta H S_z + S_z(T_{zx}I_x + T_{zy}I_y + T_{zz}I_z) \tag{6}$$

The hyperfine term may be thought of (see Fig. 7.1) as the energy of the nuclear spin in an effective magnetic field $\mathbf{H}_e$ produced by the unpaired electron, such that $\mathbf{S \cdot T \cdot I}$ is equivalent to an energy $-g_N\beta_N\mathbf{H}_e \cdot \mathbf{I}$. According to (6) the components of $\mathbf{H}_e$ must be taken to be

$$g_N\beta_N(H_{ex}, H_{ey}, H_{ez}) = (-S_z T_{zx}, -S_z T_{zy}, -S_z T_{zz}) \tag{7}$$

They therefore depend on the value of $S_z$; if we denote the fields for $m_S = +1/2$ and $m_S = -1/2$ by $\mathbf{H}_e(+)$ and $\mathbf{H}_e(-)$, their components are

$$g_N\beta_N\mathbf{H}_e(\pm) = \mp \tfrac{1}{2}(T_{zx}, T_{zy}, T_{zz}) \tag{8}$$

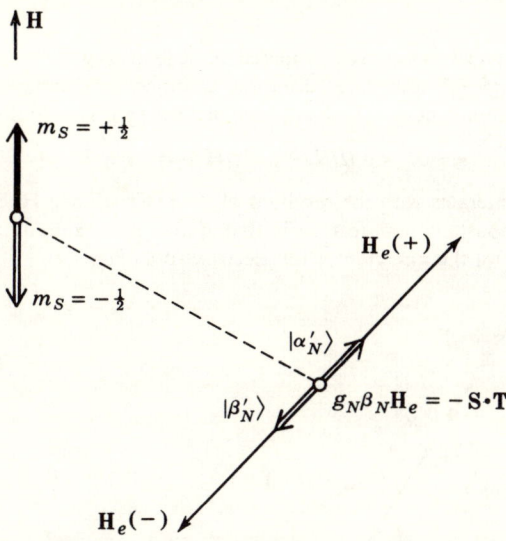

FIG. 7.1. The magnetic field at the nucleus produced by an unpaired electron.

The nuclear spin is now quantized either parallel or antiparallel to the direction of $\mathbf{H}_e$ and has two states denoted $|\alpha'_N\rangle$, $|\beta'_N\rangle$ in which the resolved spin angular momentum component along $\mathbf{H}_e$ is $+1/2$ or $-1/2$. The difference in energy between these two nuclear states is

$$g_N\beta_N|\mathbf{H}_e| = \tfrac{1}{2}\sqrt{T_{zx}^2 + T_{zy}^2 + T_{zz}^2} \tag{9}$$

We now obtain the first-order wave functions and their energies:

$$\left.\begin{matrix}|\alpha_e\alpha'_N\rangle\\|\alpha_e\beta'_N\rangle\end{matrix}\right\}\quad E = \tfrac{1}{2}g\beta H \pm \tfrac{1}{4}\sqrt{T_{zx}^2 + T_{zy}^2 + T_{zz}^2} \tag{10}$$

and

$$\left.\begin{matrix}|\beta_e\beta'_N\rangle\\|\beta_e\alpha'_N\rangle\end{matrix}\right\}\quad E = -\tfrac{1}{2}g\beta H \pm \tfrac{1}{4}\sqrt{T_{zx}^2 + T_{zy}^2 + T_{zz}^2} \tag{11}$$

This treatment assumes that $H_e$ is much larger than $H$, which is not usually true. For example a proton hyperfine splitting of 42 Mc/s corresponds to $H_e = \pm 5{,}000$ gauss and here $H_e$ is comparable with $H$. The intensities of the allowed e.s.r. transitions depend on the matrix elements of $S_x$ between these states, so we find, as usual, that the nuclear spin cannot change. The allowed transitions are $\alpha_e\alpha'_N \leftrightarrow \beta_e\alpha'_N$ and $\alpha_e\beta'_N \leftrightarrow \beta_e\beta'_N$, and we readily find that the separation $\Delta E$ between the two absorption lines is

$$(\Delta E)^2 = T_{zx}^2 + T_{zy}^2 + T_{zz}^2 \tag{12}$$

We shall soon see how this argument is modified when the magnetic field direction is not along the $z$ axis.

## 7·4 SECOND-ORDER EFFECTS

The second-order effects which we considered in the hydrogen atom, namely the mixing of the $|\alpha_e\beta_N\rangle$ and $|\beta_e\alpha_N\rangle$ spin states, are quite unimportant compared with the effects of the nuclear Zeeman energy. If this is included the spin Hamiltonian becomes

$$\mathscr{H} = g\beta H S_z - g_N\beta_N[H + H_e(\pm)]\cdot I \tag{13}$$

and the nucleus interacts with the resultant of the external field $H$ and the electronic field $H_e$. The important new feature is that both the magnitude and direction of $(H + H_e)$ depend on the direction of the electron spin (Fig. 7.2).

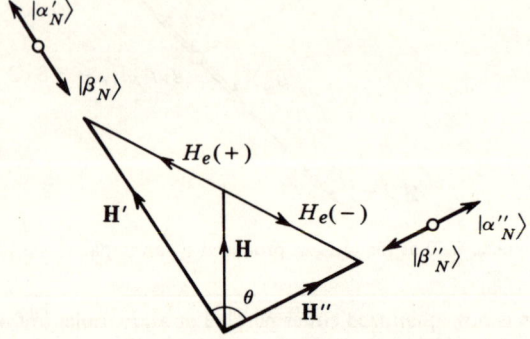

FIG. 7.2. Quantization of the nuclear spin in the combined fields $H$ and $H_e$.

First let us suppose that the electron spin is up. We shall call the effective field $\mathbf{H}'$ and describe its direction by a unit vector $\mathbf{h}'$. Then Eqs. (8) and (13) give

$$g_N\beta_N\mathbf{H}' = (-\tfrac{1}{2}T_{zx}, -\tfrac{1}{2}T_{zy}, g_N\beta_N H - \tfrac{1}{2}T_{zz}) \tag{14}$$

The two nuclear spin states $|\alpha'_N\rangle$ and $|\beta'_N\rangle$ are now quantized along $\mathbf{h}'$. The energy difference between the states $|\alpha_e\alpha'_N\rangle$ and $|\alpha_e\beta'_N\rangle$ is $g_N\beta_N|\mathbf{H}'|$ so it now depends on the strength of the external field as well as on the hyperfine tensor.

Conversely, with the electron spin down the field at the nucleus is $\mathbf{H}''$:

$$g_N\beta_N\mathbf{H}'' = (\tfrac{1}{2}T_{zx}, \tfrac{1}{2}T_{zy}, g_N\beta_N H + \tfrac{1}{2}T_{zz}) \tag{15}$$

and the corresponding spin states are $|\alpha''_N\rangle$, $|\beta''_N\rangle$, quantized along a direction $\mathbf{h}''$. To summarize the situation, the corrected wave functions and energies are:

$$\left.\begin{array}{l} |\alpha_e\alpha'_N\rangle \\ |\alpha_e\beta'_N\rangle \end{array}\right\} \quad E = \tfrac{1}{2}g\beta H \pm \tfrac{1}{2}g_N\beta_N|\mathbf{H}'| \tag{16}$$

$$\left.\begin{array}{l} |\beta_e\beta''_N\rangle \\ |\beta_e\alpha''_N\rangle \end{array}\right\} \quad E = -\tfrac{1}{2}g\beta H \pm \tfrac{1}{2}g_N\beta_N|\mathbf{H}''| \tag{17}$$

The nuclear Zeeman interaction clearly produces shifts in the absorption lines. More important, however, it causes forbidden transitions in which both *electron and nuclear* spins turn over. For example, consider the transition $\alpha_e \alpha'_N \leftrightarrow \beta_e \beta''_N$. The transition probability is proportional to the square of the matrix element

$$\langle \alpha_e \alpha'_N | S_x | \beta_e \beta''_N \rangle = \langle \alpha_e | S_x | \beta_e \rangle \langle \alpha'_N | \beta''_N \rangle \tag{18}$$

Normally the product $\langle \alpha'_N | \beta''_N \rangle$ of the nuclear spin states vanishes when the spin functions refer to a common axis of quantization. However, if the axes $\mathbf{h}'$ and $\mathbf{h}''$ are not parallel but lie at an angle $\theta$ to one another, the factor $|\langle \alpha'_N | \beta''_N \rangle|^2$ turns out to have the value $\sin^2 \theta/2$, and the transition is allowed. To compensate for this the strength of the allowed transition $\alpha_e \alpha'_N \leftrightarrow \beta_e \alpha''_N$ is reduced by a factor $\cos^2 \theta/2$.

The important new features which emerge are, that for some orientations two additional satellite lines are observed, and that detailed study of the angular variation reveals the *relative signs* of the tensor components.

## 7.5 EXPERIMENTAL DETERMINATION OF THE HYPERFINE TENSOR

In order to get the problem into proper perspective, let us first briefly consider the kind of experimental study which is performed and see qualitatively how one derives the tensor components from the experimental measurements. $\gamma$-irradiation of a single crystal of malonic acid, $CH_2(COOH)_2$, results in the formation of $CH(COOH)_2$ radicals which show a doublet splitting due to interaction of the odd electron with the C—H proton. Although there are two molecules in the unit cell of the crystal, these are related by inversion through a center of symmetry so that they give equivalent spectra at all orientations.

We commence the study by choosing three orthogonal axes $x$, $y$, $z$, which are fixed in the crystal. There is no limitation on the choice; we simply define an axis system which is easily identifiable in terms of the exterior appearance of the crystal. Usually one would choose one or more of the crystallographic axes. The crystal is then mounted with one axis ($x$, say) vertical and is either rotated about this axis, or alternatively, the magnet is rotated about the crystal. Thus one obtains a series of values for the doublet splitting as the magnetic field is rotated in the $yz$ plane. Two similar series of measurements are made with the $y$ and $z$ axes perpendicular to the field direction in turn.

From these measurements it is possible to derive values of the elements of $\mathbf{T}$ in terms of the $x,y,z$ system of axes. The final task is to find a transformation which diagonalizes the tensor. The transformation matrix defines the relative orientation of the $x,y,z$ system with respect to the radical principal axes, which we denote $X, Y, Z$. Thus it is possible to relate the orientation of the radical (which is not necessarily the same as that of an undamaged malonic acid molecule) to the crystallographic axes. We shall find in due course that the $CH(COOH)_2$ radical is planar, the unpaired electron occupying a carbon $2p_z$ orbital perpendicular to the plane. The $X, Y, Z$ axis system is defined below in Fig. 7.3, with the $X$ axis along the C—H bond, and $Z$ parallel to the axis of the $2p_z$ orbital. It does not matter whether we choose the origin of coordinates at the carbon or hydrogen nucleus, since we are only concerned with *directions* in space.

The first step, therefore, is to derive the tensor elements in terms of the chosen $x,y,z$ axes. For the sake of simplicity we shall again ignore the nuclear Zeeman energy

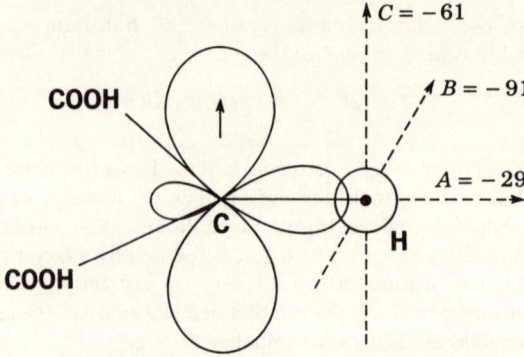

FIG. 7.3. Principal hyperfine axes of the malonic acid radical, showing the principal values $A$, $B$, $C$ (in Mc/s).

and use the first order theory (Section 7.3). For a start we need to calculate the line separation $\Delta E$ when the field $H$ is along an arbitrary direction described by three direction cosines $(l_x, l_y, l_z)$ which form the unit vector $\mathbf{I}$. The electron spin is now quantized along the $\mathbf{I}$ axis so that the effective magnetic field $\mathbf{H}_e$ acting at the nucleus is given by the equation

$$g_N \beta_N \mathbf{H}_e = -\mathbf{S} \cdot \mathbf{T} = \mp \tfrac{1}{2} \mathbf{I} \cdot \mathbf{T} \tag{19}$$

The vector $\mathbf{I} \cdot \mathbf{T}$ may also be written out in matrix notation as a row vector

$$\mathbf{I} \cdot \mathbf{T} = [l_x \ \ l_y \ \ l_z] \begin{bmatrix} T_{xx} & T_{xy} & T_{xz} \\ T_{yx} & T_{yy} & T_{yz} \\ T_{zx} & T_{zy} & T_{zz} \end{bmatrix} \tag{20}$$

We now obtain the line splitting $\Delta E = 2g_N \beta_N |\mathbf{H}_e|$, or

$$(\Delta E)^2 = (\mathbf{I} \cdot \mathbf{T})^2 = (\mathbf{I} \cdot \mathbf{T}) \cdot (\mathbf{T} \cdot \mathbf{I}) = \mathbf{I} \cdot (\mathbf{T} \cdot \mathbf{T}) \cdot \mathbf{I}$$

$$= \mathbf{I} \cdot (\mathbf{T}^2) \cdot \mathbf{I} \tag{21}$$

Equation (21) has the following meaning. $(\Delta E)^2$ is the square of the length of the row vector (20), or just the product of $\mathbf{I} \cdot \mathbf{T}$ with the column vector $\mathbf{T} \cdot \mathbf{I}$. Matrix multiplication then shows that the result involves the matrix $\mathbf{T}^2$, which is just the square of the hyperfine tensor:

$$\begin{bmatrix} (T^2)_{xx} & (T^2)_{xy} & (T^2)_{xz} \\ (T^2)_{yx} & (T^2)_{yy} & (T^2)_{yz} \\ (T^2)_{zx} & (T^2)_{zy} & (T^2)_{zz} \end{bmatrix} = \begin{bmatrix} T_{xx} & T_{xy} & T_{xz} \\ T_{yx} & T_{yy} & T_{yz} \\ T_{zx} & T_{zy} & T_{zz} \end{bmatrix}^2 \tag{22}$$

For example $(T^2)_{xy}$ is the $xy$ element of $\mathbf{T}^2$ and is not equal to $(T_{xy})^2$.

Let us suppose we mount the crystal with the $x$ axis vertical and rotate the magnetic field in the $yz$ plane, defining $\theta$ as the angle between the $z$ axis and the field direction. The direction cosines are now $(0, \sin\theta, \cos\theta)$ and Eq. (21) gives the result

$$(\Delta E)^2 = [0, \sin\theta, \cos\theta] \begin{bmatrix} (T^2)_{xx} & (T^2)_{xy} & (T^2)_{xz} \\ (T^2)_{yx} & (T^2)_{yy} & (T^2)_{yz} \\ (T^2)_{zx} & (T^2)_{zy} & (T^2)_{zz} \end{bmatrix} \begin{bmatrix} 0 \\ \sin\theta \\ \cos\theta \end{bmatrix}$$

$$= \sin^2\theta (T^2)_{yy} + 2\sin\theta\cos\theta (T^2)_{yz} + \cos^2\theta (T^2)_{zz} \tag{23}$$

Thus by measuring $\Delta E$ as a function of $\theta$ the $yy$, $yz$, and $zz$ components of the tensor $\mathbf{T}^2$ can be determined; the remaining elements are obtained by remounting the crystal and measuring $\Delta E$ in the $zx$ and $xy$ planes. All that remains is to find a transformation matrix $\mathbf{L}$ which diagonalizes $\mathbf{T}^2$, as described in Appendix D. The principal values of $\mathbf{T}^2$ are the squares $A^2$, $B^2$, $C^2$ of the principal values of $\mathbf{T}$, and the two tensors naturally have the same principal axes. It is important to note that measurements of $\Delta E$ can only determine the magnitudes of the principal components of $\mathbf{T}$ and not their signs. The matrix $\mathbf{L}$ gives the direction cosines of the three principal $X, Y, Z$ axes relative to the $x, y, z$ axes used for measurement. However one cannot determine, from the e.s.r. spectrum alone, which axis is $X$, which is $Y$, and which is $Z$. In the original study of $\gamma$-irradiated malonic acid, whose crystal structure is known, it was found that $X$, in fact, bisects the H—C—H bond angle in undamaged malonic acid molecules and that $Z$ is indeed perpendicular to the plane of the three carbon atoms. Thus provided one discounts the unlikely possibility that the radical plane differs from that of undamaged molecules by a rotation through 90°, the assignment can be made with certainty, and we shall proceed on this basis. If the crystal structure of malonic acid were not known, the experimental measurements alone would not identify the molecular symmetry axes. The principal values of the C—H proton coupling tensor of malonic acid are shown in Fig. 7.3. By a study of the second order lines it has been found that all three values have the same sign.

## 7·6  THE SIGN OF THE ELECTRON-NUCLEAR DIPOLAR COUPLING

It will be recalled from Section 7.2 that $\mathbf{T}$ consists of an isotropic part $a$ and a traceless dipolar part $\mathbf{T}'$. Depending on the absolute sign of $A$, $B$, and $C$ there are two possibilities;

$$a = +60.3 \text{ Mc/s}, \quad t'_1 = -31.3, \quad t'_2 = +30.7, \quad t'_3 = +0.7,$$
$$a = -60.3 \qquad\quad t'_1 = +31.3 \quad t'_2 = -30.7 \quad t'_3 = -0.7$$

The numerical calculation of the elements of the dipolar interaction tensor $\mathbf{T}'$ is rather complicated but it is not difficult to arrive at qualitative estimates which resolve the ambiguity. First we note that if the electron were localized at the carbon nucleus, the calculation of the dipole-dipole coupling would follow precisely the same lines as the proton-proton interaction discussed in Chapter 3, except that the Hamiltonian would contain the factor $-g\beta g_N\beta_N$ rather than $g_N^2\beta_N^2$. The present problem is rather more complicated because the electron is not localized at the carbon nucleus but is, in fact, distributed over the region of space represented by the $p$ orbital wave function. By analogy with Eq. (7) in Chapter 3, the electron-nuclear interaction Hamiltonian is

$$\mathscr{H} = -g\beta g_N\beta_N\left\{S_xI_x\left\langle\frac{r^2-3x^2}{r^5}\right\rangle + S_xI_y\left\langle\frac{-3xy}{r^5}\right\rangle + \cdots\right\} \tag{24}$$

where $\langle(r^2-3x^2)/r^5\rangle$ is an average over the whole spatial distribution of the unpaired spin; and so the principal values are

$$t'_1 = -g\beta g_N\beta_N\left\langle\frac{R^2-3X^2}{R^5}\right\rangle$$

$$t'_2 = -g\beta g_N\beta_N\left\langle\frac{R^2-3Y^2}{R^5}\right\rangle \tag{25}$$

$$t'_3 = -g\beta g_N\beta_N\left\langle\frac{R^2-3Z^2}{R^5}\right\rangle$$

To estimate the sign of $t'_3$, for example, we need only determine the sign of the quantity $\langle (3\cos^2\theta - 1)/R^3 \rangle$ where $\theta$ is the angle between $Z$ and the direction of $R$. Consider therefore the situation depicted in Fig. 7.4(c) below. The sign of the expression $(3\cos^2\theta - 1)$ in different regions of space is indicated, the dotted line indicating the cone for which $\cos^2\theta = 1/3$ ($\theta = 54°44'$). We see that the nodal surfaces bisect the regions of maximum electron density, so that $t'_3$ is expected to be very small. However if the field is along the $Y$ axis it is clear that $\theta$ is generally greater than $55°$ so that $\langle 3\cos^2\theta - 1\rangle$ is negative, and $t'_2$ is definitely negative. Similarly in Fig. 7.4(a) the electron is found mainly in the positive cone, and $t'_1$ is certainly positive.

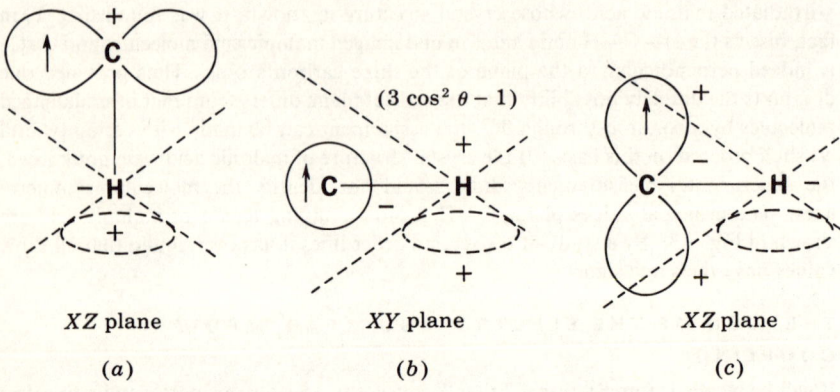

|  |  |  |
|:---:|:---:|:---:|
| *XZ* plane | *XY* plane | *XZ* plane |
| (a) | (b) | (c) |

FIG. 7.4. Dipolar hyperfine interaction in the $CH(COOH_2)$ radical: (a) field along the CH bond; (b) across the bond; (c) perpendicular to the radical plane.

We therefore conclude that all three principal values of **T** are *negative* so that the correct choice is

$$a = -60.3 \text{ Mc/s:} \qquad t'_1 = 31.3$$
$$t'_2 = -30.7$$
$$t'_3 = -0.7$$

Detailed calculations give results in satisfactory agreement with experiment. The most important conclusion is that the isotropic splitting $a$ is negative. This result makes it clear that the C—H proton lies in the nodal plane of the carbon $\pi$ orbital and confirms that the $\sigma - \pi$ exchange coupling in the C—H bond does indeed produce a negative spin density at the proton.

## 7·7 α-PROTON COUPLING TENSORS

If there are two α-protons directly bonded to the odd carbon atom they will have different hyperfine tensors and one can determine bond angles from the e.s.r. spectrum. To cite one example, the $CH_2(COOH)$ radical is also formed by γ-irradiation of malonic acid and since the odd electron now interacts with two protons, four strong lines are observed. Measurement of the angular variation of the splittings enables the two coupling tensors to be obtained in terms of the $x, y, z$ axis system, and diagonalization of these tensors yields the principal values of each tensor in terms of the local bond axes shown below (Fig. 7.5). Assuming the radical to be planar the $Z_1$ and $Z_2$ axes

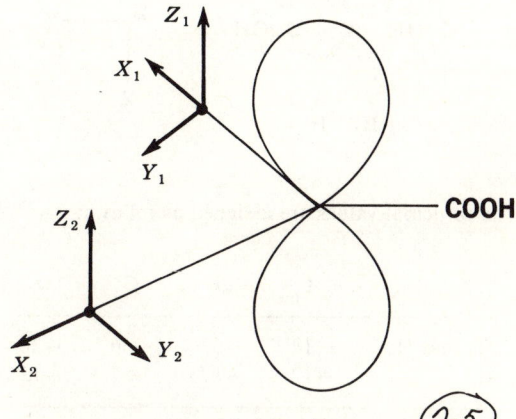

FIG. 7.5. Proton hyperfine tensors of the $CH_2(COOH)$ radical.

should be parallel; in fact the angle between them is found to be 4.4°. The angle between the $X_1$ and $X_2$ axes gives the H—C—H bond angle directly and is found to be $116 \pm 5°$. Hence within the limits of experimental error, the assumption of trigonal hybridization by the α-carbon atoms is confirmed. The experimental values for the two coupling tensors are

$$
\begin{aligned}
A_1 &= -30, & B_1 &= -55, & C_1 &= -91: & a_1 &= -62 \text{ Mc/s} \\
A_2 &= -37, & B_2 &= -59, & C_2 &= -92: & a_2 &= -63 \text{ Mc/s}
\end{aligned}
$$

where the negative signs are assumed. The principal values of the two α-proton tensors are closely similar as one would expect, and also similar to the values for the $CH(COOH)_2$ radical.

## 7·8 DELOCALIZED π-ELECTRON RADICALS

α-proton coupling tensors differ very little from one saturated hydrocarbon radical to another. If, however, the odd electron is delocalized, the principal values are reduced proportionately. One of the most interesting examples is the substituted allyl radical, $(COOH)CH=CH—\dot{C}H(COOH)$, formed by x-irradiation of glutaconic acid. The reader will recall that the allyl radical is of special interest, being the simplest π-electron radical in which negative π spin densities are expected to occur. We have already seen that it is possible to determine the sign of the isotropic coupling constant $a$ from single crystal spectra; we shall now show that orientation-dependence studies make it possible to determine the sign of the π spin density at a carbon atom and, in particular, to show that the spin density on the central carbon atom in allyl is indeed negative.

The e.s.r. spectrum indicates that the hyperfine tensors $T_1$, $T_2$, and $T_3$ of the three C—H protons have a common set of principal axes $X, Y, Z$. Furthermore two of the protons have the same principal values. This all agrees perfectly with the structural formula

$$\begin{array}{c}
\text{H}_2 \\
|
\end{array}$$

COOH   C   COOH

```
         H₂
         |
COOH     C     COOH
    \ . /   \\ /
      C      C
      |      |
      H₁     H₃
```

```
Z ──────► Y
 |
 |
 ▼
 X
```

for the radical. The principal values are assigned as follows:

|            | $A$  | $B$  | $C$  | $a$  |
|------------|------|------|------|------|
| H₁ and H₃  | −18  | −53  | −36  | −36  |
| H₂         | +12  | +17  | +7   | +12  |

The negative signs of $T_1$ and $T_3$ are assumed by analogy with the results for malonic acid; it remains for us to show that the components of $T_2$ are positive, and to identify the axes $X, Y, Z$, in terms of the radical structure.

We start by estimating the spin density on each carbon atom, and if we assume, from the results for $CH_2(COOH)$, that unit spin density on a carbon atom leads to an isotropic α-proton splitting of $-63$ Mc/s, the spin densities on the three carbon atoms are calculated to be

$$\rho_1 = \pm 0.57, \qquad \rho_2 = \pm 0.19, \qquad \rho_3 = \pm 0.57$$

There are contributions to the anisotropic splitting of $H_1$ from $\rho_1$, $\rho_2$, and $\rho_3$. However carbon atom 3 is so far away from $H_1$ that its contribution can be neglected; similarly $\rho_2$ is small and its contribution to the anisotropic splitting of $H_1$ is likely to be small. Using the results for the α-proton tensor in $CH(COOH)_2$ we expect the principal values of $T_1$ to be $-18$, $-54$, and $-36$, which are indeed very close to the measured values listed above. We deduce that for atoms 1 and 3, $X$ is along the C—H bond and $Z$ is parallel to the axis of the $p_\pi$ orbital.

The anisotropic splitting for $H_2$ does not depend solely on $\rho_2$, since, although carbon atoms 1 and 3 are further away, $\rho_1$ and $\rho_3$ are three times as large as $\rho_2$. The principal values of $T_2$ are therefore determined by summing the three contributions, the sign of the contribution being either positive or negative in the case of $\rho_2$. The calculated values are as follows:

|                  | $A$   | $B$    | $C$    |
|------------------|-------|--------|--------|
| $\rho_2$ positive | +0.9  | −19.2  | −18.4  |
| $\rho_2$ negative | +12.7 | +16.8  | +5.6   |

Recalling that the observed principal values of $T_2$ are 12, 17, and 7 Mc/s, we see that the agreement between experiment and theory is excellent provided the spin density on carbon atom 2 is taken to be negative.

# 7·9 HYPERFINE COUPLING FROM β PROTONS

The distance between an odd electron localized on the α carbon atom of a free radical and the protons attached to the next β carbon is rather large, and the anisotropic dipolar coupling is therefore small. However we saw in Chapter 6 that methyl protons may have a large contact hyperfine interaction $a$ which varies with the angle $\theta$ between the plane of the C—C and C—H bonds and the axis of the $p_z$ orbital (see Fig. 6.9). In fact

$$a = B_0 + B_2 \cos^2 \theta \tag{26}$$

The $CH_3CH(COOH)$ radical, formed by γ-irradiation of a single crystal of alanine, demonstrates this relation very nicely. At room temperature the methyl group rotates freely about the C—C bond and the three protons have identical hyperfine tensors which are almost isotropic (there is also the usual anisotropic splitting from the α proton). The spectrum at 77°K is much more complex, for the methyl protons are no longer equivalent. This means that the methyl group is either completely locked or rotates rather slowly.

Spectra taken with the field along the $c$ axis of the crystal are shown in Fig. 7.6. In this direction the α and β protons all have the same hyperfine splitting at room temperature, giving a 1:4:6:4:1 pattern. At 77° the α and $\beta_2$ protons still have equal splittings, but $\beta_1$ and $\beta_3$ are different, so there is a simple twelve-line spectrum with intensities 1:1:2:2:1:1:1:1:2:2:1:1.

Analysis of the room temperature spectrum shows that the averaged coupling tensor of the rotating methyl protons has principal values of +67.0, +67.5, and +76.5 Mc/s (isotropic part +70), with cylindrical symmetry about the C—C bond. The α proton couplings are −25.0, −89.4, and −49.8, fairly close to malonic acid, and the angle between the α $X$ principal axis and the methyl $Z$ axis is 121°. This gives a measure of the $CH_3$—C—H bond angle. The low temperature spectra taken along the three crystal axes yield the following diagonal elements for the methyl hyperfine tensors:

|            | $T_{aa}$ | $T_{bb}$ | $T_{cc}$ | Mean |
|------------|----------|----------|----------|------|
| $\beta_1$  | 116      | 118      | 128      | 120  |
| $\beta_2$  | 77       | 76       | 77       | 77   |
| $\beta_3$  | 11       | 16       | 14       | 14   |

The isotropic parts of these tensors, 120, 77, and 14 can indeed be fitted perfectly to Eq. (26) with $B_0 = +9$, $B_2 = +122$ Mc/s, and $\theta = 18°$. The average of the three splittings, 70 Mc/s, is the same as at room temperature.

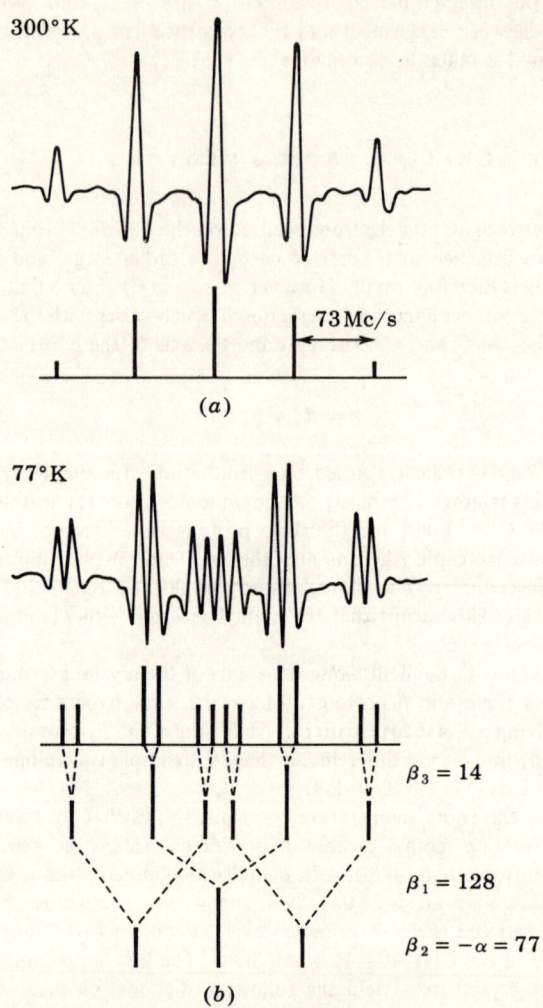

FIG. 7.6. Hyperfine structure of $CH_3CH(COOH)$ radical at (a) 300°K, and (b) 77°K. Magnetic field along the crystal $c$ axis. (The curves show the second derivatives of the resonance absorption.)

## 7·10 HYPERFINE TENSORS OF OTHER NUCLEI

### 7.10.1 Carbon 13

The *anisotropic* hyperfine tensors of $C^{13}$ and $N^{14}$ nuclei depend mainly on the unpaired electron density in the $2p$ atomic orbital of the same atom, whereas the isotropic part measures the $1s$ or $2s$ character of the odd electron.

First we note that if the unpaired electron were localized entirely in a $2p_z$ orbital the traceless dipolar hyperfine tensor would necessarily be axially symmetric, with its principal values $t'_1$, $t'_2$, $t'_3$ equal to $-(1/2)t$, $-(1/2)t$, and $+t$. As the electron is concentrated in the region where $(3 \cos^2 \theta - 1) > 0$ it is clear that $t$ is positive (see Fig. 7.7). To be more precise, if the $2p_z$ wave function takes the standard form

$$\psi = \sqrt{\frac{3}{4\pi}} \cos \theta \frac{f(r)}{r} \tag{27}$$

Eq. (25) shows us that

$$t = \frac{3}{4\pi} \int_0^\pi \int_0^{2\pi} (3 \cos^2 \theta - 1) \cos^2 \theta \sin \theta \, d\theta \, d\phi \int_0^\infty \frac{1}{r^3} f^2(r) \, dr \tag{28}$$

and we shall leave the reader, if he wishes, to obtain the formula

$$t = \frac{4}{5} g\beta g_N \beta_N \left\langle \frac{1}{r^3} \right\rangle \tag{29}$$

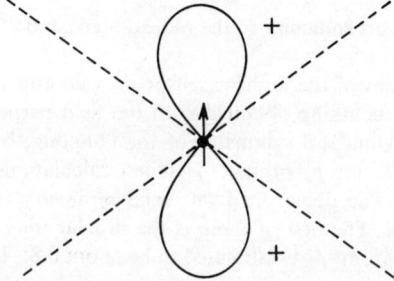

FIG. 7.7. Dipolar hyperfine coupling between a nucleus and a $2p_z$ electron on the same atom.

Analysis of the weak $C^{13}$ hyperfine structure from artificially enriched samples of malonic acid radical $CH(COOH)_2$ reveals that the principal axes of the α-proton and $C^{13}$ tensors are identical. The relative signs of the principal values are known, but the absolute signs are not; that is, either

$$a_C = +92.6, \quad t'_1 = -50.4, \quad t'_2 = -59.8, \quad t'_3 = +120.1$$

or all the signs are reversed. As we have just seen, $t'_3$ must be positive for a $2p_z$ orbital, so the signs given here are in fact correct. Hence the isotropic part $a_C$ is positive, confirming the predictions in Chapter 6. We note that there are small departures from axial symmetry which could arise from the presence of unpaired electrons on the carboxyl groups or in the C—H and C—C bonds.

### 7.10.2 Nitrogen 14

The general features of $N^{14}$ splittings in π-electron radicals are very similar to those of $C^{13}$, and again the tensors are usually close to being axially symmetric. Two examples are listed below:

|  | A | B | C |
|---|---|---|---|
| $N(SO_3)_2^{--}$ | +6 | 0 | +106 |
| $NH(SO_3)^-$ | +5.5 | +10.3 | +97.5 |

In certain cases internal motion partially averages the anisotropy. The $NH_3^+$ radical in $X$-irradiated ammonium perchlorate crystals, for example, shows an almost perfectly isotropic e.s.r. spectrum at 100°C, with $a_N = 50.7$ and $a_H = 70.0$. On lowering the temperature to 25°C restricted rotation sets in and the spectrum becomes moderately anisotropic, the principal values $A$, $B$, $C$, of the hyperfine tensors being $-73.5$, $-72.3$, $-71.6$ Mc/s for the proton and $+61.8$, $+53.3$, $+48.7$ for the nitrogen.

### 7.10.3  $F^{19}$ Splittings

Both the isotropic and anisotropic parts of fluorine interactions tend to be unusually large. A typical example is the $CHFCONH_2$ radical, formed by $\gamma$-irradiation of monofluoroacetamide. The e.s.r. spectrum shows hyperfine splitting from the $\alpha$-proton and the fluorine, the principal values of the coupling tensors being as follows:

$$a_H = -63, \qquad t'_1 = +32, \qquad t'_2 = 0, \qquad t'_3 = -33$$
$$a_F = +158, \qquad t'_1 = -203, \qquad t'_2 = -169, \qquad t'_3 = +372$$

For both tensors $Z$ is perpendicular to the radical plane and $X$ is parallel to the C—H or C—F bond.

The principal values of the fluorine tensor provide information about the C—F bond. The very large coupling obtained with the field perpendicular to the radical plane, and the near-cylindrical symmetry of the coupling, both imply considerable spin density in the fluorine $p_\pi$ orbital. Detailed calculations suggest that the spin density $\rho_F$ is $+0.12$. The departure from axial symmetry is probably due to two relatively small effects. The first of these is the dipolar interaction involving $\pi$ spin density on the carbon atom; $\rho_C$ is estimated to be about 0.8. The second contribution arises from the presence of unpaired spin in the fluorine $2p_\sigma$ orbital, resulting from spin polarization of the C—F bonding electrons.

There are therefore three main contributions to the fluorine dipolar tensor and assessment of their relative importance provides information as to the degree of carbon-fluorine double bonding.

It may be noted that $\beta$-fluorine hyperfine interactions are also quite large in the $(CO_2^-)CF_2\ CF(CO_2^-)$ radical; the coupling tensor is again close to axial symmetry, indicating delocalization of the odd electron into a fluorine $2p_\pi$ orbital.

## 7·11  RANDOMLY ORIENTED SOLIDS

Most of the early work on organic free radicals was carried out using polycrystalline or glassy samples and many interpretations were made without regard to the effects of hyperfine anisotropy; errors were therefore quite common. The development of rapid flow systems and other techniques for studying free radicals in the liquid phase has rendered much of this early work obsolete.

In general, hyperfine interactions with $\beta$ protons can usually be interpreted satisfactorily, since the lack of anisotropy simplifies the observed spectra. Internal molecular motions which result in partial averaging of anisotropic effects also make it easier to interpret the hyperfine patterns. One cannot escape the conclusion, however, that most of our extensive understanding of the e.s.r. spectra of free radicals is based on liquid phase or single crystal studies. There are, of course, many instances where single crystal studies are not possible; the study of polymers is an outstanding example.

We shall not attempt to review this field, since the spectral interpretation does not involve any new principles of spin resonance.

## PROBLEMS

**1.** Construct the $4 \times 4$ energy matrix of the first-order Hamiltonian, Eq. (6). Find its eigenvalues and eigenvectors and verify that the results agree with Eq. (9).

**2.** The resolved value $I_{z'}$ of the nuclear spin along a direction in the $xz$ plane making an angle $\theta'$ with the $z$ axis is measured by the operator $I_{z'} = I_z \cos \theta' + I_x \sin \theta'$. Write out the $2 \times 2$ matrix of $I_{z'}$ and find its eigenstates $|\alpha'\rangle$ and $|\beta'\rangle$.

**3.** In Problem 2 the angle is changed to $\theta''$, giving new states $|\alpha''\rangle$ and $|\beta''\rangle$. Prove that

$$|\langle \alpha' | \beta'' \rangle|^2 = \sin^2 \theta/2$$

where $\theta = (\theta' - \theta'')$.

**4.** A radical with axially symmetric $g$ and hyperfine tensors and a single nucleus of spin $1/2$ is placed in a magnetic field which makes an angle $\theta$ with the symmetry axis. Discuss the energy levels of the spin Hamiltonian

$$\mathcal{H} = \beta H(g_{\parallel} S_z \cos \theta + g_{\perp} S_x \sin \theta) + A(I_x S_x + I_y S_y) + C I_z S_z - g_N \beta_N H(I_z \cos \theta + I_x \sin \theta)$$

Explain any approximations which you feel you should make.

**5.** Use Eq. (29) and the observed $C^{13}$ hyperfine tensor of malonic acid to calculate the mean value of $1/r^3$ for a carbon $2p$ electron. (The value computed from a Hartree-Fock atomic wave function is 11.42 (Ångstroms)$^{-3}$.)

**6.** The radical $(COOH)\dot{C}H \, CH_2(COOH)$ is formed by $\gamma$-irradiation of succinic acid. The $\alpha$-proton coupling tensor of one of the two types of radical in the crystal unit cell was found to have components

$$\begin{bmatrix} -60 & +21 & +14 \\ +21 & -45 & -8 \\ +14 & -8 & -76 \end{bmatrix}$$

Verify that its principal values are $-92$, $-59$, and $-30$ Mc/s. Find the direction of the C—H bond.

**7.** The adipic acid radical $(COOH)\dot{C}H(CH_2)_3COOH$ shows two $\beta$-proton couplings whose isotropic parts are $+112$ and $+74$ Mc/s. Interpret this observation in terms of the ideas in Section 7.9 and estimate the twisting of the CH—CH$_2$ bond.

**8.** The microwave spectrum of NO$_2$ shows that the components of the N$^{14}$ hyperfine tensor are

$$a_N = 146.53 \text{ Mc/s}, \qquad t'_1 = -18.73, \qquad t'_2 = -19.77, \qquad t'_3 = +38.50$$

The $Z$ axis bisects the ONO bond angle and $X$ is perpendicular to the plane of the molecule. The same radical produced by irradiation of lead nitrate crystals at $77°K$ shows an axially symmetric hyperfine tensor with principal values 140, 160, 160. The principal $g$ values in the two cases are (a) $g_{zz} = 2.00199, g_{yy} = 1.991015, g_{xx} = 2.006178$, and (b) $g_{\parallel} = 2.004, g_{\perp} = 1.995$. Interpret these results in terms of molecular rotation. About which axis does the NO$_2$ radical rotate?

## SUGGESTIONS FOR FURTHER READING

Slichter: Section 7.3. Anisotropic hyperfine splitting.

Morton: *Chem. Revs.* **64**: 453 (1964). A good summary of experimental work.

McConnell, Heller, Cole, and Fessenden: *J. Am. Chem. Soc.*, **82**: 766 (1960).  Experiments on malonic acid. Analysis of the spectrum explained.

Trammell, Zeldes, and Livingston: *Phys. Rev.*, **110**: 630 (1958).  Theory of the hyperfine energy levels and local field $H_e$.

Horsfield, Morton, and Whiffen: *Mol. Phys.*, **4**: 327, 425 (1961).  $CH_2COOH$ and $CH_3CHCOOH$ radicals.

Cook, Rowlands, and Whiffen: *Mol. Phys.*, **7**: 31 (1963).  Fluorine hyperfine structure in $CHFCONH_2$.

Silverstone, Wood, and McConnell: *J. Chem. Phys.*, **41**: 2311 (1964).  Study of environment effects on $C_7H_7$ in crystalline solids.

# CHAPTER 8

# E.S.R. OF

# ORGANIC MOLECULES

# IN TRIPLET STATES

## 8·1 INTRODUCTION

A molecule in which the total spin of the electrons is one ($S = 1$) is said to be in a triplet state. This name is used because a triplet consists of three distinct sublevels or states with almost the same energy, and in an atom these sublevels can be distinguished by their spin angular momentum $S_z$ about a chosen axis, which has three possible eigenvalues $+1$, $0$, or $-1$. Just as a radical with spin 1/2 must have an odd number of electrons, so a triplet molecule must have an even number, and one often says that it must have two unpaired electrons. To understand this more clearly let us imagine the electronic structure of the molecule to be built up by adding the successive electron shells onto the bare nuclei. Each inner shell orbital is filled with two electrons of opposite spin and all the inner shell spins are paired; then comes the shell of valence electrons where some orbitals are doubly occupied by pairs of electrons, and others are singly occupied. It is the spins of these singly occupied orbitals which finally couple together in a certain way to give the total spin $S$. For example two orbitals can give rise to a singlet state ($S = 0$) or a triplet (see Appendix C).

Two factors determine whether the ground state of a molecule is a singlet or triplet, namely, the orbital energies of the valence electrons and the strengths of the electrostatic exchange interactions between electrons of parallel spin. In the ground state of an organic molecule like naphthalene, each bonding orbital holds two electrons and all the spins are paired. To uncouple a pair of electrons it is necessary to take one electron from a doubly occupied orbital and promote it to a vacant orbital of higher energy. The promotion energy is least if one excites an electron from the highest filled orbital, $\psi_a$, into the lowest empty one, $\psi_b$. The electrostatic exchange energy of the two electrons is large and negative when the spins are parallel, so that the triplet usually lies lower than the singlet excited state. The energy gap between the ground singlet state and the excited triplet depends on the difference between the orbital promotion energy $E_b - E_a$ and the exchange integral

$$K_{ab} = \iint \psi_a^* \psi_b(\mathbf{r}_1) \frac{e^2}{r_{12}} \psi_b^* \psi_a(\mathbf{r}_2) \, d\tau_1 \, d\tau_2$$

If the orbitals $a$ and $b$ are sufficiently close together the exchange energy predominates and the ground state of the molecule is a triplet. This is the case for oxygen $(O_2)$ and for several organic molecules. For example, ultraviolet irradiation of diazodiphenyl-

methane leads to the formation of diphenylmethylene which can be stabilized in solid matrices at low temperatures and has a triplet ground state.

The e.s.r. spectrum characteristic of a triplet state is only obtained if the electron spins are strongly correlated, so that the energy difference between the singlet and triplet states is larger than any magnetic interactions. For instance the spins of two hydrogen atoms one centimeter apart would clearly *not* be correlated.

If there were only exchange and electrostatic interactions between electrons the three sublevels of a triplet state would be exactly degenerate and the e.s.r. spectra would be very like those of radicals with spin 1/2. The most important single fact about triplet states is that the magnetic dipole-dipole forces between the two unpaired electrons remove the degeneracy and lead to highly anisotropic spectra in most molecules. The essential theory parallels closely the theory of nuclear dipole-dipole interactions developed in Chapter 3; only the magnitudes of the magnetic interactions are very different.

In this chapter we shall be concerned with four types of molecular system on which e.s.r. studies have been made:

1. Aromatic hydrocarbons, like naphthalene, which have phosphorescent excited triplet states possessing a long lifetime, and can be continuously excited by irradiating with ultraviolet light.
2. Organic molecules with ground triplet states.
3. Molecular complexes in solids, which have low-lying "triplet exciton" states.
4. Certain ion-radical clusters in solution.

## 8·2 ELECTRON SPIN-SPIN INTERACTION

In formulating a suitable spin Hamiltonian, we first note that there are at least four magnetic interactions to be considered, as follows:

1. The electronic Zeeman interactions.
2. The interaction of the electron spins with each other.
3. Nuclear hyperfine interactions.
4. Nuclear Zeeman interactions.

For the most part we shall be concerned with (1) and (2) only and may write the spin Hamiltonian of two electrons in a triplet state as

$$\mathscr{H} = g\beta \mathbf{H} \cdot (\mathbf{S}_1 + \mathbf{S}_2) + g^2\beta^2 \left\{ \frac{\mathbf{S}_1 \cdot \mathbf{S}_2}{r^3} - \frac{3(\mathbf{S}_1 \cdot \mathbf{r})(\mathbf{S}_2 \cdot \mathbf{r})}{r^5} \right\} \tag{1}$$

where $\mathbf{r}$ is the vector joining the two electrons. This expression closely resembles the Hamiltonian representing the dipolar interaction of two proton spins in a magnetic field. As soon as the two electrons are confined within a molecule several important points must be considered. The two spins $S_1$ and $S_2$ are parallel and form a resultant spin $S$; the isotropic $g$ value should be replaced by an anisotropic $g$ tensor; the dipolar coupling should be averaged over all possible positions and spin states of the two coupled electrons and expressed in terms of the direction of the total spin $S$. For most organic triplets the $g$ tensor anisotropy is rather small and we shall first neglect it entirely, concentrating on the spin-spin interaction.

Let us begin by assuming that both electrons are fixed in space and have some as yet unspecified triplet spin wave function. We proceed to expand the dipolar terms as follows:

$$\mathscr{H}_D = g^2\beta^2 \left\{ S_{1x}S_{2x}\left(\frac{r^2 - 3x^2}{r^5}\right) + {}_1S_yS_{2y}\left(\frac{r^2 - 3y^2}{r^5}\right) \right.$$

$$+ S_{1z}S_{2z}\left(\frac{r^2 - 3z^2}{r^5}\right) - (S_{1x}S_{2y} + S_{1y}S_{2x})\frac{3xy}{r^5}$$

$$\left. - (S_{1y}S_{2z} + S_{1z}S_{2y})\frac{3yz}{r^5} - (S_{1z}S_{2x} + S_{1x}S_{2z})\frac{3zx}{r^5} \right\}. \tag{2}$$

Since the two spins are correlated it will be convenient to express everything in terms of the total spin $S$. For example,

$$S_x^2 = (S_{1x} + S_{2x})^2$$

$$= 2S_{1x}S_{2x} + S_{1x}^2 + S_{2x}^2$$

$$= 2S_{1x}S_{2x} + \tfrac{1}{2} \tag{3}$$

since $S_{1x}^2 = S_{2x}^2 = 1/4$ (see Appendix C). Furthermore the relation

$$S_xS_y = (S_{1x} + S_{2x})(S_{1y} + S_{2y})$$

$$= (S_{1x}S_{2y} + S_{2x}S_{1y}) + S_{1x}S_{1y} + S_{2x}S_{2y}$$

$$= (S_{1x}S_{2y} + S_{2x}S_{1y}) + \tfrac{1}{2}i(S_{1z} + S_{2z}) \tag{4}$$

follows because $S_{1x}S_{1y} = \tfrac{1}{2}(iS_{1z})$. Also $S_{1y}S_{1x} = -\tfrac{1}{2}(iS_{1z})$, and one may derive an analogous expression for $S_yS_x$, obtaining

$$(S_xS_y + S_yS_x) = 2(S_{1x}S_{2y} + S_{2x}S_{1y}) \tag{5}$$

Substitution of (3) and (5) into (2) leads then to a simplified spin Hamiltonian

$$\mathscr{H}_D = \frac{1}{2}g^2\beta^2\left\{ S_x^2\left(\frac{r^2 - 3x^2}{r^5}\right) + \cdots + \cdots - (S_xS_y + S_yS_x)\frac{3xy}{r^5} - \cdots - \cdots \right\} \tag{6}$$

expressed entirely in terms of the total spin $S$. Finally one must average over all positions $r_1$ and $r_2$ of the two electrons, and the final result is an *effective spin Hamiltonian* of the type

$$\mathscr{H}_D = S \cdot D \cdot S \tag{7}$$

$D$ is a symmetric tensor called the zero-field splitting tensor and its components are given by averages over the electronic wave function:

$$D_{xx} = \frac{1}{2}g^2\beta^2\left\langle \frac{r_{12}^2 - 3x_{12}^2}{r_{12}^5} \right\rangle, \qquad D_{xy} = \frac{1}{2}g^2\beta^2\left\langle \frac{-3x_{12}y_{12}}{r_{12}^5} \right\rangle \tag{8}$$

and so on. In terms of the principal axes which diagonalize the zero-field tensor the Hamiltonian becomes

$$\mathcal{H}_D = -XS_x^2 - YS_y^2 - ZS_z^2 \tag{9}$$

where $-X$, $-Y$, and $-Z$ are the principal values. However the tensor is traceless $(X + Y + Z = 0)$ so that the zero-field splitting can be rewritten in terms of just two independent constants, which are commonly called $D$ and $E$:

$$\mathcal{H}_D = D(S_z^2 - \tfrac{1}{3}S^2) + E(S_x^2 - S_y^2) \tag{10}$$

We shall see in the next section that the spin Hamiltonian $\mathcal{H}_D$ removes the degeneracy of the three triplet wave functions even if there is no external magnetic field—hence the name "zero-field splitting." We should point out that in transition metal complexes with triplet ground states there is also a zero-field splitting, formally described by a spin Hamiltonian identical with (10). However the splitting is due mainly to spin-orbit coupling instead of spin-spin magnetic dipole forces.

## 8·3  THE TRIPLET ENERGY LEVELS

In the last section we arrived at the Hamiltonian

$$\mathcal{H} = g\beta \mathbf{H} \cdot \mathbf{S} - XS_x^2 - YS_y^2 - ZS_z^2 \tag{11}$$

We could commence the calculation of the energy levels by taking the usual triplet spin functions

$$|T_{+1}\rangle = |\alpha_1\alpha_2\rangle$$
$$|T_0\rangle \;\; = \frac{1}{\sqrt{2}}|\alpha_1\beta_2 + \beta_1\alpha_2\rangle \tag{12}$$
$$|T_{-1}\rangle = |\beta_1\beta_2\rangle$$

in which the spin component $S_z$ has the values $+1$, $0$ or $-1$. However we shall find it much easier to work in terms of three different spin functions

$$|T_x\rangle = \frac{1}{\sqrt{2}}|\beta_1\beta_2 - \alpha_1\alpha_2\rangle = \frac{1}{\sqrt{2}}|T_{-1} - T_{+1}\rangle$$

$$|T_y\rangle = \frac{i}{\sqrt{2}}|\beta_1\beta_2 + \alpha_1\alpha_2\rangle = \frac{i}{\sqrt{2}}|T_{-1} + T_{+1}\rangle \tag{13}$$

$$|T_z\rangle = \frac{1}{\sqrt{2}}|\alpha_1\beta_2 + \beta_1\alpha_2\rangle = |T_0\rangle$$

These new functions have their spin quantized along the $x$, $y$, and $z$ axes, respectively, and they therefore diagonalize the zero-field Hamiltonian.

We now calculate the matrix elements of $\mathcal{H}$ within the set of basis functions (13). The reader may verify the following formulae for the effect of the various spin operators on the new functions:

$$
\begin{aligned}
S_z|T_x\rangle &= i|T_y\rangle & S_z^2|T_x\rangle &= |T_x\rangle \\
S_z|T_y\rangle &= -i|T_x\rangle & S_z^2|T_y\rangle &= |T_y\rangle \\
S_z|T_z\rangle &= 0 & S_z^2|T_z\rangle &= 0
\end{aligned}
\tag{14}
$$

and so on. These results are obtained directly from the definitions in (13). For example,

$$S_z|T_x\rangle = \frac{1}{\sqrt{2}}(S_{1z} + S_{2z})|\beta_1\beta_2 - \alpha_1\alpha_2\rangle$$

$$= \frac{1}{\sqrt{2}}|-\beta_1\beta_2 - \alpha_1\alpha_2\rangle$$

$$= i|T_y\rangle \tag{15}$$

The Hamiltonian matrix is therefore as follows

$$
\begin{array}{c}
\langle T_x| \\
\langle T_y| \\
\langle T_z|
\end{array}
\begin{bmatrix}
X & -ig\beta H_z & ig\beta H_y \\
ig\beta H_z & Y & -ig\beta H_x \\
-ig\beta H_y & ig\beta H_x & Z
\end{bmatrix}
\tag{16}
$$

Let us now be specific and consider the case of the naphthalene triplet state with the molecular axes defined below.

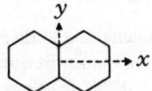

The $z$ axis is perpendicular to the plane of the molecule and we can readily obtain the eigenvalues of (16) for the field in the $z$ direction, putting $H_x = H_y = 0$. $T_z$ is then an eigenstate with energy $W = Z$ but $T_x$ and $T_y$ are mixed together. The energy matrix for these two states yields the secular determinant

$$\begin{vmatrix} X - W & -ig\beta H \\ ig\beta H & Y - W \end{vmatrix} = 0 \tag{17}$$

with eigenvalues

$$W = \tfrac{1}{2}(X + Y) \pm \sqrt{\tfrac{1}{4}(X - Y)^2 + (g\beta H)^2} \tag{18}$$

The energy levels and wave functions for a triplet state with the magnetic field along the $z$ axis may therefore be summarized as follows:

$$|\psi_x\rangle = \cos\theta|\alpha\alpha\rangle - \sin\theta|\beta\beta\rangle \qquad W = \tfrac{1}{2}(X + Y) + \tfrac{1}{2}(X - Y)\tan\theta + g\beta H$$

$$|\psi_y\rangle = \sin\theta|\alpha\alpha\rangle + \cos\theta|\beta\beta\rangle \qquad W = \tfrac{1}{2}(X + Y) - \tfrac{1}{2}(X - Y)\tan\theta - g\beta H$$

$$|\psi_z\rangle = \frac{1}{\sqrt{2}}|\alpha\beta + \beta\alpha\rangle \qquad\qquad W = Z$$

where $\tan 2\theta = (X - Y)/2g\beta H$. $(X - Y)$ is positive for naphthalene, and when $H$ is very small $\theta = \pi/4$ so that the stationary states go smoothly into $|T_x\rangle$, $|T_y\rangle$, and $|T_z\rangle$. On the other hand a strong field breaks up the spin-spin coupling; as $\theta$ approaches zero $|\psi_x\rangle$ and $|\psi_y\rangle$ approach $|\alpha\alpha\rangle$ and $|\beta\beta\rangle$, respectively, the spin being then totally quantized by the magnetic field. In Fig. 8.1 we plot the energy of the three spin levels as a function of magnetic field strength $H$.

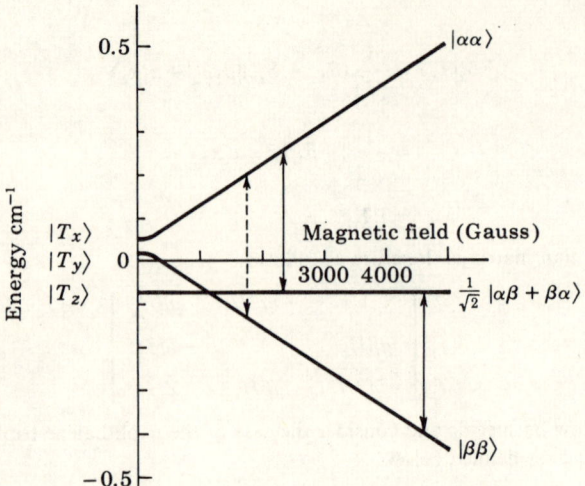

Fig. 8.1. Triplet energy levels of naphthalene in a magnetic field perpendicular to the molecular plane. Transitions are shown for a resonance frequency of 9654 Mc/s. The solid arrows are $\Delta m = \pm 1$ transitions observed with the rf field perpendicular to $H$. The dotted arrow is the $\Delta m = \pm 2$ transition observed with the rf field parallel to $H$.

If we now consider the case when the magnetic field is in the $x$ direction, we find that $|T_x\rangle$ is an eigenstate but $|T_y\rangle$ and $|T_z\rangle$ are mixed together. The new energies are readily obtained and their variation with magnetic field strength is summarized in Fig. 8.2. Clearly we can also solve for the eigenvalues when the field is in the $y$ direction, and thus complete our study of the variation of wave functions and energies with varying magnetic field.

In the lowest triplet state of naphthalene the electron resonance measurements fit the spin Hamiltonian (11) with the constants

$$X = 0.0478 \text{ cm}^{-1}$$

$$Y = 0.0196 \text{ cm}^{-1}$$

$$Z = -0.0675 \text{ cm}^{-1}$$

$$g = 2.0030$$

These values may also be converted into the alternative form (10) by using the relations

$$D = \tfrac{1}{2}(X + Y) - Z = +0.1012 \text{ cm}^{-1}$$
$$E = -\tfrac{1}{2}(X - Y) \quad = -0.0141 \text{ cm}^{-1}$$

(19)

(see Fig. 8.3). $D$ and $E$ are converted into Mc/s by multiplying by 29,979.3 ($D = 3,034$, $E = -423$). These frequencies correspond to the Zeeman energy of an electron with $g = 2.0030$ at fields of 1,082 and 151 gauss. The relative signs of $D$ and $E$ are known, but e.s.r. experiments do not give the absolute signs, except at very low temperatures.

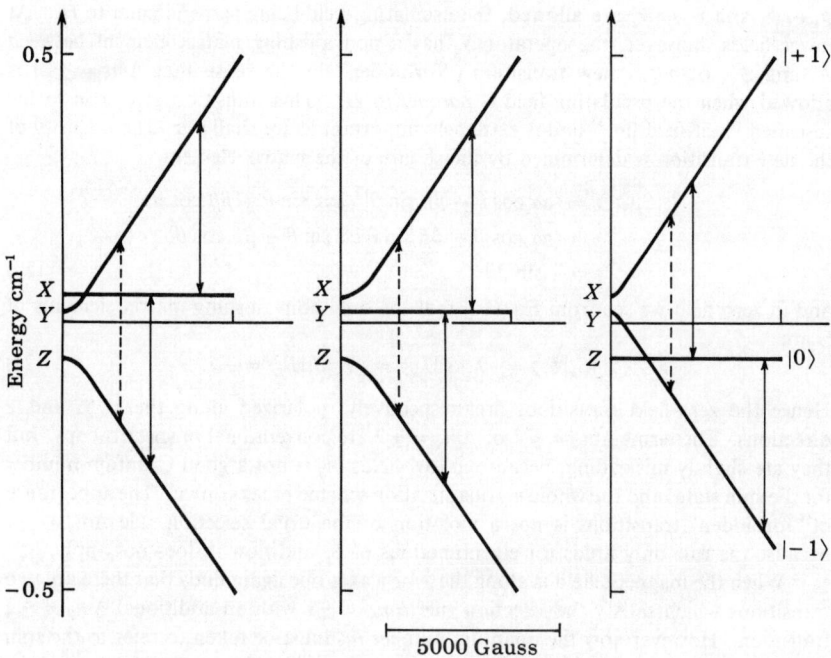

FIG. 8.2. Triplet energy levels of naphthalene. Transitions are shown for a frequency of 9600 Mc/s when the magnetic fields are, respectively, in the $x$, $y$, and $z$ directions.

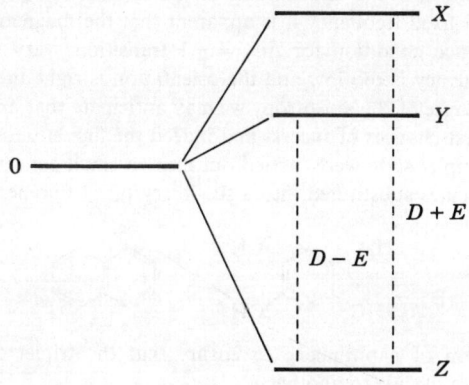

FIG. 8.3. Zero-field splitting of a triplet state.

## 8·4 TRANSITIONS WITH $\Delta m_s = 2$

We are now in a position to examine the possible transitions. At high fields when the spin is totally quantized by the magnetic field we may classify the levels according to their $m_S$ value and the usual selection rule $\Delta m_S = \pm 1$ holds. Hence the transitions

$\psi_x \leftrightarrow \psi_z$ and $\psi_y \leftrightarrow \psi_z$ are allowed, the oscillating field being perpendicular to $H_z$. At lower fields, however, the operator $S_z$ has a nonvanishing matrix element between $\psi_x$ and $\psi_y$ so that a new transition ("forbidden" in the sense that $\Delta m_S = \pm 2$) is allowed when the oscillating field is *parallel* to $H_z$. This transition gives rise to the so-called "half-field line" and is extremely important as we shall see. The intensity of the new transition is determined by the square of the matrix element

$$\langle \psi_x | S_z | \psi_y \rangle = \langle \alpha\alpha \cos \theta - \beta\beta \sin \theta | S_z | \alpha\alpha \sin \theta + \beta\beta \cos \theta \rangle$$

$$= \langle \alpha\alpha \cos \theta - \beta\beta \sin \theta | \alpha\alpha \sin \theta - \beta\beta \cos \theta \rangle$$

$$= + \sin 2\theta \tag{20}$$

and in zero field we see from Eq. (14) that the only nonvanishing matrix elements of S are

$$\langle T_y | S_x | T_z \rangle = \langle T_z | S_y | T_x \rangle = \langle T_x | S_z | T_y \rangle = -i \tag{21}$$

Hence the zero-field transitions are, respectively, polarized along the $x$, $y$, and $z$ directions. The terms $\Delta m_S = \pm 1$ or $\Delta m_S = \pm 2$ are conventional in spectroscopy, but they are slightly misleading, because in low fields $m_S$ is not a good quantum number for the spin states and the whole $m_S$ quantization scheme breaks down. The appearance of "forbidden" transitions is not a violation of the usual selection rule $\Delta m_S = \pm 1$ because the rule only holds for eigenfunctions of $S_z$ and now it does not apply.

When the magnetic field is along the $x$ or $y$ axes one again finds that there are two transitions which satisfy the selection rule $\Delta m_S = \pm 1$ with an additional $\Delta m_S = \pm 2$ transition. However now the quantum number $m_S$ must be taken to refer to the spin component resolved along the field direction rather than the $z$ axis. In general the transition between the two outermost energy levels of Fig. 8.2 is always called the $\Delta m_S = 2$ transition.

Let us now consider the implications of the information contained in Fig. 8.2. If we work with a fixed frequency it is apparent that the magnetic fields required to satisfy the resonance condition for $\Delta m_S = \pm 1$ transitions vary over a wide range. Indeed if the frequency is too low and the orientation is right there may be no resonance at all! Because of the anisotropy we may anticipate that an oriented specimen would offer the best chances of success and indeed the first successful experiments on the naphthalene triplet state were carried out using a small concentration of oriented naphthalene molecules substituted into a single crystal of durene. (One must have a

small concentration of naphthalene to ensure that the triplet excitation does not migrate from one molecule to another.)

The $\Delta m_S = \pm 2$ transition presents a rather different situation in that it is relatively isotropic. Consequently it is possible to study aromatic triplet states in low temperature glasses and this technique has been of enormous value. The precise position and shape of the $\Delta m_S = 2$ line in a glass depends upon the values of $D$ and $E$ so that they can both be determined if the line is visible, although this method is not very accurate. The $\Delta m_S = \pm 2$ transition has an intense peak at the low field edge $H_{min}$ of the spectrum, at the position

$$g\beta H_{min} = \sqrt{\tfrac{1}{4}(h\nu)^2 - \tfrac{1}{3}(D^2 + 3E^2)} \tag{22}$$

As we have already remarked, the $\Delta m_S = \pm 1$ transitions are quite anisotropic and at one time their detection in a glass specimen was considered to be unlikely. However, there is a sharp change in the number of molecules which are in resonance at those fields where **H** is parallel or nearly parallel to a principal axis of the zero-field splitting. With **H** parallel to either the $x$, $y$, or $z$ axis a pair of sharp lines is seen, with doublet splittings approximately equal to $(D - 3E)$, $(D + 3E)$, and $D$. Hence $D$ and $E$ can be determined almost as accurately as in a single crystal study.

In liquids we would expect the traceless tensor $D$ to average out to zero, but in spite of this the large fluctuating anisotropic spin-spin forces produce very strong relaxation effects and the e.s.r. line is too broad to detect. No triplet e.s.r. spectra have yet been studied successfully in solution, except in some "biradicals" where the two unpaired electrons are a long distance apart and $D$ and $E$ are small. The principal disadvantage of working with glassy samples is that the detailed nuclear hyperfine interaction cannot be studied except in a few cases. A more complete spin Hamiltonian would contain a term $\mathbf{S \cdot T \cdot I}$ for each nucleus, and ring proton hyperfine structure is better resolved in the naphthalene-durene crystals than in most glasses.

# 8·5  HYPERFINE STRUCTURE

Rather few triplet molecules show resolved proton hyperfine structure, and the analysis of the splittings is difficult, even in a single crystal, since each hyperfine tensor may have different principal axes which need not be parallel to the principal axes of $D$ and $E$.

We should expect that the isotropic hyperfine interaction in a triplet state is still proportional to the electron spin density at the nucleus. This indeed is true, but two important new effects must be considered; there are now two unpaired electrons, and their spins are coupled by the zero-field tensor.

Let us first consider a molecule with just one proton and apply a strong magnetic field along the $z$ axis to break up the spin-spin coupling, so that the electron spins are completely quantized by the field. The first-order spin Hamiltonian for the contact interaction is then

$$\mathscr{H}_c = \frac{8\pi}{3} g\beta g_N \beta_N \{S_{1z}\delta(\mathbf{r}_1) + S_{2z}\delta(\mathbf{r}_2)\}I_z \tag{23}$$

Now suppose that the two unpaired electrons with parallel spins occupy a pair of molecular orbitals $\psi_a$ and $\psi_b$. Because of the Pauli exclusion principle the spatial wave function $\Psi(\mathbf{r}_1, \mathbf{r}_2)$ must be antisymmetric in the coordinates $\mathbf{r}_1$ and $\mathbf{r}_2$, so we write

$$\Psi(\mathbf{r}_1, \mathbf{r}_2) = \frac{1}{\sqrt{2}} \{\psi_a(\mathbf{r}_1)\psi_b(\mathbf{r}_2) - \psi_b(\mathbf{r}_1)\psi_a(\mathbf{r}_2)\} \tag{24}$$

In the complete electronic wave function the space part is multiplied by one of the three spin functions $|\alpha\alpha\rangle$, $|\beta\beta\rangle$ or $(1/\sqrt{2})|\alpha\beta + \beta\alpha\rangle$ and it is straightforward to show that the three values of $\mathscr{H}_c$ are

$$\frac{1}{2}\{|\psi_a(0)|^2 + |\psi_b(0)|^2\}\frac{8\pi}{3}g\beta g_N\beta_N I_z \qquad |\alpha\alpha\rangle$$

$$0 \qquad\qquad\qquad (1/\sqrt{2})|\alpha\beta + \beta\alpha\rangle$$

$$-\frac{1}{2}\{|\psi_a(0)|^2 + |\psi_b(0)|^2\}\frac{8\pi}{3}g\beta g_N\beta_N I_z \qquad |\beta\beta\rangle \tag{25}$$

This is equivalent to a coupling $aS_zI_z$ between the nuclear spin and the *total spin $S_z$* of both electrons with a splitting constant $a$ proportional to the *normalized spin density*

$$\rho(\mathbf{r}) = \frac{1}{2}\{|\psi_a(\mathbf{r})|^2 + |\psi_b(\mathbf{r})|^2\} \tag{26}$$

at the nucleus.

The naphthalene–durene crystal illustrates this result. Here the orbitals $\psi_a$ and $\psi_b$ are the lowest antibonding and highest bonding orbitals—the same orbitals used by the unpaired electron in the negative and positive naphthalene radical ions. Because of the pairing theorem for alternant hydrocarbons the spin distribution (26) should be very similar to that in the naphthalene anion. Analysis of the e.s.r. spectrum shows that the normalized $\pi$-electron spin densities at carbons 1, 2, and 9 are $+0.219$, $+0.062$, and $-0.063$, as expected. It is important to notice that although there are two unpaired electrons the hyperfine splittings in the $\Delta m_S = 1$ transitions are not twice as large as they are in a radical with $S = 1/2$.

In a weak magnetic field the average value of $S_z$ in the states $|\psi_x\rangle$ and $|\psi_y\rangle$ is reduced from $\pm 1$ to $\pm \cos 2\theta$, with the result that the observed hyperfine splitting is smaller by a factor $\cos 2\theta$ and vanishes at zero field.

## 8·6  FURTHER STUDIES OF EXCITED TRIPLET STATES

It is not easy to grow dilute single crystal specimens which meet all the necessary conditions and very few single crystal studies have been reported. The naphthalene–durene system was the first to be investigated and all the features of its spectrum are now clearly understood.

The lowest excited triplet states of phenanthrene in diphenyl and quinoxaline in durene

have also been studied in single crystal specimens. The quinoxaline triplet spectrum shows hyperfine structure from the two ring nitrogen atoms and the values of $D$ and $E$ ($0.1007$ and $-0.0182$ cm$^{-1}$) are very similar to those of naphthalene.

The theoretical calculation of $\pi$-electron zero-field splittings in aromatic molecules is fairly complicated because both unpaired electrons can move over the whole conjugated system. An important consequence of the antisymmetry of the electronic wave function $\Psi(\mathbf{r}_1, \mathbf{r}_2)$, Eq. (24), is that the probability distribution $|\Psi(\mathbf{r}_1, \mathbf{r}_2)|^2$ becomes vanishingly small as the electrons approach one another, and there are no large contributions to $D$ and $E$ from configurations where the two electrons are close together. The values of $D$ and $E$ are not always sensitive to fine details of the spin distribution, and tell rather less about the triplet $\pi$-electron wave function than one would wish.

Some very interesting time-dependent effects have been observed in the e.s.r. spectra of excited triplet states. The lowest triplet state of benzene has been observed in a low temperature glass and the line shape indicates that the molecule no longer possesses its full hexagonal symmetry. For axially symmetric molecules like benzene and triphenylene

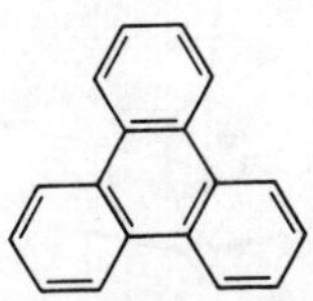

the principal symmetry axis is chosen to be the $z$ direction and since the $x$ and $y$ directions are equivalent, the zero-field splitting parameter $E$ should vanish. The shape of the half-field line for benzene, however, shows that $E$ is not zero; moreover the line shape is temperature dependent. It is thought that the molecule exists in three equivalent distorted structures

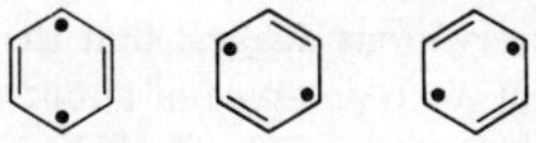

and tunnels between them at a frequency of 1,000 to 10,000 Mc/s.

Tribenzotriptycene ($a$) exhibits a temperature dependent line shape for the quite different reason that the triplet excitation migrates from one ring to another.

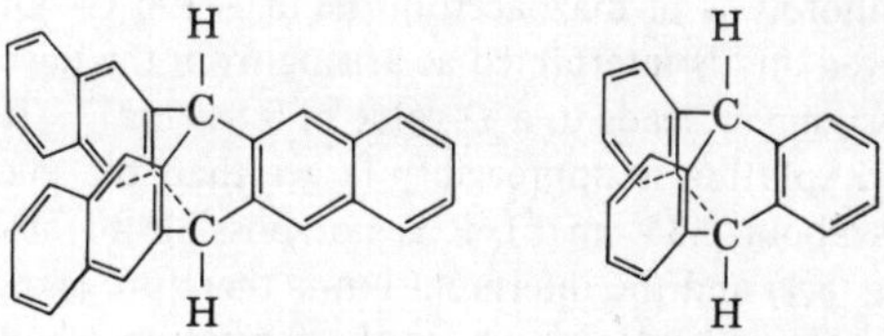

H
|
C
|
C
|
H

($a$)

H
|
C
|
C
|
H

($b$)

In triptycene itself, ($b$), the transfer rate is very fast but in tribenzotriptycene excitation transfer between the three naphthalene subsystems is slow enough to affect the line shape which, indeed, varies with temperature. At very low temperatures the excitation is localized on one ring system and the e.s.r. spectrum is virtually identical with that of naphthalene itself.

## 8.7 ORGANIC MOLECULES WITH TRIPLET GROUND STATES

### 8.7.1 Methylene and Nitrene Derivatives

Ultraviolet photolysis of diazodiphenyl methane results in the formation of diphenylmethylene which can be stabilized either in a single crystal of benzophenone or in a low temperature glass. The e.s.r. spectrum indicates that the molecule has a triplet ground state and the zero-field splitting parameters are $D = 0.4052$ cm$^{-1}$, $E = -0.0194$ cm$^{-1}$. The large value of $D$ indicates that the two unpaired electrons are mainly localized on the central carbon atom; also the anisotropic $C^{13}$ hyperfine couplings of $+41.5$, $+16.2$, and $-57.6$ Mc/s are about the same as the values calculated

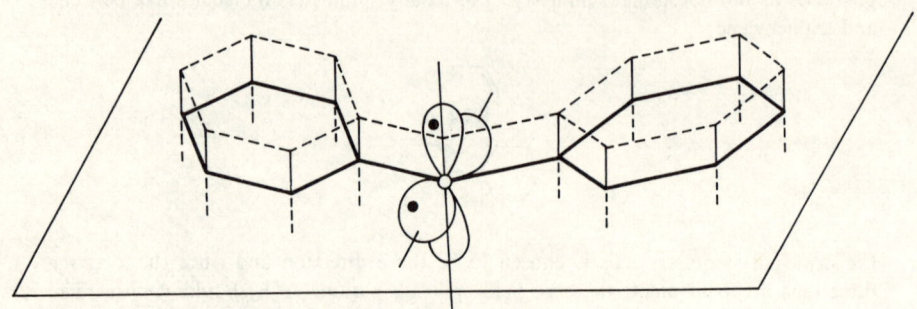

FIG. 8.4. Unpaired electrons in diphenylmethylene.

for a localized electron pair, although they lack cylindrical symmetry about the $z$ axis (Fig. 8.4). All these observations suggest that the rings lie more or less planar but the $\phi$—C—$\phi$ bond is bent, with an angle of about 140–155° (see Fig. 8.4).

The e.s.r. spectrum of methylene (CH$_2$) itself has not yet been detected but the related species $\phi$—CH and CH(C≡N) have been observed. The spectrum of phenyl-methylene in a powder shows a small hyperfine splitting of less than 15 Mc/s from the CH proton; this splitting is remarkably small and suggests that the C—C—H bond is bent, whereas methylene itself is known to be approximately linear. CH(C≡N) is formed by the photolysis of diazoacetonitrile at −196°C. Only one e.s.r. line, at high field, is observed and is interpreted as arising from the lowest energy $\Delta m_S = \pm 1$ transition. This assignment leads to a $D$ value of 0.889 cm$^{-1}$. It is worth noting that when the zero-field splitting is appreciably larger than the microwave quantum of energy ($X$ band is about 0.33 cm$^{-1}$), it is not possible to observe all the allowed transitions (see Fig. 8.1) and the interpretation is therefore less certain than usual.

Photolysis of organic azides results in the formation of nitrenes, derivatives of N—H. These also have localized triplet ground states and a growing number of them have been studied by e.s.r. Alkyl nitrenes R—N have $D$ values in the range 1.5 to 1.8 cm$^{-1}$ and very small $E$ values ($\approx 0.003$ cm$^{-1}$) which indicate that the spin distribu-tion has nearly cylindrical symmetry, with two unpaired electrons localized on the nitrogen. In contrast the $D$ value for phenyl nitrene is 0.99 cm$^{-1}$; here one unpaired electron spreads into the phenyl ring.

### 8.7.2 Triplets with One Localized Electron

In the three triplet molecules cyclopentadienylidene, indenylidene, and fluoreny-lidene, one unpaired electron is localized at the divalent carbon atom and occupies an

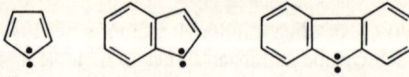

in-plane $\sigma$ orbital. The second electron occupies a $\pi$ orbital and is, to varying extents, delocalized over the ring system. The $D$ values for the three molecules are 0.4089, 0.3777, and 0.4078 cm$^{-1}$. The $E$ values are considerably smaller but by no means negligible. It has been shown that the ratio $E/D$ should be proportional to the fraction of $s$ orbital character for the in-plane $\sigma$ hybrid orbital, so that it is possible to use the measurements to estimate the apparent angle $\theta$ formed by the two occupied hybrid orbitals forming the C—C $\sigma$ bonds. $E/D$ should be zero for $\theta = 180°$ and 1 for $\theta = 90°$. The observed $E$ value for fluorenylidene is 0.028 cm$^{-1}$, and for the ratio $E/D = 0.07$ theory suggests that $\theta$ must be greater than 135°. Since it would be most unreasonable to suppose that the actual C—C—C bond angle is greater than 135°, these observations suggest that the C—C $\sigma$ bonds are "bent."

Photolysis of benzene-1,4-diazo-oxide dissolved in $p$-dichlorobenzene at low temperatures results in the formation of the triplet molecule shown below.

Although this molecule is not regularly oriented in the solid, the e.s.r. spectrum shows resolvable hyperfine structure from the two protons adjacent to the divalent carbon atom. It is possible to obtain the diagonal elements of each tensor in the molecular symmetry axes from the structure of the $\Delta m_S = \pm 1$ absorption lines (24.4, 18.5, and 29.7 Mc/s). The zero-field splitting parameters are $D = 0.3179$, $E = 0.0055$ cm$^{-1}$; the value of $D$ is determined largely by the $\pi$-electron density $\rho$ on the divalent carbon and the observed value indicates that $\rho$ is about 0.4. Comparison with the liquid phase e.s.r. spectrum of the phenoxy radical

(which has a doublet ground state) suggests that the $\pi$-electron distribution is similar in both systems. The rather large hyperfine splitting from the protons adjacent to the divalent carbon atom arises mainly from the fact that there is a large spin density in the *sigma* orbital of the carbon; this contribution dominates the $\sigma - \pi$ spin exchange effects which are normally responsible for ring proton splittings.

### 8.7.3 $\pi$-Electron Triplets

On theoretical grounds it is expected that aromatic molecules with two electrons occupying degenerate molecular $\pi$ orbitals are likely to have triplet ground states. Unfortunately the simplest of these, the benzene dinegative ion has never been prepared, but the 1,3,5-triphenyl benzene dianion ($a$) has and it does indeed give an e.s.r. spectrum characteristic of a triplet state. The pentaphenylcyclopentadienyl cation ($b$) has also been shown to have a very low-lying triplet state. However the dianion of coronene ($c$) is, on the contrary, diamagnetic.

(a)          (b)          (c)

## 8·8 TRIPLET EXCITONS

The cyano derivatives tetracyanoethylene (TCNE) and tetracyanoquinodimethane (TCNQ),

are very powerful electron acceptors and they form stable free radical salts with a variety of diamagnetic cations. Many of these salts have low electrical resistivities and remarkable e.s.r. properties. The complex $(\phi_3 PCH_3)^+(TCNQ)_2^-$, for example, has been studied in single crystal form and shows the resonance characteristic of a triplet state. The crystal structure is such that *pairwise* correlation of the electron spins of two complexes occurs, the ground state being a singlet, but the triplet being only 0.062 eV higher in energy. At moderately low temperatures the zero-field splitting of the resonance is observed and the values $D = 0.0062$ cm$^{-1}$ and $E = 0.00098$ cm$^{-1}$ are obtained. The triplet excitation is not localized on one group of molecules, however, but migrates through the crystal; one of the consequences of this migration is that hyperfine structure is not observed. As the temperature is increased, the two $\Delta m_S = \pm 1$ lines broaden and finally collapse into a single sharp line. This effect is due to increasing numbers of excitons at the higher temperatures, with the result that exciton-exciton collisions which randomize the electron spin orientation become progressively more important.

The free radical salt, Wurster's blue perchlorate, exhibits

similar effects in its e.s.r. spectrum although the reasons are quite different. At room temperature it shows the magnetic susceptibility and e.s.r. spectrum (a single sharp line) characteristic of a normal pure free radical salt, but below 186°K the susceptibility decreases rapidly. Detailed studies reveal that there is a crystal phase transition. At room temperature the Wurster's blue ions are stacked at equal distances apart in long parallel one-dimensional chains which extend throughout the crystal lattice. Below the transition point pairs of ions move together slightly to form an alternating structure (Fig. 8.5) where the two spins are coupled antiparallel, and each pair in the chain is coupled more weakly to the next pair to give an antiferromagnetic array. Resonance is observed from thermally excited triplet excitons which move freely along the chains but have difficulty in jumping from one chain to another. A small zero-field splitting is clearly visible at 50°K and can be accounted for by a spin Hamiltonian $E(S_x^2 - S_y^2)$ with $E = 212$ Mc/s.

These studies show that the magnetic and electrical properties of organic solids deserve and will receive increasing attention; the impact on the applied sciences may well be spectacular.

Virtually all organic charge-transfer complexes (as distinct from ion-radical salts) show a single e.s.r. line with an apparent $g$ value close to 2.002. Very little information on single crystal spectra is available and the interpretation is very uncertain. There

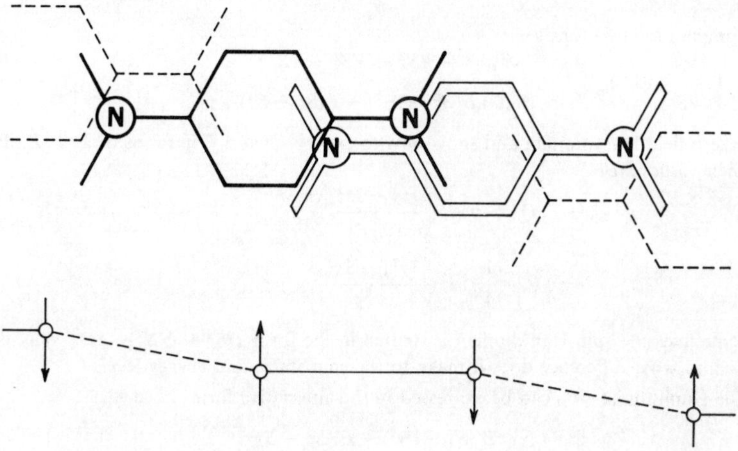

FIG. 8.5. Spin pairing in Wurster's blue perchlorate at low temperatures. The ions form one-dimensional chains with alternate strong and weak couplings and an antiferromagnetic spin arrangement.

seems little doubt that impurities or crystal defects are at least partly responsible for the resonance absorption; convincing evidence that triplet excitons are also involved has not yet been provided.

## 8·9  RADICAL-ION CLUSTERS IN SOLUTION

We have already described liquid phase studies of ion-pairing between aromatic radical anions and alkali metal cations. In some cases, there is evidence for the presence of both single and double ion pairs. An example is sodium fluorenone whose optical

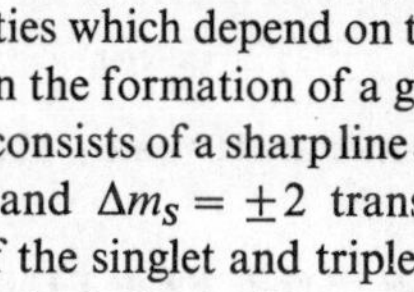

absorption in methyltetrahydrofuran at room temperature shows two peaks at 4550 and 5200 Å with relative intensities which depend on the concentration. Rapid cooling of the solution to 77°K results in the formation of a glass with almost the same optical properties. The e.s.r. spectrum consists of a sharp line at $g = 2$ due to radical monomers, together with the $\Delta m_S = \pm 1$ and $\Delta m_S = \pm 2$ transitions characteristic of a triplet state. The relative intensities of the singlet and triplet e.s.r. spectra correspond to the relative intensities of the two optical absorption peaks, and there is a small zero-field splitting $D$ whose size depends on the alkali metal, being 333, 274, and 204 Mc/s for lithium, sodium, and potassium ions.

## PROBLEMS

**1.**  Compare the two forms

$$\mathscr{H}_D = -XS_x^2 - YS_y^2 - ZS_z^2$$

$$\mathscr{H}_D = D(S_z^2 - \tfrac{1}{3}\mathbf{S}^2) + E(S_x^2 - S_y^2)$$

of the zero-field Hamiltonian and show how to express $D$ and $E$ in terms of $X, Y, Z$. Prove that their values are

$$D = \frac{3}{4}g^2\beta^2 \left\langle \frac{r_{12}^2 - 3z_{12}^2}{r_{12}^5} \right\rangle$$

$$E = \frac{1}{4}g^2\beta^2 \left\langle \frac{3y_{12}^2 - 3x_{12}^2}{r_{12}^5} \right\rangle$$

**2.**  Sometimes the spin Hamiltonian is written in the form $DS_z^2 + E(S_x^2 - S_y^2)$. Why is this wrong, and what difference does it make to the spin states and energy levels?

**3.**  The Hamiltonian $\mathscr{H}_D$ can be expressed in the alternative form

$$D'(S_x^2 - \tfrac{1}{3}\mathbf{S}^2) + E'(S_y^2 - S_z^2)$$

Express $D'$ and $E'$ in terms of the original $D$ and $E$.

**4.**  Use Eqs. (17) and (18) to calculate the wave functions $|\psi_x\rangle$, $|\psi_y\rangle$ and $|\psi_z\rangle$ for naphthalene in a field $H_z$ of 1707 gauss. Show that $\Delta m_S = \pm 2$ transitions will occur at a frequency of 9654 Mc/s, and use Eq. (20) to estimate their intensity.

**5.**  The e.s.r. spectrum of the excited triplet state of naphthalene was observed in a glass at low temperature, with the rf field parallel to the steady field, at a frequency of 9422 Mc/s. The spectrum showed an intense peak at 1551 gauss. What information about $D$ and $E$ can you deduce from this? Do your results agree with the single crystal studies?

**6.**  Estimate values of $D$ and $E$ for the following model systems:

(a) One electron localized at the origin and the other localized at a point on the $z$ axis 1.40 Å away.

(b) A network of four point atoms in the $xy$ plane, with dimensions $ac = 4$ Å, $bd = 3$ Å. The two electrons may never be on the same atom, and the six configurations $ab$, $bc$, $cd$, $da$, $ac$, $bd$ are all equally probable.

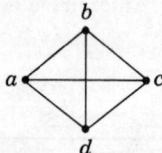

**7.**  The ion-radical salt $(Cs^+)_2(TCNQ)_3^=$ has a low-lying triplet state with energy $J = 0.16$ eV above the singlet ground state. Show that the intensity of the e.s.r. absorption should vary as

$$\frac{1}{kT} \frac{3e^{-J/kT}}{1 + 3e^{-J/kT}}$$

and estimate the ratios of the intensities at 180° and −120°C (see Chapter 1).

## SUGGESTIONS FOR FURTHER READING

Hutchison and Mangum: *J. Chem. Phys.*, **34**: 908 (1961). The first e.s.r. of an organic triplet state. Naphthalene in a single crystal of durene.

Van Der Waals and De Groot: *Mol. Phys.*, **2**: 333 (1959): *Mol. Phys.*, **6**: 545 (1963). E.S.R. of naphthalene in a glass. Triptycene.

Wasserman, Snyder, and Yager: *J. Chem. Phys.*, **41**: 1763 (1964). Triplet e.s.r. of diphenylmethylene in a glass.

Yager, Wasserman, and Cramer: *J. Chem. Phys.*, **37**: 1148 (1962). $\Delta m = \pm 1$ transitions in glasses.

Wasserman and Murray: *J. Amer. Chem. Soc.*, **86**: 4203 (1964). Photolysis of benzene-1,4-diazo-oxide.

Chesnut and Phillips: *J. Chem. Phys.*, **35**: 1002 (1961). Ion-radical salts of TCNQ.

Thomas, Keller, and McConnell: *J. Chem. Phys.*, **39**: 2321 (1963). Exciton magnetic resonance in Wurster's blue perchlorate.

# CHAPTER 9

# THEORY OF THE
# $g$ TENSOR AND THE
# E.S.R. SPECTRA OF
# INORGANIC RADICALS

## 9·1 DETERMINATION OF THE $g$ TENSOR IN CRYSTALS

We now turn to molecules in which the $g$ tensor is anisotropic, particularly inorganic radicals and transition metal complexes. The theory becomes considerably more complicated but correspondingly more information about electronic structure is obtained. The $g$ tensor is anisotropic when the electron possesses both spin and orbital angular momentum; for the moment we will take it for granted that the effective interaction between the magnetic field $H$ and the electron spin angular momentum can be represented as a tensor coupling $\beta \mathbf{H} \cdot \mathbf{g} \cdot \mathbf{S}$, although the matter is by no means trivial and we will return to it in the next section. For the present, however, we will suppose that the spin Hamiltonian has been formulated correctly and address ourselves to the experimental task of determining the principal components of the $g$ tensor.

The problem is similar to that of determining the hyperfine interaction tensor. As we saw in Section 7.5, the angular variation of the hyperfine structure yields the squares of the principal values, so that unique determination of the signs is not possible. A similar problem exists in deriving the $g$ tensor; it is not serious, however, since we know from theory that the principal values of the $g$ tensor are positive in practically every case. It will further be recalled that in deriving the hyperfine tensor we assumed complete quantization of the electron spin parallel to the magnetic field. Such an assumption would be quite out of place here; $g$ tensors are anisotropic precisely because the spin is *not* oriented exactly along the field direction.

We commence as usual by choosing a set of orthogonal axes $x$, $y$, $z$ fixed in the crystal. In general these axes are not principal directions of the $g$ tensor so that we write the spin Hamiltonian as

$$\mathscr{H} = \beta [H_x, H_y, H_z] \begin{bmatrix} g_{xx} & g_{xy} & g_{xz} \\ g_{yx} & g_{yy} & g_{yz} \\ g_{zx} & g_{zy} & g_{zz} \end{bmatrix} \begin{bmatrix} S_x \\ S_y \\ S_z \end{bmatrix} \tag{1}$$

The electron spin does not interact directly with the field **H**. Instead its energy depends on the magnitude and direction of the vector **H**·**g**. We denote the direction of **H**·**g** by a unit vector **h'** and the magnitude by $Hg'$, where

$$\mathbf{h'}Hg' = \mathbf{H}\cdot\mathbf{g} \tag{2}$$

The number $g'$ then describes the effective magnetic moment of the molecule when its spin is quantized along the direction **h'**. There are two spin states $|\alpha'\rangle$ and $|\beta'\rangle$, quantized along **h'** with energies $\pm(1/2)g'\beta H$. The separation between this pair of electron spin levels is equal to $g'\beta H$; hence from Eq. (2) we obtain

$$(\Delta E)^2 = g'^2\beta^2 H^2 = \beta^2(\mathbf{H}\cdot\mathbf{g})\cdot(\mathbf{g}\cdot\mathbf{H}) = \beta^2\mathbf{H}\cdot\mathbf{g}^2\cdot\mathbf{H} \tag{3}$$

which is similar to Eq. (12) of Section 7.3. $\mathbf{g}^2$ is the square of the g tensor. Let us now consider an experiment in which we measure the apparent $g$ value, $g'$, when the magnetic field **H** has direction cosines $l_x, l_y, l_z$ relative to the crystal axes. The components of **H** are $(Hl_x, Hl_y, Hl_z)$, so we find

$$(g')^2 = [l_x \ l_y \ l_z]\begin{bmatrix}(g^2)_{xx} & (g^2)_{xy} & (g^2)_{xz}\\(g^2)_{yx} & (g^2)_{yy} & (g^2)_{yz}\\(g^2)_{zx} & (g^2)_{zy} & (g^2)_{zz}\end{bmatrix}\begin{bmatrix}l_x\\l_y\\l_z\end{bmatrix} \tag{4}$$

Measurement of the $g'$ value at different orientations gives the elements of the tensor $\mathbf{g}^2$, which can then be transformed to principal axes. We could begin by rotating the field in the $xy$ plane, where the direction cosines are $(\cos\theta, \sin\theta, 0)$. It follows that

$$(g')^2 = (g^2)_{xx}\cos^2\theta + 2(g^2)_{xy}\sin\theta\cos\theta + (g^2)_{yy}\sin^2\theta \tag{5}$$

so that three tensor elements are measured in this experiment. Measurements in the $xz$ and $yz$ planes give the values of the other three.

In many cases the same axis system diagonalizes both the hyperfine tensor and the $g$ tensor. This would not be true for an aromatic radical, however, since the $g$ tensor would be diagonal in terms of the molecular symmetry axes but each hyperfine tensor would be diagonal in terms of its own local symmetry axes, one of which would be along the C—H bond.

# 9.2 THEORY OF THE $g$ TENSOR AND THE EFFECTIVE SPIN HAMILTONIAN

We have represented the electronic Zeeman interaction by the term in the spin Hamiltonian,

$$\mathcal{H} = \beta\mathbf{H}\cdot\mathbf{g}\cdot\mathbf{S} \tag{6}$$

where **g** is a tensor which may well be anisotropic. The reader should, however, be suspicious about Eq. (6) because we know that when the electron possesses spin angular momentum only, the $g$ tensor is isotropic and has the free electron value $g_e = 2.0023$. Hence if **g** is found to be anisotropic, **S** in Eq. (6) cannot possibly represent the true spin; whatever the symbol **S** does represent, we should perhaps distinguish it from the real spin and rewrite (6) as

$$\mathcal{H} = \beta\mathbf{H}\cdot\mathbf{g}\cdot\hat{\mathbf{S}} \tag{7}$$

reserving the uncapped symbol **S** for the true spin.

In the presence of a magnetic field there will be an interaction between $\mathbf{H}$ and the orbital angular momentum $\mathbf{L}$ as we have seen already in Chapter 4. Thus the true Hamiltonian for the magnetic interaction should be written

$$\mathscr{H} = \beta\mathbf{H}\cdot\mathbf{L} + g_e\beta\mathbf{H}\cdot\mathbf{S} \tag{8}$$

Our task is to show that the effects of this perturbing Hamiltonian may be represented by means of the "effective spin Hamiltonian" (7) in which $\hat{\mathbf{S}}$ is called the *fictitious spin*.

Let us first take a molecule having a single unpaired electron with the space wave function $\psi_0(\mathbf{r})$ whose spin may be either $\alpha$ or $\beta$. We shall calculate the Zeeman energies of the two states $|\psi_0\alpha\rangle$ and $|\psi_0\beta\rangle$ in a field $H$ by first order perturbation theory. For the $\alpha$ state, and the field along the $z$ axis, the result is

$$
\begin{aligned}
E &= \beta H\langle\psi_0\alpha|L_z + g_e S_z|\psi_0\alpha\rangle \\
&= \beta H\langle\psi_0|L_z|\psi_0\rangle + g_e\beta H\langle\alpha|S_z|\alpha\rangle \\
&= \beta H\langle L_z\rangle + \tfrac{1}{2}g_e\beta H
\end{aligned} \tag{9}
$$

However, the mean value $\langle L_z\rangle$ of the orbital angular momentum vanishes in all molecules which do not have orbitally degenerate ground states. Consequently the energy $E$ is just $(1/2)g_e\beta H$, corresponding to the free electron $g$ value $g_e$. To prove this statement we note that $L_z$ is represented by the quantum mechanical operator

$$L_z = -i\hbar\frac{\partial}{\partial\phi} \tag{10}$$

and the wave function of a nondegenerate state is always real; hence the average is

$$
\begin{aligned}
\langle L_z\rangle &= -i\hbar\int\psi_0\frac{\partial}{\partial\phi}\psi_0\,dv \\
&= -i\hbar\frac{\partial}{\partial\phi}\int\left(\frac{1}{2}\right)\psi_0\psi_0\,dv = 0
\end{aligned} \tag{11}
$$

The only way the odd electron can acquire some orbital angular momentum is through the effect of spin-orbit coupling, which can be represented in a simplified form by the expression

$$\zeta\mathbf{L}\cdot\mathbf{S} \tag{12}$$

where $\zeta$ is a constant called the spin-orbit coupling constant. This operator mixes the ground state wave functions with excited states, and by first-order perturbation theory the modified wave function $|\psi_0\alpha\rangle$ will be

$$|+\rangle = |\psi_0\alpha\rangle - \sum_n\frac{\langle n|\zeta\mathbf{L}\cdot\mathbf{S}|\psi_0\alpha\rangle}{E_n - E_0}|n\rangle \tag{13}$$

We now have to consider the various matrix elements of $\mathbf{L}\cdot\mathbf{S}$. The operator may be expressed in the form

$$\mathbf{L}\cdot\mathbf{S} = L_z S_z + \tfrac{1}{2}(L^+ S^- + L^- S^+) \tag{14}$$

where $L^+$ and $L^-$ are shift operators which operate on the space wave functions $\psi_0$ and $\psi_n$. The $L_z S_z$ part mixes $|\psi_0\alpha\rangle$ with $|\psi_n\alpha\rangle$ leaving the electron spin unchanged, but $(1/2)L^+S^-$ mixes the $\alpha$ ground state with the $\beta$ excited state $|\psi_n\beta\rangle$, and the corrected wave function becomes

$$|+\rangle = |\psi_0\alpha\rangle - \frac{1}{2}\zeta\sum_n \frac{\langle\psi_n|L_z|\psi_0\rangle}{E_n - E_0}|\psi_n\alpha\rangle$$
$$- \frac{1}{2}\zeta\sum_n \frac{\langle\psi_n|L_x + iL_y|\psi_0\rangle}{E_n - E_0}|\psi_n\beta\rangle \tag{15}$$

Spin-orbit coupling also converts the unperturbed state $|\psi_0\beta\rangle$ into a corresponding corrected state

$$|-\rangle = |\psi_0\beta\rangle + \frac{1}{2}\zeta\sum_n \frac{\langle\psi_n|L_z|\psi_0\rangle}{E_n - E_0}|\psi_n\beta\rangle$$
$$- \frac{1}{2}\zeta\sum_n \frac{\langle\psi_n|L_x - iL_y|\psi_0\rangle}{E_n - E_0}|\psi_n\alpha\rangle \tag{16}$$

Evidently $|+\rangle$ and $|-\rangle$ are no longer eigenstates of the true spin $S_z$. It is now time to give a precise definition of the fictitious spin $\hat{S}$. The operators $\hat{S}_x$, $\hat{S}_y$, $\hat{S}_z$ are defined to act on the states $|+\rangle$ and $|-\rangle$ in exactly the same way as the true spin operators $S_x$, $S_y$, $S_z$ act on the spin functions $|\alpha\rangle$ and $|\beta\rangle$; that is

$$\hat{S}_z|+\rangle = \tfrac{1}{2}|+\rangle, \qquad \hat{S}_x|+\rangle = \tfrac{1}{2}|-\rangle \tag{17}$$
$$\hat{S}_z|-\rangle = -\tfrac{1}{2}|-\rangle, \qquad \hat{S}_y|+\rangle = \tfrac{1}{2}i|-\rangle \tag{18}$$

and so on. In a magnetic field along the $z$ axis the spin Hamiltonian (7) becomes

$$\mathscr{H} = \beta H(g_{zx}\hat{S}_x + g_{zy}\hat{S}_y + g_{zz}\hat{S}_z) \tag{19}$$

and is therefore represented by the $2 \times 2$ matrix

$$\begin{array}{cc} & \begin{array}{cc} |+\rangle & \qquad\qquad |-\rangle \end{array} \\ \begin{array}{c}\langle+| \\ \langle-|\end{array} & \begin{bmatrix} \tfrac{1}{2}\beta H g_{zz} & \tfrac{1}{2}\beta H(g_{zx} - ig_{zy}) \\ \tfrac{1}{2}\beta H(g_{zx} + ig_{zy}) & -\tfrac{1}{2}\beta H g_{zz} \end{bmatrix} \end{array} \tag{20}$$

On the other hand, the true Zeeman Hamiltonian $\beta H(L_z + g_eS_z)$ has matrix elements

$$\beta H \begin{bmatrix} \langle+|L_z + g_eS_z|+\rangle & \langle+|L_z + g_eS_z|-\rangle \\ \langle-|L_z + g_eS_z|+\rangle & \langle-|L_z + g_eS_z|-\rangle \end{bmatrix} \tag{21}$$

If the spin Hamiltonian $\mathscr{H}$ has any meaning the two matrices (20) and (21) have to be identical, and the components of the $g$ tensor must be found from the equations

$$g_{zz} = 2\langle+|L_z + g_eS_z|+\rangle \tag{22}$$
$$(g_{zx} + ig_{zy}) = 2\langle-|L_z + g_eS_z|+\rangle \tag{23}$$

It is straightforward to compute $\langle+|L_z|+\rangle$ and $\langle+|S_z|+\rangle$ from the wave function (15). To first order in $\zeta$ the value of $S_z$ is still exactly $1/2$, while $L_z$ is unaffected by the excited states with $\beta$ spin. We then obtain

$$g_{zz} = g_e - 2\zeta\sum_n \frac{\langle\psi_0|L_z|\psi_n\rangle\langle\psi_n|L_z|\psi_0\rangle}{E_n - E_0} \tag{24}$$

A rather more lengthy computation of the cross terms $\langle -|L_z|+\rangle$ and $\langle -|S_z|+\rangle$ yields the off-diagonal components; a typical one is

$$g_{zx} = g_e - 2\zeta \sum_n \frac{\langle \psi_0|L_z|\psi_n\rangle\langle \psi_n|L_x|\psi_0\rangle}{E_n - E_0} \tag{25}$$

Finally we note that our $g$ tensor only applies to the ground state $\psi_0$. In general it will have quite different components in some other state $\psi_n$.

## 9·3  A SIMPLE EXAMPLE

The theory of the $g$ tensor is probably best understood in terms of a specific example. Let us therefore take the case of the hydrogen atom in a $2p$ state which is subjected to a potential field which lowers the energy of the $p_z$ orbital relative to $p_x$ and $p_y$. The unpaired electron will occupy $p_z$, $p_x$ and $p_y$ being degenerate excited orbitals which are separated from the ground state by an energy difference $\Delta E$. We have already used almost the same example in Chapter 4 to illustrate the theory of the chemical shift, and the argument here follows similar lines. Since we are concerned with orbital angular momentum, we represent the orbitals in terms of their $m_L$ values, where $m_L$ is the component of angular momentum about the $z$ axis. Since $L = 1$, $m_L$ is equal to $+1$, $0$ or $-1$, and the relationships between the orbitals $|m_L\rangle$ and $p_x$, $p_y$, $p_z$ are as follows:

$$p_0 = p_z$$

$$p_{+1} = -\frac{1}{\sqrt{2}}(p_x + ip_y)$$

$$p_{-1} = \frac{1}{\sqrt{2}}(p_x - ip_y) \tag{26}$$

The two ground state wave functions are

$$|p_0\alpha\rangle, \quad |p_0\beta\rangle \tag{27}$$

and since $L_z|p_0\rangle = 0$ we need only consider the effects of $L^+S^-$ and $L^-S^+$. The relevant matrix elements are

$$\langle m_L + 1|L^+|m_L\rangle = \langle m_L|L^-|m_L + 1\rangle = \sqrt{L(L+1) - m_L(m_L+1)} \tag{28}$$

which reduce to the relations

$$L^+|p_0\rangle = \sqrt{2}|p_{+1}\rangle$$
$$L^-|p_0\rangle = \sqrt{2}|p_{-1}\rangle \tag{29}$$

Substituting into (15) and (16) we see that the two modified wave functions are

$$|+\rangle = |p_0\alpha\rangle - \frac{\zeta|p_{+1}\beta\rangle}{\Delta E\sqrt{2}}$$

$$|-\rangle = |p_0\beta\rangle - \frac{\zeta|p_{-1}\alpha\rangle}{\Delta E\sqrt{2}} \tag{30}$$

and it is very easy to calculate all the matrix elements of $(L_z + g_e S_z)$. The g tensor has axial symmetry about the z axis, with $g_{zz}$ unchanged from the free-electron value. However the perpendicular components $g_{xx}$ and $g_{yy}$ do alter, and Eq. (24) gives

$$g_{xx} = 2.0023 - \frac{2\zeta}{\Delta E} \{ \langle p_0|L_x|p_{+1} \rangle \langle p_{+1}|L_x|p_0 \rangle$$

$$+ \langle p_0|L_x|p_{-1} \rangle \langle p_{-1}|L_x|p_0 \rangle \}$$

$$= 2.0023 - \frac{2\zeta}{\Delta E} \tag{31}$$

The value of $\zeta$ is usually obtained from the analysis of atomic spectra, and in many cases $\Delta E$ can be deduced from the electronic spectrum.

# 9·4  THE g TENSOR IN MOLECULES

The calculation of the g tensor is a difficult task, but it can be simplified if certain approximations are made. The first is that the electrons in a molecule move independently in a set of molecular orbitals. All but one of the occupied orbitals contain two paired electrons, which do not contribute to the g tensor, and the odd electron is in a molecular orbital which we label $\psi_0$. If this were the only difference between a molecule and a hydrogen atom the components of the g tensor would still be given to a good approximation, by Eqs. (24) and (25). The orbitals $\psi_n$ would now be all the filled and unfilled molecular orbitals, and $E_n$ would be the orbital energy.

However, the spin-orbit coupling operator for an electron in a molecule no longer has the simple form $\zeta \, \mathbf{L} \cdot \mathbf{S}$, because the electron moves over several atoms with different values of $\zeta$, and we must look more deeply into the origin of the coupling.

The electrons in an *atom* are usually assumed to be bound in a radial force field with a certain potential energy $V(r)$. It may then be shown that the electronic Hamiltonian contains a term

$$\xi(r)\mathbf{L} \cdot \mathbf{S}$$

$$\xi(r) = \frac{eh^2}{2m^2c^2} \left( \frac{1}{r} \frac{\partial V}{\partial r} \right) \tag{32}$$

The simpler form $\zeta \, \mathbf{L} \cdot \mathbf{S}$ is derived for an electron in a particular atomic orbital $\phi_k(\mathbf{r})$ on atom $k$ by averaging over the radial probability distribution:

$$\zeta_k = \int \phi_k^* \xi_k(r_k) \phi_k \, dv \tag{33}$$

where $\xi_k(r_k)$ refers to the atom on which the orbital is centered, and decreases rapidly to zero at small distances from the nucleus. The molecular orbitals are usually expressed as linear combinations of atomic orbitals

$$\psi_0 = \sum_k C_{ko} \phi_k$$

$$\psi_n = \sum_k C_{kn} \phi_k \tag{34}$$

and the full spin orbit Hamiltonian can be written approximately as

$$\mathscr{H}_{LS} = \sum_k \xi_k(r_k)\mathbf{L}_k \cdot \mathbf{S} \tag{35}$$

$L_k$ being the orbital angular momentum about the center of the $k$th atom. If we neglect the overlap of atomic orbitals on different atoms it is now possible to show that Eq. (24) for the $g$ tensor reduces to a simplified form

$$g_{zz} = 2.0023 - 2 \sum_n \sum_{k,j} \frac{\langle \psi_0 | \zeta_k L_{zk} \delta_k | \psi_n \rangle \langle \psi_n | L_{zj} \delta_j | \psi_0 \rangle}{E_n - E_0} \tag{36}$$

Here the sum runs over all pairs of atoms $k, j$, and the symbol $\delta_k$ means that when the $L_{zk}$ operator inside the bracket acts on some atomic orbital it gives zero unless the orbital belongs to atom $k$. The sign of $(E_n - E_0)$ changes as $\psi_n$ changes from an occupied to an unoccupied orbital, so that, *other things being equal*, the filled and empty orbitals give opposite contributions to the $g$ value. Equation (36) looks a little complicated but it is quite easy to use in practice, as we now proceed to show.

The values of $\zeta$ for several atoms are given in Table 9.1.

TABLE 9.1.   Spin-Orbit Coupling Constants $\zeta$ for Atoms with $p$ Electrons (cm$^{-1}$)

| | | | |
|---|---|---|---|
| B | 11 | Na | 11 |
| C | 28 | P | 299 |
| N | 76 | S | 382 |
| O | 151 | Cl | 586 |
| F | 270 | Br | 2460 |

# 9·5  THE CO$_2^-$ RADICAL

### 9.5.1 Experimental Results

Irradiation of a single crystal of sodium formate, $Na^+ HCO_2^-$, with $\gamma$ rays results in loss of a hydrogen atom to form oriented $CO_2^-$ radicals. The $g$ tensor is found to be anisotropic and anisotropic hyperfine structure from $C^{13}$ is also observed. The spin Hamiltonian may therefore be written in the usual way

$$\mathscr{H} = \beta \mathbf{H} \cdot \mathbf{g} \cdot \mathbf{S} + \mathbf{S} \cdot \mathbf{T} \cdot \mathbf{I} \tag{37}$$

in which $\mathbf{S}$ is really the fictitious spin. The principal values of the two tensors are found to be as follows.

$$\begin{aligned}
\mathbf{T}: \quad & t'_1 = -32 \text{ Mc/s}, \quad a_C = +468 \\
& t'_2 = -46 \\
& t'_3 = +78
\end{aligned}$$

$$\begin{aligned}
\mathbf{g}: \quad & g_{xx} = 2.0032, \quad \text{average } g = 2.0006 \\
& g_{yy} = 1.9975 \\
& g_{zz} = 2.0014
\end{aligned}$$

Since $CO_2^-$ is isoelectronic with $NO_2$ it is expected to have a bent structure with $C_{2v}$ symmetry (Fig. 9.1). We shall now discuss the molecular orbitals of the bent radical and show that they provide a consistent interpretation of the e.s.r. spectra. The axis system is shown in Fig. 9.1. The $x$ axis is perpendicular to the plane of the molecule and the $z$ axis bisects the O—C—O bond angle.

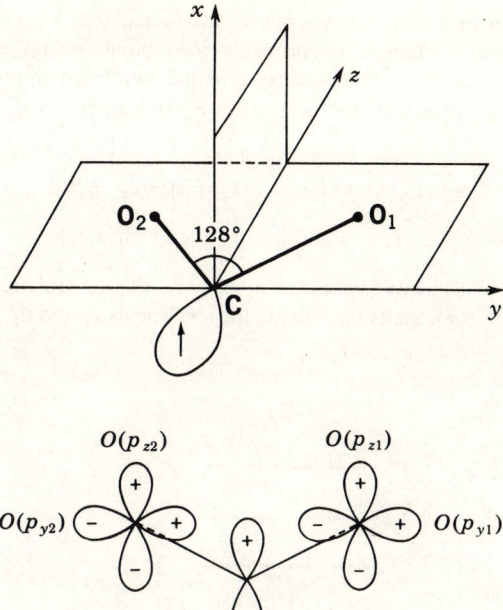

FIG. 9.1. Top: structure of the CO$_2^-$ radical. Bottom: definitions of the atomic orbitals.

### 9.5.2 Molecular Orbitals

The radical has two reflection planes of symmetry, the $xz$ and $yz$ planes, and each molecular orbital has a definite symmetry, being either symmetric or antisymmetric in each plane. The four types of orbital in the group $C_{2v}$ are usually denoted $a_1$, $a_2$, $b_1$, and $b_2$ according to the scheme below:

|        | $xz$ | $yz$ |
|--------|:----:|:----:|
| $a_1$  |  +   |  +   |
| $a_2$  |  −   |  −   |
| $b_1$  |  +   |  −   |
| $b_2$  |  −   |  +   |

Each of the three atoms has a filled shell of $1s$ electrons, while the $2s$ and $2p$ functions are available to form bonds. As a first step towards the construction of molecular orbitals we combine the carbon and oxygen $2s$ and $2p$ functions into twelve group orbitals which have a definite symmetry, as can be seen from Fig. 9.1. There

| $a_1(\sigma)$ | $a_2(\pi)$ | $b_1(\pi)$ | $b_2(\sigma)$ |
|---------------|-----------------------|------------------|----------------------|
| C(s)          | O($p_{x1} - p_{x2}$)  | C($p_x$)         | C($p_y$)             |
| C($p_z$)      |                       | O($p_{x1} + p_{x2}$) | O($s_1 - s_2$)   |
| O($s_1 + s_2$) |                      |                  | O($p_{y1} + p_{y2}$) |
| O($p_{y1} - p_{y2}$) |                |                  | O($p_{z1} - p_{z2}$) |
| O($p_{z1} + p_{z2}$) |                |                  |                      |

are therefore five $a_1$ *molecular* orbitals, one $a_2$, two $b_1$, and four $b_2$. The orbital energies are not known with certainty, but probably correspond roughly to the scheme of Fig. 9.2. The orbitals of each symmetry type are numbered in order of increasing energy, and the unpaired seventeenth electron is expected to occupy the antibonding orbital

$$\psi(4a_1) = c_1 C(s) + c_2 C(p_z) + c_3 O(p_{z1} + p_{z2})$$

$$+ c_4 O(p_{y1} - p_{y2}) + c_5 O(s_1 + s_2) \tag{38}$$

The oxygen $2s$ orbitals lie well below the others, and so $c_5$ is probably very small. We shall now use the e.s.r. results to estimate the coefficients $c_1$ and $c_2$.

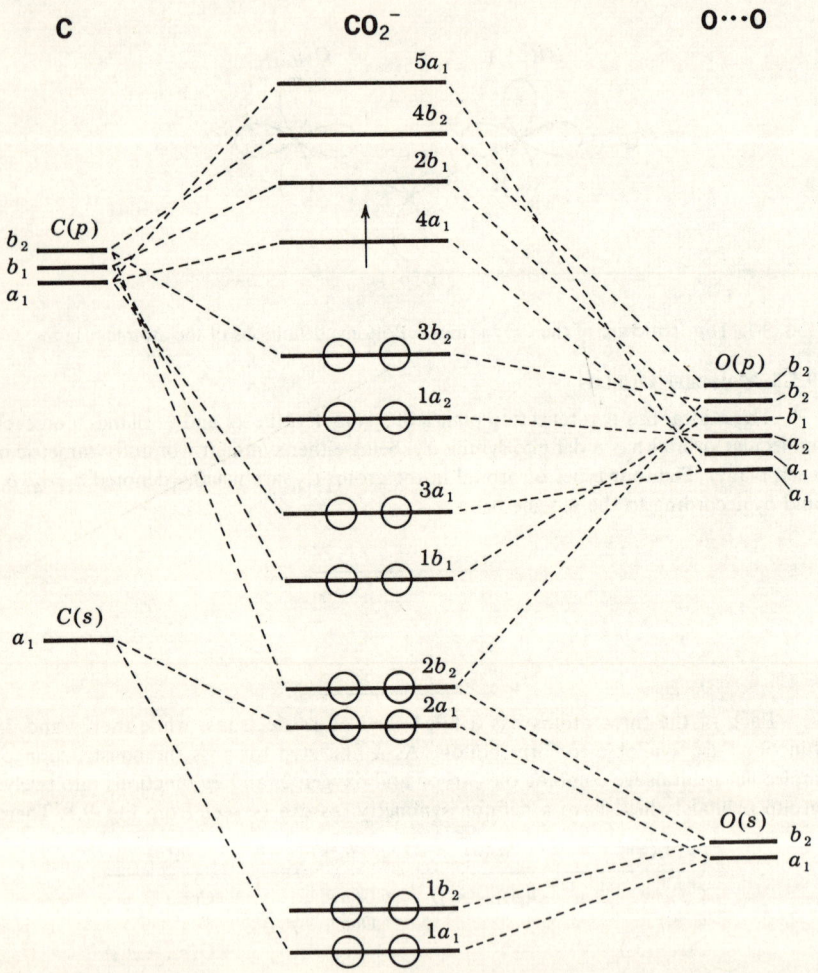

FIG. 9.2. Molecular orbital energy level scheme for $CO_2^-$.

### 9.5.3 Interpretation of the $T$ and $g$ Tensors

We begin with the $C^{13}$ hyperfine splitting. The large isotropic part, $a_C = 468$ Mc/s is due to unpaired spin density in the carbon $2s$ orbital. Since it is known that the isotropic splitting for a Hartree $2s$ atomic orbital is about 3330 Mc/s we have $|c_1|^2 = 468/3330 = 0.140$, or $c_1 = 0.374$. The anisotropic part $T'$ is nearly cylindrical about the $z$ axis, and clearly derives mainly from the carbon $2p_z$ electron. In fact we can reconstruct the tensor components $(-32, -46, +78)$ by superposing two cylindrical tensors $(-42, -41, +83)$ and $(+10, -5, -5)$ oriented along the $z$ and $x$ axes. The $(-42, -41, +83)$ part should be compared with the splittings of $(-50, -70, +120)$ from the localized unpaired electron in malonic acid; and it suggests for $c_2$ the value $|c_2|^2 = 83/120 = 0.692$, or $c_2 = \pm0.832$. The residual $(+10, -5, -5)$ anisotropy has the wrong sign to come from electron density in the oxygen $p_z$ orbitals. It arises instead from a small positive spin density of about 0.08 in carbon $2p_x$.

We therefore conclude that the unpaired electron is mainly localized in a carbon $sp^2$ hybrid orbital

$$\psi(4a_1) = 0.374\, C(s) - 0.832\, C(p_z) + \cdots \tag{39}$$

which points along the negative $z$ axis (Fig. 9.1). We shall soon see that $c_3$ is about 0.397, and since our estimates give a normalization constant

$$c_1^2 + c_2^2 + 2c_3^2 = 1.1 \tag{40}$$

greater than one, it is clear that $c_4$ and $c_5$ must be rather small.

Interpretation of the $g$ tensor depends on the properties of the excited states. The deviations of the principal values from $g_e = 2.0023$ are

$$\Delta g_{xx} = +0.0009$$
$$\Delta g_{yy} = -0.0048$$
$$\Delta g_{zz} = -0.0009$$

and by far the largest departure from the free spin value is observed when the magnetic field is in the $y$ direction. Now the angular momentum operator $L_y$ acts on the $p$ orbitals to give

$$L_y|p_y\rangle = 0, \qquad L_y|p_z\rangle = i|p_x\rangle, \qquad L_y|p_x\rangle = -i|p_z\rangle \tag{41}$$

so that it mixes the modified $p_z$ orbital $\psi(4a_1)$ with the lowest excited $p_x$ orbital $\psi(2b_1)$, which has the form

$$d_2\, C(p_x) + d_3\, O(p_{x1} + p_{x2}) \tag{42}$$

The excitation $4a_1 \to 2b_1$ corresponds to the lowest excited state of $CO_2^-$, and the lowest energy absorption band in the electronic spectrum has been measured, giving an excitation energy $\Delta E$ of 29,000 cm.$^{-1}$. We now make use of Eq. (36) to calculate $\Delta g_{yy}$, substituting $L_y$ for $L_z$. The matrix elements become

$$\sum_k \langle \psi_0 | \zeta_k L_{yk} \delta_k | \psi_n \rangle = -i(\zeta_C c_2 d_2 + 2\zeta_O c_3 d_3)$$

$$\sum_j \langle \psi_n | L_{yj} | \psi_0 \rangle = +i(c_2 d_2 + 2 c_3 d_3) \tag{43}$$

Here $\zeta_C$ and $\zeta_O$ are the spin-orbit constants of the two atoms. One therefore obtains the result

$$\Delta g_{yy} = \frac{-2(\zeta_C c_2 d_2 + 2\zeta_O c_3 d_3)(c_2 d_2 + 2c_3 d_3)}{\Delta E} \tag{44}$$

The orbital coefficients $d_2$ and $d_3$ are likely to be similar to those for the isoelectronic $NO_2$ radical; these have been calculated to be $d_2 = +0.798$ and $d_3 = -0.426$. Substitution of the experimental value of $\Delta g_{yy}$ into Eq. (44) leads to an estimate of $c_3$; $c_3 = +0.397$. Accordingly our best estimate of the odd orbital is

$$\psi(4a_1) = 0.374\ C(s) - 0.832\ C(p_z) + 0.397\ O(p_{z1} + p_{z2}) \tag{45}$$

This, of course, is rather rough, but it is probably reasonable. For instance, a rather crude molecular orbital calculation for $NO_2$ gives similar results:

$$\psi(4a_1) = 0.360\ N(s) - 0.584\ N(p_z) + 0.786\ O(p_{z1} + p_{z2}) \tag{46}$$
$$+ 0.112\ O(p_{y1} - p_{y2})$$

Clearly it is difficult to obtain an accurate description of the molecular orbital containing the unpaired electron, since there are a number of parameters of uncertain value. Nevertheless one obtains a good qualitative description and the e.s.r. method is probably far preferable to other methods, for example, those based on the estimation of oscillator strengths of optical absorption bands.

## 9·6  OTHER INORGANIC RADICALS

The study of $CO_2^-$ described in the last section is typical of many similar investigations of inorganic radicals. Many other radicals or radical ions of the structure $XO_2$ or $XO_3$ have been studied extensively, as well as simple diatomic radicals like $N_2^-$ or OH. The oxyanion radicals often show hyperfine structure from the central atom, but none from oxygen because the natural abundance of $O^{17}$ is too low. Some of the more interesting ones are listed in Table 9.2.

The anisotropic part of the hyperfine interaction arises predominantly from spin density in a $p$ orbital on the central atom; in particular, if the anisotropic part of the tensor is axially-symmetric, one can estimate the central $p$ orbital character in the molecular orbital of the odd electron quite accurately.

The isotropic part of the hyperfine structure gives a precise value of the unpaired spin density $\rho(0)$ at the central nucleus. However it is dangerous to assume that $\rho(0)$ gives an accurate measure of the $2s$ character of the odd orbital, unless the splitting is very large, for we have seen that in many radicals an electron in a $2p$ orbital polarizes the bonds and has a large isotropic splitting (e.g., malonic acid in Chapter 7). Estimates of both the $s$ and $p$ orbital character are sometimes used to deduce bond angles, but this also is rather suspect, owing to the possibility that a radical may have bent bonds.

As we have seen, the $g$ tensor depends on both the energy and nature of nearby excited states. In order to calculate its components accurately, one requires knowledge of the ground and excited state wave functions, in addition to the values of the excitation energies and spin-orbit coupling constants. Often one can only hope to show that the $g$ tensor is consistent with a molecular orbital scheme; this, nevertheless, is a worthwhile achievement.

TABLE 9.2.    $g$ Tensors and Hyperfine Coupling Constants (Mc/s) of Inorganic Radicals[a]

| Radical | $g_{xx}$ | $g_{yy}$ | $g_{zz}$ | Nucleus | $a$ | $t'_1$ | $t'_2$ | $t'_3$ | Notes |
|---|---|---|---|---|---|---|---|---|---|
| $N_2^-$ | 1.984 | 2.001 | 2.001 | $N^{14}$ | ± 13 | ∓ 2 | ∓ 2 | ± 5 | $KN_3$ |
| $F_2^-$ | 2.0230 | 2.0230 | 2.0031 | $F^{19}$ | ± 883 | ∓718 | ∓718 | ±1437 | LiF |
| OH | 2.008 | 2.008 | 2.013 | $H^1$ | ± 116 | ∓ 17 | ∓ 17 | ± 34 | $H_2O$ |
| XeF | 2.1264 | 2.1264 | 1.9740 | $F^{19}$ | ±1230 | ∓704 | ∓704 | ±1407 | $XeF_4$ |
|  |  |  |  | $Xe^{129}$ | ±1605 | ∓381 | ∓381 | ± 763 |  |
| $CO_2^-$ | 2.0032 | 1.9975 | 2.0014 | $C^{13}$ | + 468 | − 32 | − 46 | + 78 | $NaHCO_2$ |
| $NO_2$ | 2.006178 | 1.991015 | 2.00199 | $N^{14}$ | + 146.53 | − 18.73 | − 19.77 | + 38.50 | Microwave spectrum |
| $ClO_2$ | 2.0036 | 2.0183 | 2.0088 | $Cl^{35}$ | + 42 | +162 | − 86 | − 76 | $KClO_4$ |
| $F_3^-$ | — | — | — | $F^{19}$ | +1455 | −740 | +1691 | − 951 | LiF |
|  |  |  |  | $2F^{19}$ | + 561 | −337 | +499 | − 163 |  |
| $CO_3^-$ | 2.0086 | 2.0184 | 2.0066 | $C^{13}$ | − 31 | + 3 | − 3 | − 7 | $KHCO_3$ |
| $SO_3^-$ | 2.004 | 2.004 | 2.004 | $S^{33}$ | + 353 | − 37 | − 39 | + 75 | $NH_4^+SO_3^-$ |
| $ClO_3^-$ | 2.008 | 2.008 | 2.007 | $Cl^{35}$ | + 358 | − 35 | − 35 | + 71 | $NH_4^+ClO_4$ |
| $HPO_2^-$ | 2.0019 | 2.0035 | 2.0037 | $P^{31}$ | +1385 | +314 | −157 | −157 | $NH_4^+H_2PO_2^-$ |
|  |  |  |  | $H^1$ | + 230 | − 3 | + 8 | − 6 |  |

[a] For diatomic radicals, the $z$ axis is parallel to the bond. For $XO_2$, $z$ bisects the bond angle and $x$ is perpendicular to the molecular plane. For $XO_3$, $z$ is the vertical symmetry axis. $HPO_2^-$, $SO_3^-$, and $ClO_3^-$ are pyramidal.

## PROBLEMS

**1.** $XeF_4$ forms a monoclinic crystal with the following angles between the crystal axes: $\mathbf{a} \cdot \mathbf{b} = 90°$, $\mathbf{a} \cdot \mathbf{c} = 99 \cdot 6°$, $\mathbf{b} \cdot \mathbf{c} = 90°$. When the crystal is irradiated to form XeF radicals, it is found that the $g$ tensor has axial symmetry with $g_{\parallel} = 1.9740$, $g_{\perp} = 2.1264$. The symmetry axis lies in the $ac$ crystal plane, bisecting the $ac$ angle.

(a) Calculate the $g_{aa}$, $g_{ab}$, and $g_{bb}$ components of the $g$ tensor.

(b) Electron resonance is observed at 9,070 Mc/s, rotating the magnetic field in the $ab$ plane. Calculate the maximum and minimum resonance fields $H$. At what angle is the field midway between these values? (Ignore hyperfine structure.)

**2.** A radical in a single crystal has the spin Hamiltonian (in crystal axes)

$$\mathscr{H} = \beta[g_{\parallel} H_z S_z + g_{\perp}(H_x S_x + H_y S_y)]$$

The average $g$ value is $g = \frac{1}{3}(g_{\parallel} + 2g_{\perp})$ and $\Delta g = (g_{\parallel} - g_{\perp})$ is small. Derive an expression for the resonance field $H$ at fixed frequency $\nu$ when the field makes an angle $\alpha$ with the $z$ axis.

**3.** The crystal in Problem 2 is mounted with its $(1, 1, 1)$ axis across the magnetic field and the e.s.r. spectrum is observed while rotating the crystal about this axis. Sketch the variation of $H$ with angle and describe how you would determine the value of $\Delta g$.

**4.** The spin Hamiltonian of $Xe^{132}F^{19}$ in a field along the molecular axis has the form $\mathscr{H} = \beta g_{\parallel} H S_z + A S_z I_z + \frac{1}{2}B(S^+ I^- + S^- I^+)$. Here $I$ refers to the $F^{19}$ spin as $Xe^{132}$ has none. Set up the $4 \times 4$ energy matrix (see Chapter 2) and work out the energy levels to second order. How large is the splitting between the two allowed transitions at 9070 Mc/s? Use the data in Table 9.2.

**5.** A C—H fragment which is part of a localized $\pi$-electron radical has its unpaired electron in the carbon $2p_z$ orbital. The (C—H) bond has bonding and antibonding orbitals of the form $(1/\sqrt{2})(c \pm h)$, where $c$ is a carbon $sp^2$ hybrid

$$c = \frac{1}{\sqrt{3}} [C(2s) + \sqrt{2} \, C(2p_x)]$$

and $h$ is hydrogen $1s$. The bonding orbital contains two electrons. Use the method of Section 9.3 to obtain a theoretical expression for the $g$ tensor. Explain how you could calculate the $g$ tensor of the methyl radical.

The *average* $g$ value of $CH_3$ is 2.00268 compared with 2.00232 for a free electron. What do you expect the values of $g_{\parallel}$ and $g_{\perp}$ to be?

**6.** The molecular orbitals of $ClO_2$ are probably fairly similar to those of $CO_2^-$. Draw a diagram similar to Fig. 9.2 and discuss the most likely form of the odd orbital. Does the anisotropic hyperfine tensor support your assignment? Why is $g_{yy}$ now much *higher* than the free-electron value?

**7.** According to Fig. 9.2 the odd electron in $F_3^-$ (assuming it is bent) could occupy either the $5a_1$ or $4b_2$ orbital. Use the anisotropic hyperfine tensors (Table 9.2) to decide between these possibilities. Theoretical estimates of $F^{19}$ hyperfine constants are that $a = 47,900$ Mc/s for a $2s$ electron, and the anisotropic components are $-1515$, $-1515$, $+3030$ for a $2p_z$ electron. Estimate the orbital coefficients of the odd electron.

**8.** In XeF the odd electron is believed to occupy a combination of $Xe(5p_z)$ and $F(2p_z)$ orbitals. Theoretical estimates of the $Xe^{129}$ anisotropic tensor for a $5p_z$ orbital are $-1052$, $-1052$, $+2104$ Mc/s. (For $F^{19}$ see Problem 7.) Discuss the e.s.r. data for XeF and the light they throw on the electronic structure of the radical.

## SUGGESTIONS FOR FURTHER READING

Griffith: Chapter 5. Theory of the fictitious spin and the spin Hamiltonian.

Slichter: Chapter 7. Spin-orbit coupling.

Pryce: *Proc. Phys. Soc.* (London), **A63**: 25 (1950). The theory of the $g$ tensor in more general form.

Ballhausen and Gray: *Molecular Orbital Theory* (New York: W. A. Benjamin, Inc., 1964). Chapter 6. Orbitals of $NO_2$ and other small molecules.

Morton: *Chem. Revs.* **64**: 453 (1964). Review including many inorganic radicals.

Ovenall and Whiffen: *Mol. Phys.*, **4**: 135 (1961). E.S.R. spectrum of $CO_2^-$ and its interpretation.

Symons: *Advances in Physical Organic Chemistry*, **1**: 283 (1963). A useful review with much information about inorganic radicals.

Morton and Falconer: *J. Chem. Phys.*, **39**: 427 (1963). E.S.R. spectrum of the XeF radical analyzed and interpreted.

# CHAPTER 10

# E.S.R. OF TRANSITION METAL IONS AND COMPLEXES

## 10·1 INTRODUCTION

The theory of the electronic structure of transition metal ions is both satisfying and successful. In large measure the successes are due to the comprehensive and precise results of thousands of electron spin resonance studies. The main ideas underlying the theory are quite simple, but their full elaboration in each of the special cases that arise has become a highly sophisticated and technical pursuit. We shall not attempt to review the whole subject. Instead we shall deal only with that part of the theory which relates directly to e.s.r. measurements and we shall pick out one or two examples which illustrate the principal phenomena. However, it is first necessary to give a quick sketch of the electronic structure of the compounds.

TABLE 10.1. Free Ions of the Transition Metals

| Number of $d$ Electrons | Ions | Spin Orbit Constant $\zeta$ (cm$^{-1}$) | Spin Arrangement of Free Ion | Total Spin $S$ |
|---|---|---|---|---|
| $3d^1$ | Titanium$^{+++}$ | 154 | ↑ – – – – | 1/2 |
| $3d^2$ | Vanadium$^{+++}$ | 209 | ↑ ↑ – – – | 1 |
| $3d^3$ | Vanadium$^{++}$ | 167 | ↑ ↑ ↑ – – | 3/2 |
| | Chromium$^{+++}$ | 273 | | |
| $3d^4$ | Chromium$^{++}$ | 230 | ↑ ↑ ↑ ↑ – | 2 |
| | Manganese$^{+++}$ | 352 | | |
| $3d^5$ | Manganese$^{++}$ | 347 | ↑ ↑ ↑ ↑ ↑ | 5/2 |
| | Iron$^{+++}$ | — | | |
| $3d^6$ | Iron$^{++}$ | 410 | ↑ ↑ ↑ ↑ 0 | 2 |
| | Cobalt$^{+++}$ | — | | |
| $3d^7$ | Cobalt$^{++}$ | 533 | ↑ ↑ ↑ 0 0 | 3/2 |
| $3d^8$ | Nickel$^{++}$ | 649 | ↑ ↑ 0 0 0 | 1 |
| $3d^9$ | Copper$^{++}$ | 829 | ↑ 0 0 0 0 | 1/2 |

0·2 **ENERGY LEVELS OF THE $d$ ELECTRONS**

### 10.2.1 The $d$ Orbitals of a Free Ion

We are interested in ions of the first transition series which have $n$ $3d$ electrons in their valence shell outside an argon core. Although the metal atoms $M$ possess one or two $4s$ electrons the positive charges on the metal *ions* $M^{++}$ or $M^{+++}$ make the $d$ orbitals relatively more stable, and so the ions adopt a $(3d)^n$ electron configuration, as shown in Table 10.1.

The five $d$ orbitals of a free ion have precisely the same energy, and may be classified by their orbital angular momenta $\mathbf{L}^2$ and $L_z$. For $L = 2$ the possible values of the $L_z$ quantum number $m_L$ are 2, 1, 0, $-1$, $-2$, and the orbitals have the form

$$\sqrt{\frac{15}{8\pi}} \times f(r) \times \begin{cases} \dfrac{1}{4} \sin^2 \theta \, e^{2i\phi} = |2\rangle \\[2mm] -\sin \theta \cos \theta \, e^{i\phi} = |1\rangle \\[2mm] \dfrac{1}{\sqrt{6}} (3 \cos^2 \theta - 1) = |0\rangle \\[2mm] +\sin \theta \cos \theta \, e^{-i\phi} = |-1\rangle \\[2mm] \dfrac{1}{4} \sin^2 \theta \, e^{-2i\phi} = |-2\rangle \end{cases} \tag{1}$$

The electronic energy levels of the free ion are determined by the interplay of three interactions:

1. Coulomb repulsion between the $d$ electrons, which makes the energy depend on the combined *total angular momentum*, $\mathbf{L}$, of all the $d$ electrons.

2. Exchange forces between $d$ electrons of the same spin, which lower the energies of configurations with *high total spin*, $S$.

3. Spin-orbit coupling $\zeta \, \mathbf{L} \cdot \mathbf{S}$ which couples $\mathbf{L}$ and $\mathbf{S}$ together into a resultant total angular momentum vector, $\mathbf{J}$.

The Coulomb and exchange energies of $1{,}000 - 10{,}000$ cm$^{-1}$ are generally larger than spin-orbit coupling energies, since $\zeta$ is less than $1{,}000$ cm$^{-1}$ (Table 10.1), and the ground state electron configuration of the free ion will have the highest total spin consistent with the Pauli exclusion principle. Thus for the $d^1$ to $d^5$ ions each $d$ electron goes into a different orbital and all the spins are parallel, giving $S$ values ranging from $1/2$ to $5/2$. From $d^6$ to $d^9$ it becomes necessary to fill one or more $d$ orbitals with two paired electrons and the number of unpaired spins falls progressively as shown in the last column of Table 10.1.

Let us now consider an octahedral complex where the metal ion is surrounded by six negative ions or other groups which form bonds with it. These "ligands" introduce a new factor into the energy level scheme, namely

4. The "ligand-field splitting" of the $d$ orbitals which removes the degeneracy of the $d$ orbitals and mixes together orbitals with different $L_z$ quantum numbers.

The new electronic states are best discussed in terms of the real $d$ orbitals, which are not eigenfunctions of $L_z$, but conform to the octahedral symmetry of the complex.

The real orbitals are formed by taking the real and imaginary parts of the complex orbitals in Eq. (1), and have the following angular variation:

$$(d_{z^2}, d_{x^2-y^2}) = \sqrt{\frac{15}{16\pi}}\left(\frac{3z^2 - r^2}{r^2\sqrt{3}}, \frac{x^2 - y^2}{r^2}\right)$$

$$(d_{xy}, d_{yz}, d_{xz}) = \sqrt{\frac{15}{16\pi}}\left(\frac{2xy}{r^2}, \frac{2yz}{r^2}, \frac{2xz}{r^2}\right)$$

(2)

They are illustrated in Fig. 10.1.

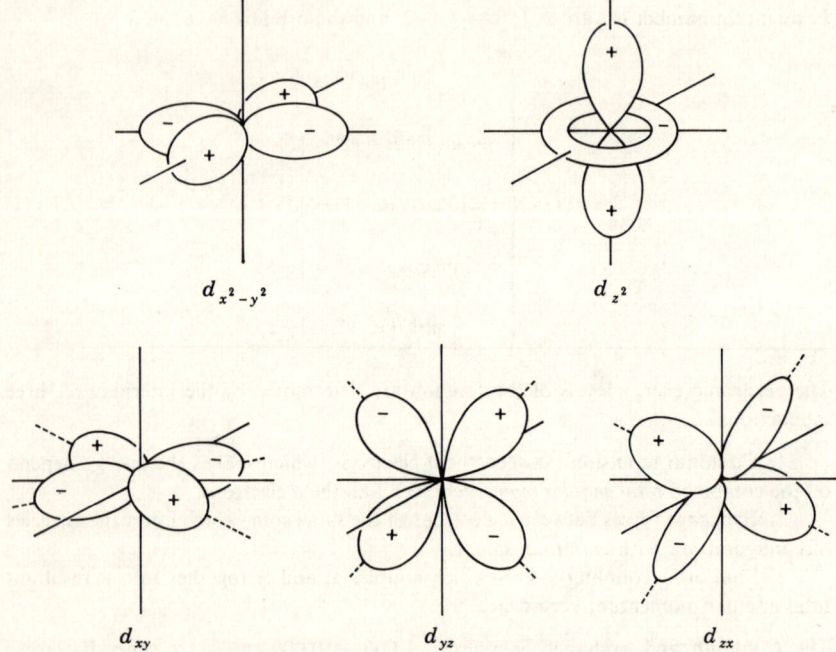

$$d_{x^2-y^2} \qquad\qquad d_{z^2}$$

$$d_{xy} \qquad\qquad d_{yz} \qquad\qquad d_{zx}$$

FIG. 10.1. Real forms of the $d$ orbitals.

## 10.2.2 The Ligand-Field Splitting

The ligands are usually negative ions, or neutral molecules with a prominent lone pair of electrons. Typical examples are

$$I^-, \ Cl^-, \ F^-, \ H_2O, \ NH_3, \ NO_2^-, \ CN^-$$

There are also double-ended chelating ligands like ethylenediamine and the acetylacetonate ion which attach themselves to the metal ion in two places so that three ligands form an approximately octahedral complex. The ligands affect the $d$ electrons in two distinct ways. One is through the electrostatic field of the negative charges; the other is by covalent bonding with the ligand orbitals. Both these effects alter the energies of the $d$ orbitals in approximately the same manner. Those $d$ orbitals which have a large electron density near the ligands are repelled by the ligands and have a higher energy; those which avoid the ligands are not pushed up so high. Thus in an octahedral complex $d_{z^2}$ and $d_{x^2-y^2}$ are both raised by an equal amount while $d_{xy}$, $d_{yz}$, and $d_{xz}$ are relatively lower. The energy gap between the two types of orbital is called the ligand field splitting, $\Delta$. Figure 10.2 shows the energy level scheme for this and other types of symmetry.

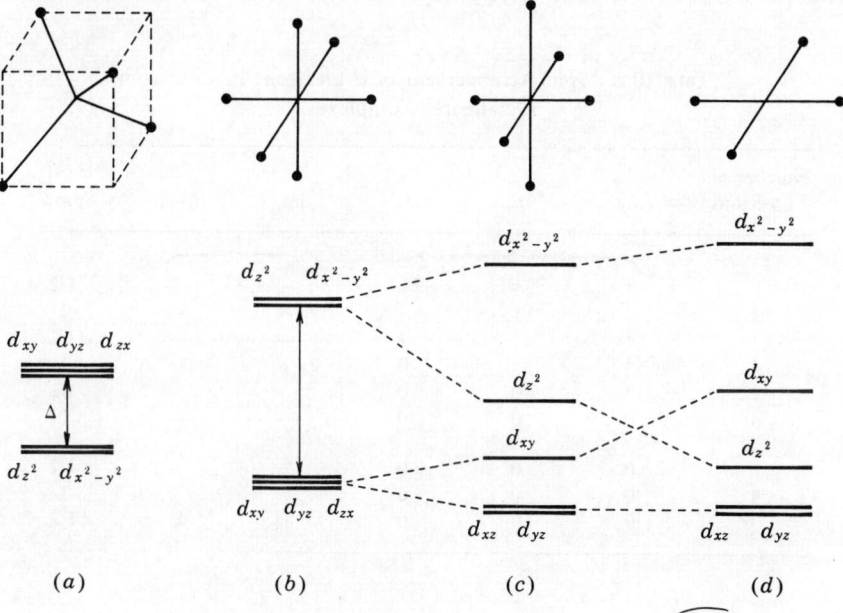

FIG. 10.2. Ligand field splitting of the $d$ orbitals. (a) Tetrahedral. (b) Octahedral. (c) Octahedron with tetragonal distortion. (d) Square planar.

The size of the splitting varies from 10,000 to 30,000 cm$^{-1}$ and our list of ligands above is arranged in a sequence, the "spectrochemical series" with small splittings for I$^-$ and Cl$^-$ on the left, large ones for CN$^-$ on the right. For example, $\Delta$ values for Cr$^{+++}$ with various ligands are shown below.

Ligand Field Splittings for
Cr$^{+++}$ Ion (cm$^{-1}$)

| | |
|---|---|
| CN$^-$ | 26,300 |
| ethylenediamine | 21,900 |
| NH$_3$ | 21,600 |
| H$_2$O | 17,400 |
| Cl$^-$ | 13,600 |

Two distinct situations are possible, depending on the size of $\Delta$. If $\Delta$ is much smaller than the Coulomb and exchange energies of the electrons in the free ion, the electron configurations will be just like the ones shown in Table 10.1; a complex like $(Fe^{+++})$ $(F^-)_6$ has five unpaired $d$ electrons with a total spin $S = 5/2$. This is called the weak field, or high spin case, and applies to ions where the exchange energies of the electrons are more important than the other interactions. The electrons prefer to go into the configuration with the largest number of parallel spins and the highest exchange energy. On the other hand, if $\Delta$ is very large, as it is in $(Fe^{+++})(CN^-)_6$, the electrons first fill up the low-lying $d_{xy}, d_{xz}, d_{yz}$ orbitals before they begin to occupy $d_{z^2}$ or $d_{x^2-y^2}$. This is the strong field or low spin case. The first three electrons can each go into different orbitals of the lower set and have parallel spins, just as they do in a high spin complex. However, the next three go into the same orbitals and pair up with the

TABLE 10.2.  Spin Arrangement of $d$ Electrons in Low Spin Octahedral Complexes

| Number of $d$ Electrons | $d_{xy}$ | $d_{yz}$ | $d_{xz}$ | $d_{z^2}$ | $d_{x^2-y^2}$ | Spin $S$ |
|---|---|---|---|---|---|---|
| 1 | ↑ | — | — | — | — | 1/2 |
| 2 | ↑ | ↑ | — | — | — | 1 |
| 3 | ↑ | ↑ | ↑ | — | — | 3/2 |
| 4 | ↑ | ↑ | 0 | — | — | 1 |
| 5 | ↑ | 0 | 0 | — | — | 1/2 |
| 6 | 0 | 0 | 0 | — | — | 0 |
| 7 | 0 | 0 | 0 | ↑ | — | 1/2 |
| 8 | 0 | 0 | 0 | ↑ | ↑ | 1 |
| 9 | 0 | 0 | 0 | ↑ | 0 | 1/2 |

electrons which are already there. Thus $Fe^{+++}$ would only have spin 1/2, as shown in Table 10.2. Any further electrons then go into the $d_{z^2}$ or $d_{x^2-y^2}$ orbitals and the spin reaches a second maximum value of 1.

### 10.2.3 Regular and Distorted Complexes

The atomic orbitals of the metal ion must be considerably distorted by the ligands, and their energies change in a complicated way. The most important point, however, is to decide which orbitals remain degenerate and which do not. The degeneracy is completely determined by the symmetries of the atomic orbitals and the complex. For example, the $d_{xy}$, $d_{yz}$, and $d_{xz}$ orbitals *must* inevitably be degenerate in a regular octahedral or tetrahedral environment; $d_{xz}$ and $d_{yz}$ are degenerate in a square planar complex, and so on. Chemists often use the notation of group theory to describe the symmetry, denoting the $d_{xz}$ and $d_{yz}$ orbitals of the planar complex as being of symmetry $e_g$, for example. For our purposes these symbols are merely labels and it is not necessary to understand precisely what they mean. However, for convenience, the symmetry notation is summarized in Table 10.3 and the most important point to note is that the ligands still allow some degeneracy in the $d$ orbital scheme. Orbitals of type $t$ in the octahedron are always triply degenerate while $e$ orbitals are doubly degenerate. $a$ and $b$ orbitals are nondegenerate.

TABLE 10.3.    Symmetry Notation for Transition Metal
Orbitals in Complexes

| Atomic Orbital | Octahedron $O_h$ | Tetrahedron $T_d$ | Square Planar $D_{4h}$ |
|---|---|---|---|
| $s$ | $a_{1g}$ | $a_1$ | $a_{1g}$ |
| $p_x$ | | | $e_u$ |
| $p_y$ | $t_{1u}$ | $t_2$ | |
| $p_z$ | | | $a_{2u}$ |
| $d_{xy}$ | | | $b_{2g}$ |
| $d_{xz}$ | $t_{2g}$ | $t_2$ | |
| $d_{yz}$ | | | $e_g$ |
| $d_{z^2}$ | $e_g$ | $e$ | $a_{1g}$ |
| $d_{x^2-y^2}$ | | | $b_{1g}$ |
| $f_{xyz}$ | $a_{2u}$ | $a_2$ | $a_{2u}$ |

Very few complexes in the solid state actually possess regular symmetry and the classification into octahedral, tetrahedral, square planar, etc., is only approximate. There are several reasons for this which we shall not explore here. In electron resonance investigations it is convenient to classify the complex according to the nearest "regular" symmetry group and then to improve the description by introducing an appropriate distortion. The commonest distortion in tetrahedral complexes is the so-called tetragonal distortion corresponding to a flattening of the tetrahedron, as indicated by the arrows in Fig. 10.3. As a result of this distortion the degeneracy of the $d$ orbitals is further removed as shown in Fig. 10.5.

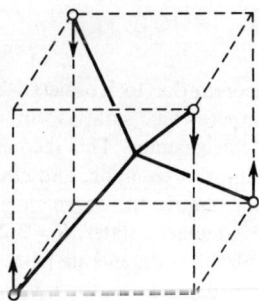

FIG. 10.3. Tetrahedral complex with a tetragonal distortion.

We have already seen in Fig. 10.2(c) that octahedral complexes can undergo a tetragonal distortion into a structure with four short metal-ligand bonds and two long ones. Octahedral complexes also frequently show a trigonal distortion (Fig. 10.4), corresponding to a slight compression or elongation along a three-fold axis. Electron resonance studies are able to detect these distortions quite easily, as we shall soon see.

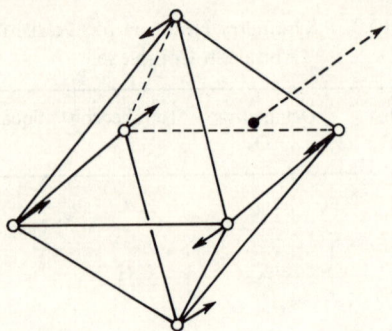

Fig. 10.4. Octahedral complex with a trigonal compression along the (1, 1, 1) direction.

## 10·3 GENERAL FEATURES OF THE E.S.R. SPECTRA

Many of the main ideas have been encountered already. We will begin with a discussion of the $g$ tensor in ions with spin 1/2, and as we shall see, the theory is actually simpler than it is for inorganic free radicals like $CO_2^-$. This is because the unpaired electrons in transition metal complexes are largely localized on the metal atom; also the spin-orbit coupling in a metal atom is much greater than in most ligand atoms, so that contributions to the $g$ tensor arising from electron delocalization can nearly always be neglected.

For metal ions with triplet ground states the general features of the e.s.r. spectra are similar to those described for organic molecules in Chapter 8. However, the most important mechanism of zero-field splitting involves spin-orbit mixing of ground and excited electronic states, and this can produce splittings of 10 cm$^{-1}$ or more. We shall also meet cases where the spin is greater than 1; here, too, the e.s.r. spectra are dominated by zero-field splittings arising from spin-orbit coupling.

## 10·4 KRAMERS' THEOREM

There is a most important theorem due to Kramers which states that in the absence of external magnetic fields the electronic states of any molecule *with an odd number of electrons* are at least doubly degenerate. This theorem implies that all the complicated ligand-field splitting, spin-orbit coupling, and electron spin-spin interactions in transition metal ions can never remove the degeneracy of a state with spin $S = 1/2$. Kramers' theorem also applies to quartet states, $S = 3/2$, where the zero-field splitting produces two "Kramers' doublets" or degenerate pairs of states with $m_S = \pm 3/2$ and $m_S = \pm 1/2$. Kramers' theorem depends on the invariance of the electronic Hamiltonian under the symmetry operation of "time reversal" which reverses the spins and momenta of all the electrons, and it does not apply to molecules in a magnetic field.

# 10·5 THE $g$ TENSOR IN IONS WITH $S = 1/2$

### 10.5.1 The $Ti^{+++}$ Ion in a Tetrahedral Complex

The trivalent titanium ion possesses one $d$ electron, and in the absence of a distortion it will occupy either the $d_{z^2}$ or $d_{x^2-y^2}$ orbitals shown in Fig. 10.2(a). The $g$ value of a regular tetrahedral complex is necessarily isotropic, so it is more interesting to consider the effect of a tetragonal distortion (Fig. 10.3).

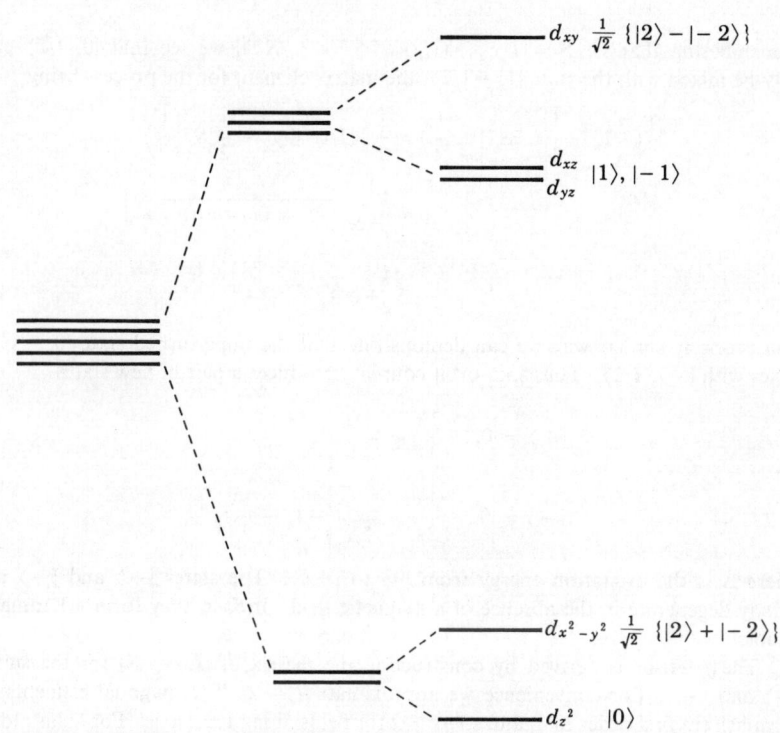

FIG. 10.5. $d$ orbitals in a tetrahedral complex with a tetragonal distortion.

The $d$ orbital energy level scheme is shown in Fig. 10.5. Since we are dealing with angular momentum, the familiar real forms of the $d$ orbitals are inconvenient. Instead it is better to go back to the $|m_L\rangle$ form used in Eq. (1). Each of the $d_{z^2}$ or $d_{x^2-y^2}$ orbitals is a linear combination of the orbitals $|0\rangle$, $|2\rangle$, and $|-2\rangle$ with a zero average value for the angular momentum $\langle L_z \rangle$, and if they have a different energy from the $d_{xy}$, $d_{yz}$, and $d_{xz}$ orbitals, the angular momentum is completely quenched by the ligand field. If this were a complete description of the system we would expect the complex to have the isotropic free electron $g$ value. However, the spin-orbit coupling operator $\zeta \mathbf{L} \cdot \mathbf{S}$ mixes the ground state functions with excited states and generates some orbital angular momentum. In the ground state the unpaired electron occupies $d_{z^2}$, or $|0\rangle$, and the two spin states will be labeled by their $m_L$, $m_S$ quantum numbers

$$\left|0, \frac{1}{2}\right\rangle \qquad \left|0, -\frac{1}{2}\right\rangle \tag{3}$$

According to first-order perturbation theory spin-orbit coupling changes the $|0, 1/2\rangle$ state into

$$|+\rangle = \left|0, \frac{1}{2}\right\rangle - \sum_{m_L, m_S} |m_L m_S\rangle \frac{\left\langle m_L m_S \left| \zeta \mathbf{L} \cdot \mathbf{S} \right| 0, \frac{1}{2}\right\rangle}{E(m_L m_S) - E\left(0, \frac{1}{2}\right)} \tag{4}$$

Remembering that $\mathbf{L} \cdot \mathbf{S} = [L_z S_z + (1/2)(L^+ S^- + L^- S^+)]$ we see that $|0, 1/2\rangle$ can only be mixed with the state $|1, -1/2\rangle$, the matrix element for the process being

$$\frac{1}{2}\zeta\left\langle 1, -\frac{1}{2}\left| L^+ S^- \right| 0, \frac{1}{2}\right\rangle = \frac{1}{2}\zeta\langle 1|L^+|0\rangle\left\langle -\frac{1}{2}\left| S^- \right| \frac{1}{2}\right\rangle$$

$$= \frac{1}{2}\zeta\sqrt{L(L+1) - m_L(m_L + 1)}$$

$$= \frac{1}{2}\zeta\sqrt{6} \tag{5}$$

In a precisely similar way we can demonstrate that the unperturbed state $|0, -1/2\rangle$ mixes with $|-1, 1/2\rangle$. Thus spin-orbit coupling produces a pair of new states

$$|+\rangle = \left|0, +\frac{1}{2}\right\rangle - \frac{\zeta\sqrt{6}}{2\Delta}\left|1, -\frac{1}{2}\right\rangle$$

$$|-\rangle = \left|0, -\frac{1}{2}\right\rangle - \frac{\zeta\sqrt{6}}{2\Delta}\left|-1, +\frac{1}{2}\right\rangle \tag{6}$$

where $\Delta$ is the excitation energy from $|0\rangle$ to $|\pm 1\rangle$. The states $|+\rangle$ and $|-\rangle$ are strictly degenerate in the absence of a magnetic field. In fact, they form a Kramers' doublet.

The $g$ tensor is derived by constructing the matrix of $(\mathbf{L} + g_e \mathbf{S})$ for the states $|+\rangle$ and $|-\rangle$. For convenience we approximate $g_e = 2$. The diagonal elements of $L_z$ vanish (to first order in $\zeta$) and so $g_\parallel = 2$ for fields along the $z$ axis. For fields along the $x$ axis the matrix of $S_x$ is simple, but we also require the matrix of $L_x$. This involves terms like

$$\left\langle 0, -\frac{1}{2}\left| L_x \right| 1, -\frac{1}{2}\right\rangle = \left\langle 0, +\frac{1}{2}\left| L_x \right| -1, +\frac{1}{2}\right\rangle = \frac{1}{2}\sqrt{6} \tag{7}$$

and we find the complete matrix

$$(L_x + 2S_x) = \begin{array}{cc} & \begin{array}{cc} |+\rangle & \qquad\qquad |-\rangle \end{array} \\ \begin{array}{c} \langle +| \\ \\ \langle -| \end{array} & \left[ \begin{array}{cc} 0 & 1 - \left(\dfrac{3\zeta}{\Delta}\right) \\ \\ 1 - \left(\dfrac{3\zeta}{\Delta}\right) & 0 \end{array} \right] \end{array} \tag{8}$$

Exactly the same result is obtained for the field in the $y$ direction and we describe the situation by saying that $g_\perp = 2\{1 - (3\zeta/\Delta)\}$. It follows from these results and the discussion in Section 9.3 that the spin Hamiltonian is

$$\mathcal{H} = \beta\{g_{\parallel}H_z\hat{S}_z + g_\perp(H_x\hat{S}_x + H_y\hat{S}_y)\} \tag{9}$$

where $\hat{S}_x$, $\hat{S}_y$, $\hat{S}_z$ are components of the fictitious spin $\hat{S}$. The states $|+\rangle$ and $|-\rangle$ are eigenfunctions of $\hat{S}_z$ with eigenvalues $+1/2$ and $-1/2$, respectively.

Not very many tetrahedral $d^1$ complexes are known. The manganate ion, $MnO_4^{2-}$, is a good example but the ion is rather strongly distorted in the crystal and the above treatment cannot be applied satisfactorily; all of the principal values of the $g$ tensor are less than 2. Another is $VCl_4$, but its e.s.r. spectrum has not yet been studied in detail. We shall see in due course that the theory for octahedral $d^9$ complexes is similar to that for tetrahedral $d^1$. The major point to be noted is that the states which are mixed by spin-orbit coupling are separated from each other by an amount comparable with the ligand field splitting (10,000 to 20,000 cm$^{-1}$) and the $g$ tensor anisotropy is therefore rather small compared with that encountered in complexes with one or two unpaired electrons in the $t_2$ orbitals.

### 10.5.2 The Ti$^{3+}$ Ion in an Octahedral Complex·

We begin by considering a trigonal distortion of the type shown in Fig. 10.4. If we take the axis of the distortion as a new axis of quantization the $d$ orbitals indicated in Fig. 10.6 take the following form:

$$|e^+\rangle = \sqrt{\frac{2}{3}}|1\rangle - \sqrt{\frac{1}{3}}|-2\rangle$$
$$|e^-\rangle = \sqrt{\frac{2}{3}}|-1\rangle + \sqrt{\frac{1}{3}}|2\rangle \tag{10}$$

$$|t^+\rangle = \sqrt{\frac{1}{3}}|1\rangle + \sqrt{\frac{2}{3}}|-2\rangle$$
$$|t^-\rangle = \sqrt{\frac{1}{3}}|-1\rangle \quad \sqrt{\frac{2}{3}}|2\rangle \tag{11}$$

$$|t_0\rangle = |0\rangle$$

The major ligand field splitting is $\Delta$, and the subsidiary splitting of the $t_{2g}$ orbitals due to the distortion is $\delta$. The most stable orbital is $d_{z^2}$ or $|t_0\rangle$.

Spin-orbit coupling mixes the ground state functions $|t_0, 1/2\rangle$, $|t_0, -1/2\rangle$ with the degenerate states immediately above, and since $\delta$ may be comparable with the spin-orbit coupling energy, we will not use perturbation theory but derive exact solutions. It is readily shown that the operator $\zeta\mathbf{L}\cdot\mathbf{S}$ mixes the ground state $|t_0, 1/2\rangle$ with $|t^+, -1/2\rangle$ and similarly mixes $|t_0, -1/2\rangle$ with $|t^-, +1/2\rangle$; but it does not mix $|t^+\rangle$ with $|t^-\rangle$ because the $m_L$ values of the $d$ functions which form these orbitals differ by

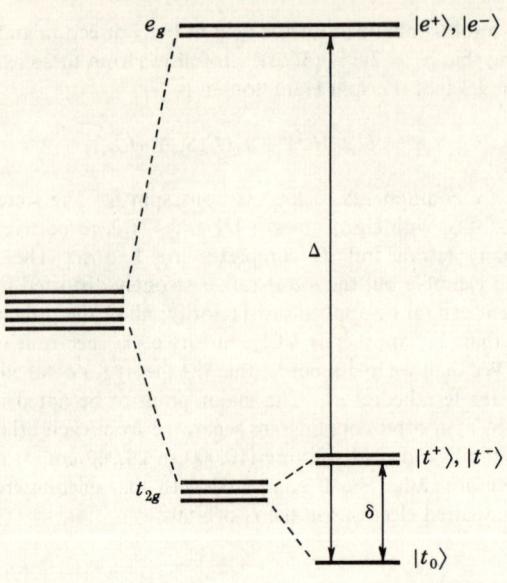

FIG. 10.6. Orbitals for an octahedral complex with a trigonal distortion (Symmetry $D_{3h}$).

more than one. The off-diagonal matrix element for $|t_0, 1/2\rangle$ and $|t^+, -1/2\rangle$ is

$$\left\langle t^+, -\frac{1}{2}\middle|\zeta\mathbf{L}\cdot\mathbf{S}\middle|t_0, \frac{1}{2}\right\rangle = \frac{1}{2}\zeta\langle t^+|L^+|t_0\rangle\left\langle -\frac{1}{2}\middle|S^-\middle|\frac{1}{2}\right\rangle$$

$$= \frac{1}{2}\zeta\left[\sqrt{\frac{1}{3}}\langle 1|L^+|0\rangle + \sqrt{\frac{2}{3}}\langle -2|L^-|0\rangle\right]$$

$$= \zeta/\sqrt{2} \qquad (12)$$

as we see from Eq. (5), and so the $2 \times 2$ matrix of the spin-orbit coupling and the trigonal distortion energies becomes

$$
\begin{array}{cc}
 & \left|t_0, \dfrac{1}{2}\right\rangle \qquad \left|t^+, -\dfrac{1}{2}\right\rangle \\
\begin{array}{c}\left\langle t_0, \dfrac{1}{2}\right| \\[2mm] \left\langle t^+, -\dfrac{1}{2}\right|\end{array} &
\begin{bmatrix} 0 & \dfrac{\zeta}{\sqrt{2}} \\[3mm] \dfrac{\zeta}{\sqrt{2}} & \delta + \dfrac{1}{2}\zeta \end{bmatrix}
\end{array}
\qquad (13)
$$

There is an identical matrix for the $|t_0, -1/2\rangle$ and $|t^-, 1/2\rangle$ states. The exact wave functions are

$$|+\rangle = \cos\theta\left|t_0, \frac{1}{2}\right\rangle - \sin\theta\left|t^+, -\frac{1}{2}\right\rangle$$

$$|-\rangle = \cos\theta\left|t_0, -\frac{1}{2}\right\rangle + \sin\theta\left|t^-, \frac{1}{2}\right\rangle$$

$$(14)$$

and form a Kramers' doublet, $\theta$ being defined so that

$$\tan 2\theta = \frac{\zeta}{(\zeta + 2\delta)\sqrt{2}} \tag{15}$$

The reader may care to calculate the matrix elements of $(\mathbf{L} + 2\mathbf{S})$ and verify that the principal components of the $g$ tensor are

$$g_{\parallel} = \frac{3(\zeta + 2\delta)}{\sqrt{(\zeta + 2\delta)^2 + 8\zeta^2}} - 1$$

$$g_{\perp} = \frac{(2\delta - 3\zeta)}{\sqrt{(\zeta + 2\delta)^2 + 8\zeta^2}} + 1 \tag{16}$$

Spin-orbit coupling also mixes the $|t\rangle$ with the $|e^{\pm}\rangle$ orbitals, but this is unimportant if $\delta \ll \Delta$.

It will be clear that the $g$ tensor in $d^1$ octahedral complexes is considerably more anisotropic than in tetrahedral complexes, simply because the energy difference between the states which are mixed by spin-orbit coupling is so much smaller. In general, if the $t_{2g}$ shell in a distorted octahedral or tetrahedral complex is incompletely filled (but not if it is exactly half-filled), the $g$ tensor is much more anisotropic than if the odd electron is in an $e$ orbital.

One example is the $(Ti^{+++})(H_2O)_6$ ion in $CsTi(SO_4)_2 \cdot 12H_2O$, where electron resonance spectra are observed with $g_{\parallel} = 1.25$, $g_{\perp} = 1.14$. These values can be fitted rather roughly to the theoretical formula after making allowance for covalent bonding with the water molecules. The chelated complex titanium trisacetylacetonate (Fig. 10.7) also has trigonal symmetry, but now $\delta$ is much larger than $\zeta$, and the spin-orbit coupling with the higher $|e^{\pm}\rangle$ levels must be included. Here the theory gives

$$g_{\parallel} = 2$$

$$g_{\perp} = 2 - \frac{2\zeta}{\delta} - \frac{4\zeta}{\Delta} \tag{17}$$

and the experimental values $g_{\parallel} = 2.00$ and $g_{\perp} = 1.93$ can be interpreted successfully by taking $\delta = 7,500 \text{ cm}^{-1}$, if $\Delta = 20,000 \text{ cm}^{-1}$, and $\zeta = 150 \text{ cm}^{-1}$.

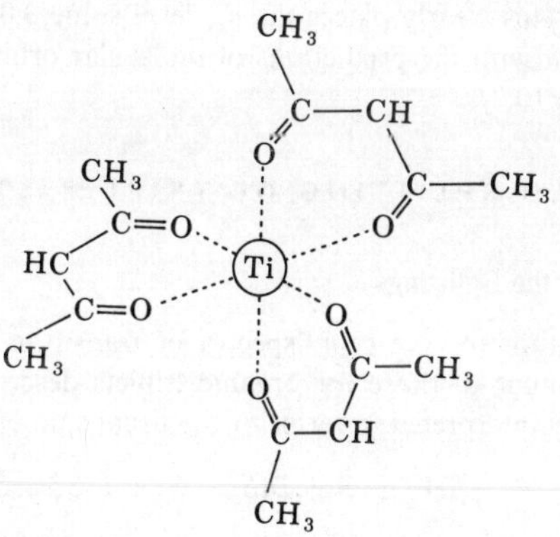

FIG. 10.7. Titanium trisacetylacetonate.

We have already mentioned that the $g$-tensor theories for $d^1$ tetrahedral and $d^9$ octahedral complexes are very similar. In the tetrahedral $Ti^{+++}$ complex the odd electron occupies the degenerate $e$ orbitals; in an octahedral $d^9$ complex there is a "hole" in the $e$ orbitals (i.e., one electron missing). The theory developed in Section 10.5.1 would be applicable to an octahedral copper complex in which there was a compression of two ligands along the $z$ axis; it is only necessary to change the *sign* of the spin-orbit coupling constant to account for the fact that we are considering a "hole" rather than an electron. In practice octahedral complexes of copper invariably show a tetragonal distortion corresponding to elongation along a four-fold symmetry axis $z$ [Fig. 10.2(c)]. Such a distortion lowers the energy of $d_{z^2}$ so that the hole is in $d_{x^2-y^2}$. It is then relatively easy to show that, with neglect of second-order terms, the principal components of the $g$ tensor are

$$g_\parallel = 2\left(1 + \frac{4\zeta}{\Delta_1}\right)$$

$$g_\perp = 2\left(1 + \frac{\zeta}{\Delta_2}\right)$$

(18)

where $\Delta_1$ is the separation between $d_{x^2-y^2}$ and $d_{xy}$ and $\Delta_2$ is the separation between $d_{x^2-y^2}$ and $d_{xz}$, $d_{yz}$. An extreme case of a strong tetragonal perturbation is copper phthalocyanine

where it is found that $g_\parallel = 2.165$ and $g_\perp = 2.045$. Since $\zeta = 829$ cm$^{-1}$ one may infer, after making a small correction to allow for covalent bonding, that $\Delta_1 = 31,700$ cm$^{-1}$ $\Delta_2 = 29,000$ cm$^{-1}$. This clearly places the $d_{xy}$ level some 3,000 cm$^{-1}$ below the $d_{xz}$, $d_{yz}$ pair, in agreement with the predictions of molecular orbital theory (but contrary to the scheme of Fig. 10.2).

## 10·6 THE ZERO-FIELD SPLITTING OF TRIPLET STATES

### 10.6.1 The Origin of the Splittings

The general features of the e.s.r. spectra of transition metal ions with triplet ground states are similar to those for organic triplets described in Chapter 8. The spectra are invariably interpreted in terms of the usual spin Hamiltonian

$$\mathscr{H} = \beta \mathbf{H} \cdot \mathbf{g} \cdot \hat{\mathbf{S}} + D(\hat{S}_z^2 - \tfrac{2}{3}) + E(\hat{S}_x^2 - \hat{S}_y^2)$$

(19)

with additional terms involving nuclear spin if necessary. However, we saw that the mechanism of the zero-field splitting in organic molecules involves the dipolar interaction of the electron spins; in contrast, the zero-field splitting in metal ions usually arises because of spin-orbit coupling and in some cases can be too large to permit the observation of electron resonance absorption.

The spin-orbit coupling leads to an indirect electron spin-spin coupling in very much the same way as the contact hyperfine interaction $a\mathbf{I} \cdot \mathbf{S}$ gives indirect nuclear spin-spin coupling in diamagnetic molecules (see Section 5.5). According to second-order perturbation theory the electronic energy contains a term

$$-\zeta^2 \sum_n \frac{\langle 0|\mathbf{L} \cdot \mathbf{S}|n\rangle\langle n|\mathbf{L} \cdot \mathbf{S}|0\rangle}{E_n - E_0} \qquad (20)$$

and if we assume that the space and spin parts of the electronic wave functions are independent of one another in all the states $|0\rangle$ and $|n\rangle$ it is readily seen that (20) reduces to the form

$$\mathscr{H} = \hat{\mathbf{S}} \cdot \mathbf{D} \cdot \hat{\mathbf{S}} \qquad (21)$$

where $\mathbf{D}$ is a spin-spin coupling tensor. The components of this tensor are

$$D_{ik} = -\zeta^2 \sum_n \frac{\langle \psi_0|L_i|\psi_n\rangle\langle \psi_n|L_k|\psi_0\rangle}{E_n - E_0} \qquad i, k = x, y, z \qquad (22)$$

Here it has been assumed that the excited state $|n\rangle$ is formed by promoting an electron out of a certain molecular orbital $\psi_0$ into an excited orbital $\psi_n$, and now $E_n$, $E_0$ are the orbital energies. Apart from a numerical factor the same tensor $\mathbf{D}$ appears in the theory of the $g$ factor in Chapter 9.

We now describe in detail how one calculates the zero-field splitting in a typical $d^2$ ion. We choose to examine the $V^{+++}$ ion in a tetragonally distorted octahedral complex.

### 10.6.2 Zero-Field Splitting in the $V^{+++}$ Ion

The ground state electron configuration of the undistorted ion is constructed by assigning two electrons to the triply degenerate $t_2$ orbitals $d_{xy}$, $d_{xz}$, and $d_{yz}$, according to Fig. 10.2(b). The $(t_2)^2$ electron configuration gives rise to several electronic states since there are three ways of putting two electrons in the same orbital (with opposed spins) and three ways of arranging them in different orbitals, where the spins may be either parallel or antiparallel. These states differ in energy because of the electron repulsions, but the lowest is a triplet with threefold orbital degeneracy and symmetry $T_1$. The electrons occupy two of the $d_{xy}$, $d_{yz}$, $d_{xz}$ orbitals and since each orbital configuration can take up three spin orientations there are nine degenerate states. The lowest excited states in the $(t_2)^2$ configuration are a group of singlets ($E$ and $T_2$) about 9,000 cm$^{-1}$ above the ground state and an isolated singlet of $A_1$ symmetry about 21,000 cm$^{-1}$ above. We shall ignore these upper states and consider the effect of a tetragonal distortion on the degenerate triplet ground term.

Elongation of the octahedron along the $z$ axis shifts the $d_{xz}$ and $d_{yz}$ orbitals below $d_{xy}$, as shown in Fig. 10.2(c). So far as the electronic states are concerned, the $(d_{xz})(d_{yz})$ triplet is clearly lowest, while the $(d_{xy})(d_{xz})$ and $(d_{xy})(d_{yz})$ triplets are degenerate, with a slightly higher energy $\delta$, as we see from Fig. 10.8. Thus there are nine states which are written as Slater determinants as follows:

$$A_2: \quad \psi_1 = \|(xz)^\alpha(yz)^\alpha\|$$

$$\psi_2 = (1/\sqrt{2})\{\|(xz)^\alpha(yz)^\beta\| + \|(xz)^\beta(yz)^\alpha\|\} \qquad (23)$$

$$\psi_3 = \|(xz)^\beta(yz)^\beta\|$$

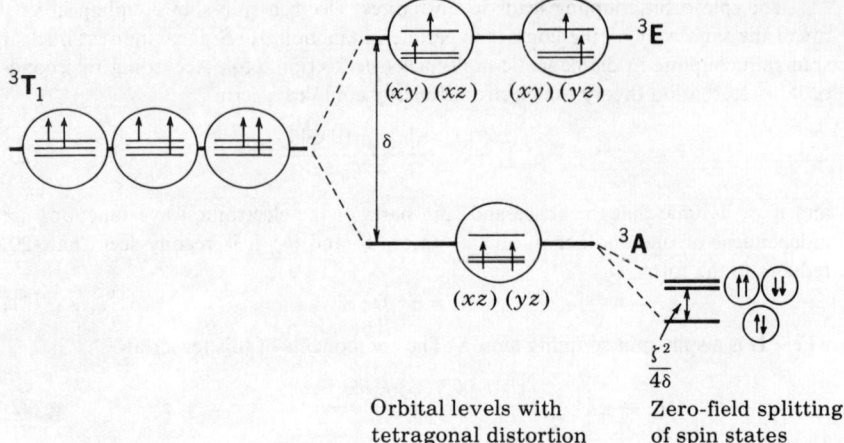

FIG. 10.8. Splitting of the octahedral $(d)^2$ ground state by a tetragonal distortion and spin-orbit coupling.

$$E: \quad \psi_4 = \|(xy)^\alpha(xz)^\alpha\|$$

$$\psi_5 = (1/\sqrt{2})\{\|(xy)^\alpha(xz)^\beta\| + \|(xy)^\beta(xz)^\alpha\|\}$$

$$\psi_6 = \|(xy)^\beta(xz)^\beta\|$$

$$\psi_7 = \|(xy)^\alpha(yz)^\alpha\| \tag{24}$$

$$\psi_8 = (1/\sqrt{2})\{\|(xy)^\alpha(yz)^\beta\| + \|(xy)^\beta(yz)^\alpha\|\}$$

$$\psi_9 = \|(xy)^\beta(yz)^\beta\|$$

The symbol $\|(xz)^\alpha(yz)^\alpha\|$ stands for a normalized determinant, as described in Section 6.4.2. The splitting $\delta$ between $A_2$ and $E$ will be in the range 100 to 1000 cm$^{-1}$ depending on the magnitude of the distortion.

At this stage of the calculation, the three spin components of the $A_2$ ground state are still degenerate. This degeneracy is, however, removed by spin-orbit coupling which mixes the $A_2$ state with the $E$ state. Our final problem therefore is to calculate modified $^3A_2$ wave functions and energies, treating the spin-orbit mixing with the $E$ states by second-order perturbation theory.

The spin-orbit coupling operator for two electrons takes the form

$$\mathscr{H}_{LS} = \zeta(\mathbf{L}_1 \cdot \mathbf{S}_1 + \mathbf{L}_2 \cdot \mathbf{S}_2) \tag{25}$$

and we shall first consider its effect on $\psi_1$. For this it is necessary to consult Table 10.4 which lists the effect of the angular momentum operators on the $t_{2g}$ orbitals.

The $L_z$ operator converts the spatial configuration $(xz)(yz)$ into $(yz)^2$ or $(xz)^2$, which both belong to a singlet excited state, so we now turn to $L^+S^-$ and $L^-S^+$. The spin of $\psi_1$ already has its maximum value $m_S = 1$ and $L^-S^+$ therefore gives zero.

TABLE 10.4.   Effect of the Angular Momentum Operators
on the $d$ Orbitals

| Orbital | $L_x\psi$ | $L_y\psi$ | $L_z\psi$ |
|---|---|---|---|
| $t$ { $(yz)$ | $-2i(y^2-z^2)$ | $i(xy)$ | $-i(xz)$ |
| $(xz)$ | $-i(xy)$ | $-2i(z^2-x^2)$ | $i(yz)$ |
| $(xy)$ | $i(xz)$ | $-i(yz)$ | $-2i(x^2-y^2)$ |
| $e$ { $(z^2)$ | $-i\sqrt{3}(yz)$ | $i\sqrt{3}(xz)$ | $0$ |
| $(x^2-y^2)$ | $-(yz)$ | $-i(xz)$ | $2i(xy)$ |

Thus we have to consider whether the $L^+S^-$ terms mix $\psi_1$ with the states $\psi_5$ and $\psi_8$ which have $m_S = 0$. Table 10.4 shows that

$$\langle (xy)^\beta(xz)^\alpha|\tfrac{1}{2}(L_1^+S_1^- + L_2^+S_2^-)|(xz)^\alpha(yz)^\alpha\rangle$$

$$= -\langle xy|\tfrac{1}{2}L^+|yz\rangle = \tfrac{1}{2} \tag{26}$$

while the corresponding matrix element with $\langle(xy)^\alpha(xz)^\beta|$ vanishes because the spin functions of the $d_{xz}$ electron are orthogonal. Hence the required matrix element $\langle\psi_5|\mathscr{H}_{LS}|\psi_1\rangle$ is equal to $\zeta/2\sqrt{2}$. In a similar fashion we construct the full $3\times3$ energy matrix for $\psi_1$, $\psi_5$, and $\psi_8$

$$\begin{bmatrix} 0 & \dfrac{\zeta}{2\sqrt{2}} & \dfrac{i\zeta}{2\sqrt{2}} \\[2ex] \dfrac{\zeta}{2\sqrt{2}} & \delta & 0 \\[2ex] \dfrac{-i\zeta}{2\sqrt{2}} & 0 & \delta \end{bmatrix} \tag{27}$$

The second-order perturbed energy of $\psi_1$ is therefore lowered to

$$E_1 = E_0 - \frac{\zeta^2}{4\delta} \tag{28}$$

It is easily seen that $\psi_3$ also mixes with $\psi_5$ and $\psi_8$, having its energy lowered by an equal amount. Next we consider $\psi_2$, with $m_S = 0$, and calculate how it becomes mixed with $\psi_4$, $\psi_6$, $\psi_7$, and $\psi_9$. The energy matrix is

$$\begin{bmatrix} 0 & \dfrac{-\zeta}{2\sqrt{2}} & \dfrac{\zeta}{2\sqrt{2}} & \dfrac{i\zeta}{2\sqrt{2}} & \dfrac{i\zeta}{2\sqrt{2}} \\[2ex] \dfrac{-\zeta}{2\sqrt{2}} & \delta & 0 & \dfrac{-i\zeta}{2} & 0 \\[2ex] \dfrac{\zeta}{2\sqrt{2}} & 0 & \delta & 0 & \dfrac{i\zeta}{2} \\[2ex] \dfrac{-i\zeta}{2\sqrt{2}} & \dfrac{i\zeta}{2} & 0 & \delta & 0 \\[2ex] \dfrac{-i\zeta}{2\sqrt{2}} & 0 & \dfrac{-i\zeta}{2} & 0 & \delta \end{bmatrix} \tag{29}$$

and the perturbed energy is

$$E_2 = E_0 - \frac{\zeta^2}{2\delta} \tag{30}$$

These results are illustrated in Fig. 10.8; the states which we describe as $m_S = \pm 1$ remain degenerate because of the axial symmetry of the complex but there is a zero-field splitting between these states and that with $m_S = 0$, equal to $\zeta^2/4\delta$. Notice that $m_S$ is, in fact, no longer a good quantum number since spin-orbit coupling mixes $\psi_1(m_S = 1)$ with $\psi_5$ and $\psi_8$, both of which have $m_S = 0$.

### 10.6.3 The Spin Hamiltonian

We now wish to show that the zero-field splitting in our tetragonally distorted $V^{+++}$ complex can be represented by the term $D\hat{S}_z^2$ in the spin Hamiltonian; $E$ is zero in this example because of axial symmetry. Let us call the three perturbed states $\psi_1$, $\psi_2$, and $\psi_3$ $|+\rangle$, $|0\rangle$, and $|-\rangle$, and define a fictitious spin $\hat{S} = 1$ such that these three states are eigenfunctions of $\hat{S}_z$ with eigenvalues of 1, 0, and $-1$ equal to those of the parent unperturbed states. The matrix of the operator $D\hat{S}_z^2$ within the states $|+\rangle$, $|0\rangle$, $|-\rangle$ is

$$\begin{bmatrix} D & & \\ & 0 & \\ & & D \end{bmatrix} \tag{31}$$

and it follows that if $D = \zeta^2/4\delta$ the zero-field splitting calculated in 10.6.2 is successfully represented by the fictitious spin Hamiltonian. The extension to ions without axial symmetry follows quite readily.

### 10.6.4 E.S.R. Measurements of $D$

Octahedral $V^{+++}$ complexes generally exhibit a trigonal distortion and we have treated the case of a tetragonal distortion only because it is slightly simpler. The results are quite similar in the two cases. In a typical example $\delta$ might be approximately 500 cm$^{-1}$, and with $\zeta$ of the order 70 cm$^{-1}$ the zero-field splitting would be close to 2.5 cm$^{-1}$. This is relatively large and it would not be easy to observe the $\Delta m_S = \pm 1$ transitions using 3 cm wavelength radiation. For the $V^{+++}$ ion in $Al_2O_3$, where the environment is trigonal, the zero-field splitting is actually found to be 8 cm$^{-1}$, and $\zeta$ is estimated to be 70 cm$^{-1}$

The zero-field splittings in triplet states are large whenever the separation between the ground state and the nearest excited state is exceptionally small, as in the example described above. However, in the tetrahedral $FeO_4^{2-}$ ion which has the electron configuration $(e)^2$, the zero-field splitting is only 0.1 cm$^{-1}$; this is because the nearest excited states involve promotion of an electron from $e$ to $t_2$ and the energy difference is at least 10,000 cm$^{-1}$. A similar situation arises in distorted octahedral $Ni^{++}$ complexes where the ground $d^8$ electron configuration is $(t_{2g})^6(e_g)^2$. Spin-orbit coupling can only mix the triplet $A_2$ ground state with high excited states formed by promoting an electron from $t_{2g}$ to $e_g$. One interesting case is $(Ni^{++})6H_2O$ in crystals of $NiSiF_6.6H_2O$. This ion has a trigonal distortion and at 90°K the zero-field splitting is $-0.17$ cm$^{-1}$. $D$ is highly sensitive to alterations of temperature or pressure because the value depends on the size of the distortion and it changes sign at high pressures. It must be noted that the spin-orbit coupling constant increases from 209 cm$^{-1}$ for

$V^{+++}$ to 649 cm$^{-1}$ for Ni$^{++}$ and $D$ is proportional to $\zeta^2$, so that the large energy difference between states mixed by the spin-orbit coupling is partly offset by the larger values of $\zeta$.

# 0·7 IONS WITH SPIN $S$ GREATER THAN ONE

### 10.7.1 Quartet States

An ion with three unpaired electrons and a total spin $S = 3/2$ is said to be in a quartet state, with four components in which the $m_S$ quantum number is allowed to take the values $m_S = +3/2, +1/2, -1/2, -3/2$. However, these four states are not usually degenerate, even in zero field. The principal mechanism of the splitting again involves spin-orbit coupling combined with deviations from regular symmetry, and the e.s.r. spectra are usually interpreted in terms of the spin Hamiltonian

$$\mathscr{H} = \beta \mathbf{H} \cdot \mathbf{g} \cdot \hat{\mathbf{S}} + D(\hat{S}_z^2 - \tfrac{5}{4}) + E(\hat{S}_x^2 - \hat{S}_y^2) \qquad (32)$$

with $E = 0$ if the ion is axially symmetric. One might have expected to find third-order terms like $\hat{S}_x^3$, $\hat{S}_y^3$, etc., in the spin Hamiltonian, but it can be shown that such terms reduce either to a constant or to the second-order terms containing $\hat{S}_x^2$, $\hat{S}_y^2$, and $\hat{S}_z^2$.

We can predict the principal features of the e.s.r. spectrum by calculating the matrix of $\mathscr{H}$ within the set of four states in $|m_S\rangle$ form, $|3/2\rangle, |1/2\rangle, |-1/2\rangle, |-3/2\rangle$. It is best to begin with zero magnetic field, when the matrix reduces to the simple form

$$\begin{bmatrix} D & 0 & \sqrt{3}E & 0 \\ 0 & -D & 0 & \sqrt{3}E \\ \sqrt{3}E & 0 & -D & 0 \\ 0 & \sqrt{3}E & 0 & D \end{bmatrix} \qquad (33)$$

which splits into two identical $2 \times 2$ matrices

$$\begin{bmatrix} D & \sqrt{3}E \\ \sqrt{3}E & -D \end{bmatrix} \qquad (34)$$

The zero-field levels thus consist of a pair of Kramers' doublets with energies

$$W = \pm \sqrt{D^2 + 3E^2} \qquad (35)$$

A magnetic field along the $z$ direction removes the degeneracy completely. The spin Hamiltonian matrix now becomes

$$\begin{bmatrix} D + \tfrac{3}{2}g\beta H & 0 & \sqrt{3}E & 0 \\ 0 & -D + \tfrac{1}{2}g\beta H & 0 & \sqrt{3}E \\ \sqrt{3}E & 0 & -D - \tfrac{1}{2}g\beta H & 0 \\ 0 & \sqrt{3}E & 0 & D - \tfrac{3}{2}g\beta H \end{bmatrix} \qquad (36)$$

and the energy level diagram shown in Fig. 10.9 is obtained. There are three allowed $\Delta m_S = \pm 1$ transitions, and provided the zero-field splitting is not too large, three absorption lines are observable. If the zero-field splitting is much greater than the energy $h\nu$, one transition $(+1/2 \leftrightarrow +3/2)$ will not be accessible and another $(-1/2 \leftrightarrow$

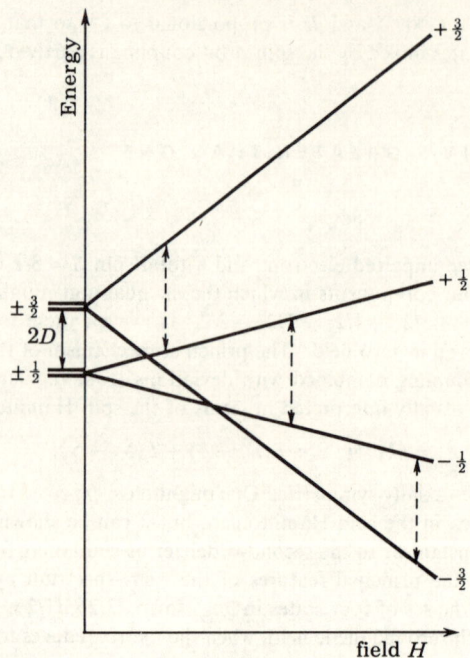

FIG. 10.9. Zero-field splitting and Zeeman splitting of a quartet state with axial symmetry.

$-3/2$) will only be observed in very high magnetic fields. The transition between $+1/2$ and $-1/2$ is always observable, no matter how large the zero-field splitting.

As in triplet states, the size of the zero-field splitting depends upon the energy separation between the ground and nearest excited state coupled to it by the spin-orbit operator. Distorted octahedral $Cr^{+++}$ complexes usually show small zero-field splittings since the ground electron configuration is $(t_{2g})^3$ and the nearest excited states involve promotion of an electron from the $t_{2g}$ orbitals to the $e_g$. A small distortion from octahedral symmetry splits this excited orbital triplet state and the splitting reacts back through the spin-orbit coupling to remove the degeneracy of the ground spin states. For example, in trigonally distorted chelated trisethylenediamine ($Cr^{+++}$) chloride the e.s.r. spectrum is described by the spin Hamiltonian (32), having $D = \pm 0.0413$ cm$^{-1}$, $E = 0$, and an isotropic $g$ value of 1.9871.

A rather different situation arises in octahedral $Co^{++}$ complexes. The strong field $d^7$ configuration is $(t_{2g})^5(e_g)^2$ and the presence of low-lying excited states results in a very large zero-field splitting. As a result only one transition is observed and the $g$ value is highly anisotropic. In contrast to this, tetrahedral $Co^{++}$ complexes (like the $CoCl_4^{2-}$ ion) have the electron configuration $(e)^4(t_2)^3$ and the situation is similar to that found for octahedral $Cr^{+++}$ complexes, although the zero-field splitting is rather larger.

### 10.7.2 Quintet and Sextet States

An ion with four unpaired electrons has spin $S = 2$ and $m_S$ can therefore take the values $+2, +1, 0, -1, -2$. In fields of low symmetry the degeneracy of these five spin states is completely removed and the resonance spectrum is quite complicated. The spectrum of the high spin chromous ion $(Cr^{++})6H_2O$ in $CrSO_4.5H_2O$ is analyzed in terms of the Hamiltonian

$$\mathcal{H} = \beta g_{\parallel} H_z \hat{S}_z + \beta g_{\perp}(H_x \hat{S}_x + H_y \hat{S}_y)$$
$$+ D(\hat{S}_z^2 - 2) + E(\hat{S}_x^2 - \hat{S}_y^2) \tag{37}$$

with $\hat{S} = 2$, $g_{\parallel} = 1.95$, and $g_{\perp} = 1.99$. The zero-field splittings $|D| = 2.24$ cm$^{-1}$, $|E| = 0.10$ cm$^{-1}$ come from a tetragonal distortion of the water octahedron with a further small asymmetric distortion. The detailed interpretation of these results is complicated since spin-orbit coupling mixes the ground state with a number of excited states, and we shall not explore the details. Notice that since there is no Kramers' degeneracy for an ion with $S = 2$, resonance absorption may be impossible if the zero-field splitting is large enough.

High spin ions with five unpaired electrons ($S = 5/2$) occupy a unique position since, in both octahedral and tetrahedral complexes, two electrons occupy the doubly degenerate $e$ orbitals and three electrons occupy the triply degenerate $t_2$ orbitals. The ground state is therefore orbitally nondegenerate. Since all excited states involve promotion of an electron from $e$ to $t_2$, or from $t_2$ to $e$, spin-orbit coupling is expected to be unimportant and the zero-field splitting should be rather small. The spin degeneracy is nevertheless removed even in an undistorted complex, and higher order spin-orbit perturbations or direct electron dipole spin-spin couplings (Chapter 8) both contribute a new term to the spin Hamiltonian, which becomes

$$\mathcal{H} = g\beta \mathbf{H} \cdot \hat{\mathbf{S}} + \frac{1}{6} a \left[ \hat{S}_x^4 + \hat{S}_y^4 + \hat{S}_z^4 - \frac{707}{16} \right] \tag{38}$$

The sextet splits into a quartet and a doublet with a separation equal to $3a$. In distorted complexes additional $D$ and $E$ terms also occur. As a result the six-fold spin degeneracy is partly removed, leaving the three Kramers' doublets shown in Fig. 10.10. In complex ions like $Mn^{++}(H_2O)_6$ and $Fe^{+++}(H_2O)_6$ the zero-field splitting is small and five $\Delta m_S = \pm 1$ transitions are observable. $a$ is about $10^{-4} - 10^{-2}$ cm$^{-1}$ and $D \approx 10^{-2}$ cm$^{-1}$.

In complexes with strong fields of tetragonal symmetry, the zero-field splitting between the Kramers' doublets may be very large. The most interesting and important examples are undoubtedly ferrihaemoglobin and its derivatives. Here the spin Hamiltonian for the sextet state is approximately

$$\mathcal{H} = 2\beta \mathbf{H} \cdot \mathbf{S} + a(S_x^4 + S_y^4 + S_z^4) + bS_z^2 + cS_z^4 \tag{39}$$

and the lowest Kramers' doublet $|\pm 1/2\rangle$ is separated from the other doublets by 5 cm$^{-1}$. Only the single resonance line from the $(+1/2 \leftrightarrow -1/2)$ transition is observed. The absorption line is highly anisotropic with an apparent $g$ value of 2 when the magnetic field is parallel to the tetragonal axis and 6 when the field is in the plane of the porphyrin ring. These results are readily understood if we calculate the matrices of $2S_z$ and $2S_x$ within the pair of states $|\pm 1/2\rangle$ of $S = 5/2$. $2S_z$ is diagonal with elements $\pm 1$ while $2S_x$ has an off-diagonal matrix element

$$2\left\langle \frac{1}{2} \Big| S_x \Big| -\frac{1}{2} \right\rangle = \sqrt{S(S+1) - m_S(m_S + 1)} \quad \left( S = \frac{5}{2} \right) \tag{40}$$

$$= \sqrt{\frac{35}{4} + \frac{1}{4}} = 3$$

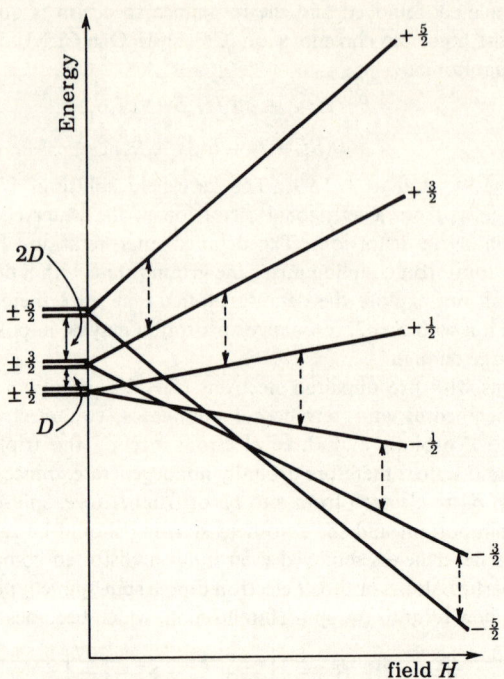

FIG. 10.10. Zero-field splitting between the three Kramers' doublets of a sextet state ($S = 5/2$).

and the two matrices are therefore

$$2S_z = \begin{bmatrix} 1 & 0 \\ 0 & -1 \end{bmatrix} \qquad 2S_x = \begin{bmatrix} 0 & 3 \\ 3 & 0 \end{bmatrix} \tag{41}$$

Clearly the results can be interpreted in terms of the Hamiltonian

$$\mathcal{H} = \beta g_{\parallel} H_z \hat{S}_z + \beta g_{\perp}(H_x \hat{S}_x + H_y \hat{S}_y)$$

for a *fictitious spin* $\hat{S}$ *of a half* having $g_{\parallel} = 2$ and $g_{\perp} = 6$.

### 10.7.3 Summary

Enough has been said to indicate the more important features which arise in ions with spin greater than one. The multiplicity of lines which arises because of zero-field splitting is called the *fine* structure of the spectrum; very often additional hyperfine structure due to electron-nuclear interactions is also observed. The most important point to note is that in ions with an odd number of unpaired electrons, the zero-field splitting results in a set of Kramers' doublets, and since the lowest level is at least doubly-degenerate in spin, resonance absorption is always possible.

## 10·8 HYPERFINE SPLITTING FROM THE METAL NUCLEUS

Many of the $3d$ transition metal atoms possess magnetic nuclei and metal hyperfine structure is often seen. A nucleus with spin $I$ has $2I + 1$ allowed orientations with respect to the direction of an applied field so that the hyperfine multiplet consists of $2I + 1$ equally intense, equally spaced lines. The most important magnetic nuclei are listed in Table 10.5. In copper, the two isotopes have almost identical magnetic moments so that a single multiplet of four lines is usually observed.

TABLE 10.5.   Nuclear Spins of the Transition Metals

| Nucleus | Abundance % | I | $g_N$ | $Q(10^{-24} \text{ cm}^2)$ |
|---|---|---|---|---|
| $Ti^{47}$ | 7.32 | 5/2 | $-0.3153$ | — |
| $Ti^{49}$ | 5.46 | 7/2 | $-0.3154$ | — |
| $V^{51}$ | 99.8 | 7/2 | 1.471 | $+0.25$ |
| $Cr^{53}$ | 9.55 | 3/2 | $-0.3163$ | — |
| $Mn^{55}$ | 100 | 5/2 | 1.387 | $+0.3$ |
| $Fe^{57}$ | 2.21 | 1/2 | $<0.10$ | 0 |
| $Co^{59}$ | 100 | 7/2 | 1.328 | $+0.5$ |
| $Cu^{63}$ | 69.1 | 3/2 | 1.484 | $-0.16$ |
| $Cu^{65}$ | 30.9 | 3/2 | 1.590 | $-0.15$ |

The hyperfine interaction is described by the usual term in the spin Hamiltonian, $S \cdot T \cdot I$, and the tensor $T$ may be isotropic or anisotropic in different cases. The isotropic hyperfine interaction arises as usual from the Fermi contact interaction but the origin of this interaction is not immediately obvious, since we have so far regarded the magnetic electrons as occupying $d$ orbitals which, of course, have nodes at the nucleus. The reader will recall that a similar problem arose when we were discussing the proton hyperfine interaction in aromatic radicals. The explanation is that electron exchange terms in the full Hamiltonian mix the $(3s)^2(3d^n)$ ground configuration with excited configurations of the type $(3s)(3d^n)(4s)$, so that the necessary unpaired spin density comes into the $s$ atomic orbitals. It is found experimentally that the $s$ electron density at the nucleus is remarkably constant throughout the first transition series and only changes appreciably when the magnetic electrons spread out on to the ligands.

There are two main contributions to the anisotropic part of the hyperfine tensor. The first is the electron-nuclear dipolar part which, though similar in principle to that described in Chapter 7, takes a rather more complex form in the transition metal ions. The second contribution arises from the coupling between the partly quenched electronic orbital angular momentum and the nuclear spin; its magnitude therefore depends upon the importance of spin-orbit coupling, and complexes which have anisotropic $g$ values invariably show anisotropic hyperfine coupling.

## 10·9   COVALENT BONDING AND LIGAND HYPERFINE STRUCTURE

In addition to metal hyperfine structure, many complexes show extra splitting from the ligand nuclei. This is partly due to covalent bonding between the central atom $d$ electrons and the ligand orbitals although direct dipole-dipole coupling is also important. The ligands can form both $\sigma$ and $\pi$ bonds, and the type of bonding is severely limited by symmetry requirements.

An undistorted octahedral complex is the simplest to understand. Let us take $(Mn^{++})(F^-)_6$ which has five $d$ electrons. As Fig. 10.11 shows, the $2s$ and $2p$ orbitals of the six fluorine atoms form molecular orbitals which have the same symmetry as each of the five $d$ orbitals, and are capable of forming covalent bonds. The $d_{z^2}$ and $d_{x^2-y^2}$ or $e_g$ orbitals only form $\sigma$ bonds, whereas the $t_{2g}$ orbitals only form $\pi$ bonds. The molecular symmetry orbitals in Fig. 10.11 take the forms shown below:

$$\sigma(4s) = \frac{1}{\sqrt{6}}(\sigma_1 + \sigma_2 + \sigma_3 + \sigma_4 + \sigma_5 + \sigma_6) \qquad a_{1g}$$

$$\sigma(4p_z) = \frac{1}{\sqrt{2}}(\sigma_5 - \sigma_6) \qquad t_{1u}$$

$$\pi(4p_z) = \frac{1}{2}(z_1 + z_2 + z_3 + z_4) \qquad t_{1u}$$

(42)

$$\sigma(d_{z^2}) = \frac{1}{\sqrt{6}}(-\sigma_1 - \sigma_2 - \sigma_3 - \sigma_4 + 2\sigma_5 + 2\sigma_6) \qquad e_g$$

$$\sigma(d_{x^2-y^2}) = \frac{1}{2}(\sigma_1 - \sigma_2 + \sigma_3 - \sigma_4) \qquad e_g$$

$$\pi(d_{xy}) = \frac{1}{2}(y_1 + x_2 - y_3 - x_4) \qquad t_{2g}$$

Here the symbol $\pi(4p_z)$, for example, means a $\pi$ molecular orbital which has the same symmetry as the metal $4p_z$ orbital; $y_1$ and $z_1$ are the $p_y$ and $p_z$ orbitals of ligand number one, and $\sigma_1$ is a $\sigma$ orbital of the same ligand ($2s$ or $2p_x$). The molecular orbital scheme which results from the combination of metal and ligand orbitals is shown in Fig. 10.12. First we notice that all the bonding orbitals are filled and contain two electrons each. Of these the $e_g$, $t_{1u}$, and $t_{2g}$ ones are mainly localized on the fluorine atoms. Next comes a pair of fluorine nonbonding orbital levels, $t_{1g}$ and $t_{2u}$, which are filled. The main feature of interest is the position of the five $t_{2g}$ and $e_g$ orbitals which hold the unpaired electrons. Both orbitals are raised in energy by covalent bonding and have the usual ligand field splitting $\Delta$. The higher antibonding orbitals are irrelevant. Notice that the final picture is much the same as in the simple electrostatic theory where ligands are supposed to be replaced by negative point charges surrounding the metal ion. The symmetries, spacings, and principal features of the orbitals which contain the *unpaired electrons* are all similar.

The bonding in a tetrahedral complex is rather different, for now the metal $e$ orbitals form $\pi$ bonds and the $t$ orbitals form mainly $\sigma$ bonds.

We are now able to attack the problem of ligand hyperfine structure in $MnF_2$. It is partly caused by the unpaired $d$ electrons entering the $s$ and $p$ orbitals of the ligands. For example, in the complex the free ion $d_{xy}$ orbital will combine with the ligand $\pi(d_{xy})$ orbital in Eq. (42), giving the molecular orbital

$$\psi_{xy} = d_{xy} \cos \theta + \pi(d_{xy}) \sin \theta \qquad (43)$$

and so on. The unpaired electrons in the fluorine $2p_\pi$ orbitals will then produce anisotropic hyperfine structure (see Chapter 9). Unpaired electrons in the fluorine $\sigma$ orbitals have both $s$ and $p$ character and give isotropic hyperfine splitting in addition.

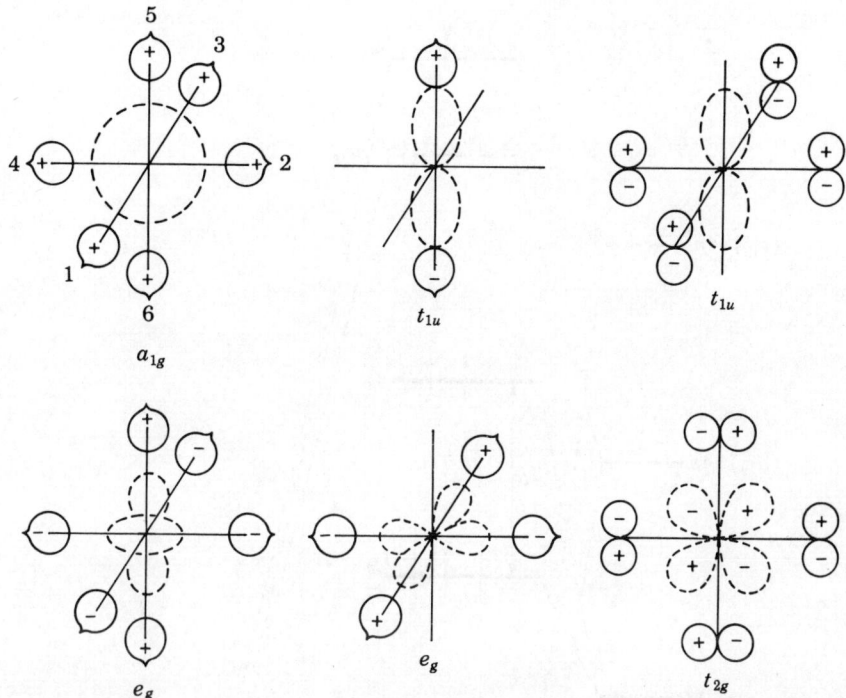

FIG. 10.11. Molecular orbitals formed from ligand $s$ and $p$ orbitals are chosen to match the octahedral symmetry of the metal ion atomic orbitals. Ligand $\sigma$ orbitals are either a $2s$ function or the $2p$ function which points towards metal.

The hyperfine structure of $Mn^{++}$ as an impurity in diamagnetic $ZnF_2$ crystals has been analyzed completely. Here each $Mn^{++}$ ion is surrounded by a distorted octahedron of fluorine atoms, as shown in Fig. 10.13. There are three pairs of equivalent fluorines $F_1F_3$, $F_2F_4$, and $F_5F_6$ with three different hyperfine tensors; $T_1$ and $T_2$ have the same principal values, while $T_5$ has slightly different ones and is diagonal in the $xyz$ axis system. Each tensor can be split into several parts. Thus $T_5$ has an isotropic contact part $a_5 = 49.6$ Mc/s, an anisotropic part $(-10.4, -10.4, +20.8)$ with axial symmetry about the MnF bond caused by $\sigma$ bonding and by long-range dipolar interaction with the $d$ electron spins, and a small anisotropic part $(+1.06, -0.53, -0.53)$ caused by asymmetry of the $\pi$ bond. The $F^{19}$ hyperfine constants given in Chapter 9 can now be used to estimate the amount of covalent bonding. The $Mn^{55}$ nucleus also gives a hyperfine splitting with axial symmetry about the $y$ axis; $A = C = -276$, $B = -271$ Mc/s.

The degree of electron delocalization may often be estimated from two other effects. It follows that if the magnetic electrons are not confined to the metal atom, the spin-orbit coupling is reduced so that the $g$ tensor is less anisotropic than it otherwise would be. It also follows that interaction with the metal nucleus is reduced, leading to smaller metal hyperfine splitting.

Ligand hyperfine structure has been analyzed in many complexes, and taken

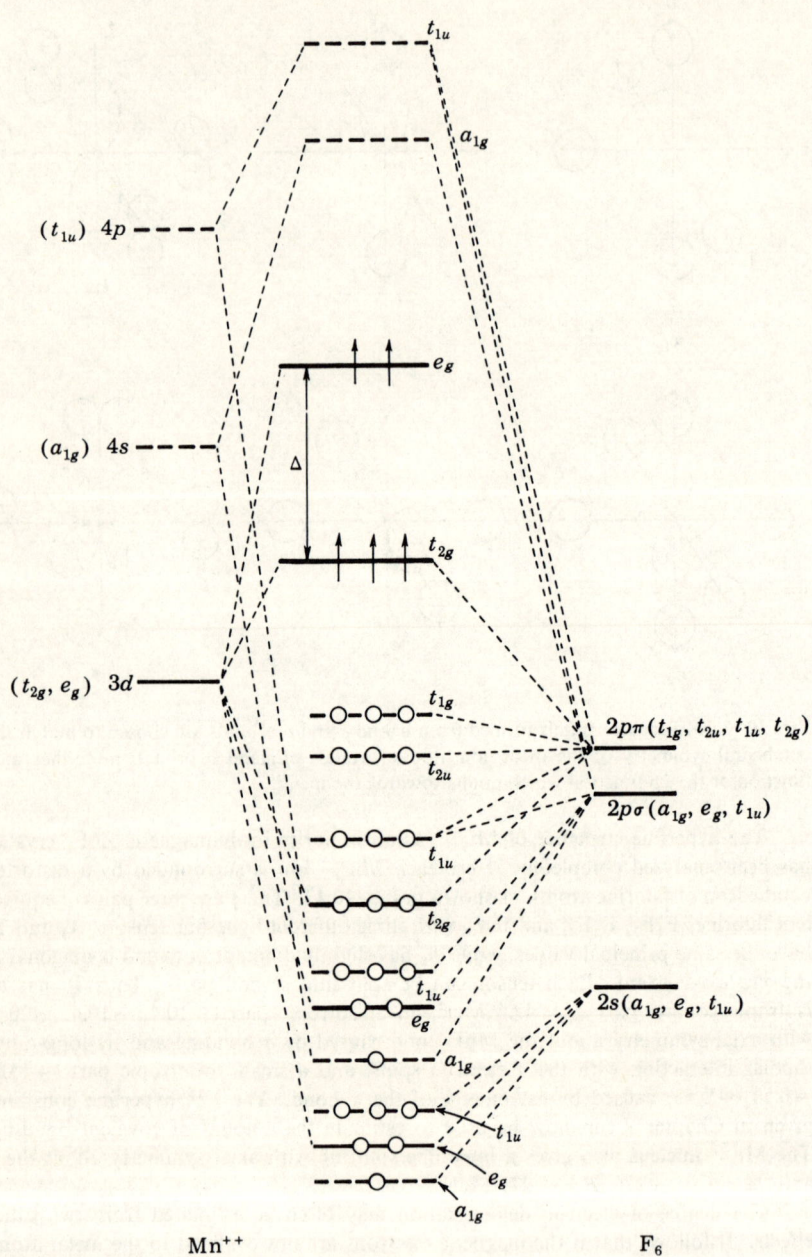

FIG. 10.12. Molecular orbital scheme for octahedral $(Mn^{++})(F^-)_6$.

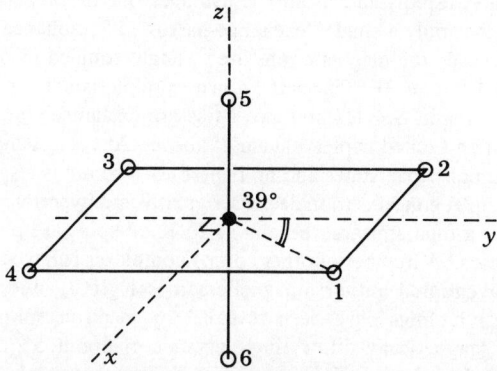

FIG. 10.13. Distorted $(F^-)_6$ octahedron surrounding the $Mn^{++}$ ion in a $ZnF_2$ crystal.

with the principal values of the $g$ and zero-field splitting tensors, it helps to provide a detailed description of the electronic wave functions.

Finally, we should say a word or two about nuclear quadrupole effects because they are sometimes quite pronounced in metal ion spectra. Most of the metal nuclei with spins also have sizeable quadrupole moments. The electric field gradient from the surrounding ligand ions or dipoles competes with the magnetic hyperfine interaction in determining the orientation of the nuclear spin. The result is that the $2I + 1$ nuclear levels are not equally spaced and the separations between adjacent lines in the hyperfine multiplet differ slightly. Also, since $m_I$ is no longer a good quantum number additional weak hyperfine lines may appear.

## 10 ELECTRON EXCHANGE COUPLING

We have already mentioned the need to study magnetically dilute crystals if narrow resonance lines are to be observed. Usually one hopes to find a diamagnetic compound which forms large single crystals easily and has a similar crystal structure. The paramagnetic ion is then introduced into the host lattice at a low concentration of 1 in 1,000 or even less. Alternatively metal ions can be introduced into single crystals of materials like MgO, $Al_2O_3$, $CaF_2$, etc.

There are two reasons why it is essential to study magnetically dilute crystals. First it is necessary to reduce the dipole-dipole coupling between the magnetic moments of different paramagnetic complexes, which may give resolvable line splittings but more often produces broadening so that detail is lost.

The second effect which occurs in magnetically concentrated crystals is quite different and much more important. If the odd electron wave functions on different complex ions overlap, the unpaired electron spins are coupled by exchange forces. The exchange energy takes the form

$$\mathscr{H} = J\mathbf{S}_1 \cdot \mathbf{S}_2 \tag{44}$$

We have already discussed exchange coupling between organic radicals in Chapter 8 and seen that it can produce an antiparallel spin arrangement in Wurster's blue perchlorate. Exchange coupling allows the electrons on neighboring lattice sites to exchange spin states rapidly and usually it averages out the hyperfine structure and fine structure so that only a single "exchange-narrowed" resonance line remains.

In several crystals the magnetic ions are strongly coupled in pairs (compare the water protons in Chapter 3). The best known example is copper acetate where the $d^9$ $Cu^{++}$ ion has a spin $S = 1/2$ and two spins are exchange coupled into a singlet ground state with an excited triplet 300 $cm^{-1}$ above. At room temperature there is a high concentration of triplet states and no hyperfine structure is resolved, but on cooling to 77°K the triplet concentration decreases greatly and hyperfine splitting from two equivalent copper atoms appears. Finally at 20°K all spins are paired and the e.s.r. spectrum disappears. A number of other copper complexes behave in the same way.

A more conventional antiferromagnetic salt is $K_2IrCl_6$, and important details of the exchange couplings have been revealed by electron spin resonance. It is possible to grow magnetically dilute single crystals of about 5% $K_2IrCl_6$ in a diamagnetic $K_2PtCl_6$ host lattice. There is a fairly high concentration of paramagnetic ion-pairs occupying adjacent lattice sites. Each $d^5$ $Ir^{4+}$ ion has $S = 1/2$ and the pairs give a triplet e.s.r. spectrum with a small zero-field splitting. As the temperature falls the spectrum weakens as the thermal population of the triplet state becomes lower, and one estimates that $J$ is about 7.5 $cm^{-1}$. This information is of great value in understanding the antiferromagnetism of pure $K_2IrCl_6$ itself. The exchange coupling between the ions is not due to direct overlap of their $d$ orbitals. Instead the spin is transmitted through the intervening chlorine ligands and the result is called a *super-exchange* coupling. It is specially important in salts like $K_2IrCl_6$ because the ligand orbitals combine effectively with the metal $d$ orbitals.

# 10·11    THE RARE-EARTH IONS

The theory which we have described in this chapter is appropriate for metal ions of the 3$d$, 4$d$, or 5$d$ transition series. The main differences arise from the fact that electron delocalization effects and spin-orbit coupling are increasingly important for the heavier atoms. The general procedure has been to start with the appropriate atomic electron configuration for the free ion, see how the orbital degeneracy is partly removed by the field of the ligands, and then introduce the spin-orbit coupling and small distortions by means of perturbation theory.

The situation is quite different for the rare-earth ions. The magnetic electrons occupy 4$f$ orbitals which are effectively shielded from the electrostatic or bonding effects of ligands. The general procedure in interpreting e.s.r. spectra consists of two stages. First, one takes the 4$f$ electrons of the free ion and couples their orbital angular momenta into a resultant $L$, and their spins into a resultant $S$ to obtain the electron configuration of the ion without spin-orbit coupling. Spin-orbit coupling is rather strong ($\zeta = 640$– $2,940$ $cm^{-1}$) and couples $L$ and $S$ into widely spaced multiplets with different values of the total angular momentum $J$. After this one treats the ligand field effects by perturbation theory. Here we shall illustrate the most important features by taking one simple example, the $Ce^{+++}$ ion.

This ion has a single 4$f$ electron ($L = 3$, $S = 1/2$) outside a rare gas core and surrounded by filled 5$s$ and 5$p$ shells. The two possible $J$ values are $(L + S) = 7/2$ or $(L - S) = 5/2$, giving ${}^2F_{7/2}$ and ${}^2F_{5/2}$ electron configurations which have different energies because of spin-orbit coupling. To calculate these energies, we rewrite the operator $\zeta \mathbf{L} \cdot \mathbf{S}$ as

$$\zeta \mathbf{L} \cdot \mathbf{S} = \tfrac{1}{2} \zeta [(\mathbf{L} + \mathbf{S})^2 - \mathbf{L}^2 - \mathbf{S}^2]$$

$$= \tfrac{1}{2} \zeta [J^2 - L^2 - S^2] \qquad (45)$$

$$= \tfrac{1}{2} \zeta [J(J + 1) - L(L + 1) - S(S + 1)]$$

Here $L$, $S$, and $J$ are all good quantum numbers so that it is correct to replace $J^2$ by its eigenvalue $J(J + 1)$, and so on. The ground state of the free ion is $^2F_{5/2}$ and lies 2,250 cm$^{-1}$ below the $^2F_{7/2}$ state.

To a first approximation the ligand field merely splits the sextet ($J = 5/2$) level into three Kramers' doublets, and in a field of trigonal symmetry the doublets are eigenfunctions of $J_z = L_z + S_z$ with $m_J$ values $\pm 1/2$, $\pm 3/2$, $\pm 5/2$. The lowest pair is $\pm 1/2$, and the spin-orbit coupled wave functions in $m_L m_S$ form are

$$|+\rangle = \sqrt{\tfrac{3}{7}} \left| 0, \tfrac{1}{2} \right\rangle - \sqrt{\tfrac{4}{7}} \left| 1, -\tfrac{1}{2} \right\rangle$$

$$|-\rangle = \sqrt{\tfrac{4}{7}} \left| -1, \tfrac{1}{2} \right\rangle - \sqrt{\tfrac{3}{7}} \left| 0, -\tfrac{1}{2} \right\rangle \qquad (46)$$

We can calculate the $g$ values from the matrices of $(L_z + 2S_z)$ and $(L_x + 2S_x)$ just as we did for ferrihaemoglobin. The result is that $g_\| = 6/7 = 0.857$ and $g_\perp = 18/7 = 2.571$.

Experimentally it is found that a diluted single crystal of Ce$^{+++}$ in lanthanum ethyl sulphate gives two resonances; one has $g_\| = 0.955$, $g_\perp = 2.185$, and the other arises from a doublet about 3 cm$^{-1}$ higher, having $g_\| = 3.72$, $g_\perp = 0.2$. The $g$ values calculated for the $m_J = \pm 5/2$ doublet are $g_\| = 30/7$ and $g_\perp = 0$, while for $m_J = \pm 3/2$ they are $18/7$ and $0$. The situation is summarized in the energy level diagram shown in Fig. 10.14; the $m_J = \pm 3/2$ doublet is apparently too high in energy to be appreciably populated at the temperatures of measurement.

It should be emphasized that we have dealt with the simplest possible case and applied only the most elementary theory to it. In a more accurate treatment one would allow for mixing of levels with different $J$ brought about by the crystal field.

## 10.12   SPIN-LATTICE RELAXATION

We conclude by mentioning one effect which has important experimental consequences. For ions with strong spin-orbit coupling, the spin system is strongly coupled to the lattice vibrations and the spin relaxation time is therefore very small at high temperatures. This may mean that the e.s.r. lines are too broad to be detectable even at 300°K and it is necessary to quench the lattice vibrations by working at extremely low temperatures. Contrasting examples are the Mn$^{++}$(H$_2$O)$_6$ ion ($S = 5/2$) which gives narrow lines at 300°K and the low spin Mn$^{++}$(CN$^-$)$_6$ ion ($S = 1/2$) in which spin-orbit coupling is important and makes it necessary to work at 20°K or less.

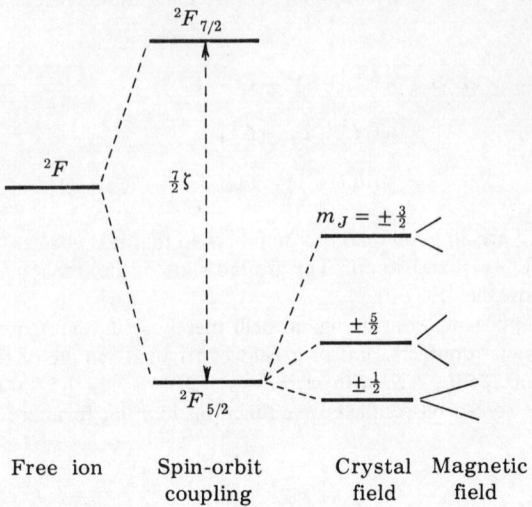

FIG. 10.14. Energy levels of $Ce^{+++}$ ion in a lanthanum ethyl sulphate crystal.

## PROBLEMS

**1.** The electron resonance spectrum of the tetragonally distorted $MnO_4^{2-}$ ion has $g_{\parallel} = 1.94$ and $g_{\perp} = 1.97$. Assuming a value for the spin-orbit coupling constant for $Mn^{++}$ of 200 cm$^{-1}$, predict the frequency of the optical absorption band arising from the $e \to t_2$ transition.

**2.** Prove Eq. (16) for the principal $g$ values in a trigonally distorted octahedral $d^1$ complex.

**3.** Equations (18) give the principal components of the $g$ tensor for an octahedral copper complex with four short and two long bonds. Would these expressions be different for a copper complex with four long and two short bonds?

**4.** Make use of the $|m_L\rangle$ forms of the $d$ orbitals given in Fig. 10.5 to prove the results of Table 10.4.

**5.** In some tetragonal $Cr^{+++}$ complexes the zero-field splitting is very large and the electron resonance spectrum may be described in terms of a fictitious spin of $1/2$, rather than the real spin of $3/2$. Neglecting nuclear effects, write down the appropriate spin Hamiltonian and predict the values of $g_{\parallel}$ and $g_{\perp}$.

**6.** Neglecting the electron Zeeman interaction, use Eq. (38) to verify the statement that a sextet spin state in a cubic complex splits into a quartet and a doublet. Express the splitting in terms of the parameter $a$.

**7.** Using Eq. (44) for the exchange interaction between two spins $S_1$ and $S_2$, prove that in the case of a pair of $Cu^{++}$ ions, the singlet and triplet states have energies $-3J/4$ and $+J/4$.

Suppose you have a complex in which there is an exchange coupling of the form (44) between a $Cu^{++}$ ion and a $Ni^{++}$ ion, each of which is in an octahedral environment. Calculate the energies of the possible spin states in terms of $J$. What sort of electron resonance spectrum might you observe?

**8.** Predict the main features of the electron resonance spectra of the ions $Fe^+$, $Ni^+$, $Co^+$, and $Cr^+$ in sites of cubic symmetry. Compare your predictions with the observations of Hayes, *Disc. Faraday Soc.*, **26**: 58 (1958).

# SUGGESTIONS FOR FURTHER READING

Orgel: Chapters 1–4. Elementary description of ligand field splitting and general features of transition metal complexes.

Coulson: Chapter 10. Introduction to ligand field theory.

Griffith: Chapter 12. Electron resonance results summarized.

Ballhausen: Chapter 6. Theory of e.s.r. in transition metal complexes.

Bleaney and Stevens: *Repts. Prog. in Phys.*, **16**: 108 (1953). Review on electron spin resonance in crystals.

Bowers and Owen: *Repts. Prog. in Phys.*, **18**: 304 (1955). A supplement to Bleaney and Stevens' article.

Low: *Paramagnetic Resonance in Solids* (New York: Academic Press Inc., 1960). A review of experimental work.

Gibson, Ingram, and Schonland: *Discussions Faraday Soc.*, **26**: 72 (1958). Haemoglobin.

Clogston, Gordon, Jaccarino, Peter, and Walker: *Phys. Rev.*, **117**: 1222 (1960). Hyperfine structure of $(Mn^{++})(F^-)_6$.

McGarvey: *J. Chem Phys.*, **41**: 3743 (1964). The spin Hamiltonian for chromium complexes.

Heine: *Group Theory in Quantum Mechanics* (London: Pergamon Press, 1960). Page 148. Kramers' theorem.

# CHAPTER 11

# SPIN RELAXATION

## 11·1 INTRODUCTION

The central theme of our book so far has been the spin Hamiltonian. All magnetic resonance spectra have been analyzed in terms of perfectly sharp resonant transitions between spin energy levels, which are the stationary states of a definite and fixed Hamiltonian. This is a very useful approach but it is also quite unrealistic, because every molecule interacts with its surroundings and these interactions limit the lifetimes of the spin states, broadening the energy levels. In fact we saw even in Chapter 1 that relaxation is essential for the success of a spin resonance experiment; the absorption line becomes completely saturated unless the spin system can give up its excess Zeeman energy to the surrounding "lattice." It is now time to develop the subject of spin relaxation more fully and make a detailed study of the coupling between the spin system and its environment.

## 11·2 BLOCH'S EQUATIONS

Before enquiring into the origins of relaxation at the molecular level, let us first look again at the resonance experiment in a rather general way, taking a macroscopic view of the situation. Temporarily we shall abandon all thoughts of spin wave functions and energy levels and consider the bulk magnetic moment **M** of a large assembly of spins at a certain temperature. They could be either electron or nuclear spins, but to fix our ideas let us think about the relaxation of the proton spins in a cubic centimeter of water at room temperature, assuming first that there is no external magnetic field.

Imagine that the spins have been aligned so that $N_\alpha$ of them are in the $\alpha$ spin state and $N_\beta$ in the $\beta$ state. The $z$ component of **M** is then

$$M_z = \gamma_N \hbar (N_\alpha - N_\beta) = \gamma_N \hbar n \tag{1}$$

where $\gamma_N$ is the magnetogyric ratio and $n = N_\alpha - N_\beta$. As we saw in Section 1.4 the population difference $n$ decays exponentially to equilibrium because of spin-lattice relaxation, and the rate of change is

$$\frac{dn}{dt} = -\frac{n}{T_1} \tag{2}$$

Since the magnetic moment $M_z$ is proportional to $n$, it obeys the equation

$$\frac{dM_z}{dt} = -\frac{M_z}{T_1} \tag{3}$$

and decays to zero with a characteristic time $T_1$. There is of course no physical distinction between the $z$ direction and any other direction in the absence of a field, so $M_x$ and $M_y$ also decay to zero at the same rate as $M_z$.

The introduction of a steady magnetic field $H_0$ along the $z$ axis alters the situation in several ways. First $M_z$ no longer vanishes in thermal equilibrium, but tends to approach a steady value $M_0$, proportional to the static magnetic susceptibility $\chi_0$ of all the $N$ spins.

$$M_0 = \gamma_N \hbar n_0 = \chi_0 H_0 \tag{4}$$

$$\chi_0 = \frac{N\gamma_N^2 \hbar^2 I(I+1)}{3\,kT} \tag{5}$$

According to Eq. (24) of Section 1.4 the relaxation of $M_z$ now obeys the equation

$$\frac{dM_z}{dt} = -\frac{(M_z - M_0)}{T_1} \tag{6}$$

Further, although the transverse components $M_x$, $M_y$ of the magnetic vector still decay exponentially to zero, the decay time is generally different from $T_1$ and it is necessary to introduce another *transverse relaxation time* $T_2$ for these components, such that

$$\frac{dM_x}{dt} = -\frac{M_x}{T_2}$$

$$\frac{dM_y}{dt} = -\frac{M_y}{T_2} \tag{7}$$

The reader may well wonder why $T_1$ and $T_2$ should not be equal. A short answer to this question is that longitudinal and transverse relaxation depend on different processes within the spin system. Changes in $M_x$ and $M_y$ do not alter the total Zeeman energy of the nuclear spins, whereas changes of $M_z$ need an exchange of Zeeman energy with the "lattice."

To complete our macroscopic description of spins in the magnetic field we must allow for the *spin angular momentum* of the nucleus, which makes the nuclear moment behave like a gyroscope rather than a bar magnet. Instead of simply lining up along the field direction the nuclear moment $\mu_N$ precesses round it, and although the nuclear spin states are quantized it happens that the total bulk magnetization $\mathbf{M}$ behaves precisely as though the nuclei obeyed classical mechanics. Thus the magnetic field produces a couple of strength $\mathbf{G} = \mu_N \times \mathbf{H}$ tending to twist $\mu_N$ into line with $\mathbf{H}$, and this torque forces the spin vector to precess round a cone, making a constant angle with the field (Fig. 11.1). The rate of change of spin angular momentum for each nucleus is

$$\frac{d}{dt}(\hbar \mathbf{I}) = \mathbf{G} = \gamma_N \hbar(\mathbf{I} \times \mathbf{H}) \tag{8}$$

so the classical equation of motion is

$$\frac{d\mathbf{I}}{dt} = \gamma_N(\mathbf{I} \times \mathbf{H}) \tag{9}$$

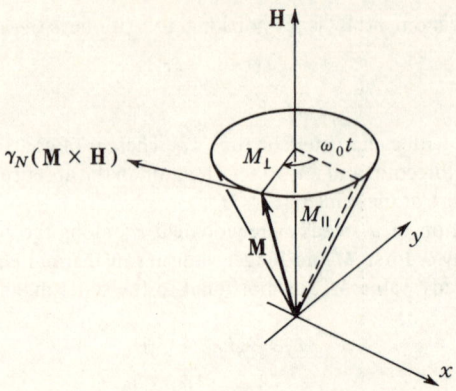

FIG. 11.1. Precession of nuclear magnetic moments about a steady magnetic field.

The bulk magnetic moment **M** is the sum of all the individual nuclear moments $\boldsymbol{\mu}_N$ and so it behaves similarly

$$\frac{d\mathbf{M}}{dt} = \gamma_N(\mathbf{M} \times \mathbf{H}) \tag{10}$$

As an illustration let us solve Eq. (10) for the precession of the spin system about a steady magnetic field $\mathbf{H}_0$ along the $z$ axis (we ignore relaxation for the moment). The equations reduce to

$$\frac{dM_x}{dt} = \omega_0 M_y$$

$$\frac{dM_y}{dt} = -\omega_0 M_x \tag{11}$$

$$\frac{dM_z}{dt} = 0$$

where $\omega_0$ is the nuclear resonance frequency ($\omega_0 = \gamma_N H_0$). A typical solution is

$$\mathbf{M} \equiv (M_x, M_y, M_z) = (M_\perp \cos \omega_0 t, -M_\perp \sin \omega_0 t, M_\parallel) \tag{12}$$

Here the tip of the spin vector moves clockwise round a cone at a uniform angular velocity $\omega_0$. $M_\parallel$ and $M_\perp$ represent the magnitudes of the longitudinal and transverse components of **M**. They remain constant as relaxation has been neglected. This motion is called the Larmor precession of the spin, and $\omega_0$ is known as the Larmor frequency. The direction of precession depends on the sign of the magnetogyric ratio; for electrons $\gamma$ is negative and the spins process anticlockwise about $\mathbf{H}_0$.

To conclude, we must combine the effects of relaxation and Larmor precession in (6), (7), and (10) into a new set of equations for **M**. These are called the Bloch equations:

$$\frac{dM_x}{dt} = \omega_0 M_y - \frac{M_x}{T_2}$$

$$\frac{dM_y}{dt} = -\omega_0 M_x - \frac{M_y}{T_2} \tag{13}$$

$$\frac{dM_z}{dt} = -\frac{(M_z - M_0)}{T_1}$$

The spin performs a damped precession in which the rotating transverse components of **M** decay to zero with a characteristic time $T_2$, while $M_z$ relaxes towards its equilibrium value $M_0$ with a decay time $T_1$.

# 11·3 THE LORENTZ LINE SHAPE

Besides providing a clear description of $T_1$ and $T_2$ Bloch's equations give a very satisfying macroscopic explanation of magnetic resonance absorption and predict a definite kind of line shape, the Lorentz line which we have already described.

First it is clear that a radio-frequency field $H_1$ which oscillates at the Larmor frequency will produce a torque on the nuclear moments which synchronizes with the free spin precession, and so induces a large oscillating magnetic moment in the sample. Conversely, a field which oscillates at some other frequency soon falls out of step with the spin and only produces small effects. Thus the possibility of observing resonance depends on the Larmor precession of the spins.

To study the resonance in detail let us take an assembly of nuclear spins in a steady field $H_0$ and apply a circularly polarized rf field $H_1$ which rotates clockwise within the $xy$ plane in the same sense as the Larmor precession, having a uniform angular velocity $\omega$ (not necessarily equal to the resonance frequency $\omega_0$). Bloch's equations become

$$\frac{d\mathbf{M}}{dt} = \gamma_N(\mathbf{M} \times \mathbf{H_0}) + \gamma_N(\mathbf{M} \times \mathbf{H_1}) - \frac{(\mathbf{i}M_x + \mathbf{j}M_y)}{T_2} - \frac{\mathbf{k}(M_z - M_0)}{T_1} \qquad (14)$$

where **i**, **j**, **k** are unit vectors along the $x$, $y$, $z$ axes and

$$\mathbf{H_1} = H_1(\mathbf{i}\cos\omega t - \mathbf{j}\sin\omega t) \qquad (15)$$

It is helpful to view the motion in a new coordinate system which rotates with $H_1$ at an angular velocity $\omega$ about the $z$ axis. In this system (Fig. 11.2) the moving axes are denoted $x'$, $y'$ and $z'$. The $z'$ axis is chosen parallel to the steady field $H_0$, while $x'$ is

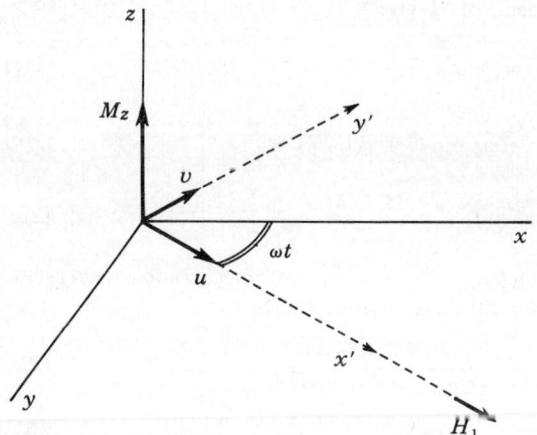

Fig. 11.2. The rotating coordinate system.

in the direction of $\mathbf{H}_1$. We also define $u$ and $v$ to be the transverse components of the magnetic moment along the $x'$ and $y'$ directions, so that in the new system the vectors $\mathbf{M}$, $\mathbf{H}_1$, and $\boldsymbol{\omega}$ are written

$$\mathbf{M}' = \mathbf{i}'u + \mathbf{j}'v + \mathbf{k}'M_z$$

$$\mathbf{H}'_1 = \mathbf{i}'H_1 \tag{16}$$

$$\boldsymbol{\omega}' = -\omega\mathbf{k}'$$

To an observer in the rotating frame the unit vectors $\mathbf{i}'$, $\mathbf{j}'$, and $\mathbf{k}'$ appear to be fixed, so that the *apparent* rate of change of $\mathbf{M}'$ is

$$\frac{\delta\mathbf{M}'}{\delta t} = \mathbf{i}'\frac{du}{dt} + \mathbf{j}'\frac{dv}{dt} + \mathbf{k}'\frac{dM_z}{dt} \tag{17}$$

However, a fixed observer sees a set of unit vectors rotating at the angular velocity $\boldsymbol{\omega}'$ and changing; for example, the rate of change of $\mathbf{i}'$ is

$$\frac{d\mathbf{i}'}{dt} = \boldsymbol{\omega}' \times \mathbf{i}' \tag{18}$$

and for $\mathbf{M}'$ itself he calculates

$$\frac{d\mathbf{M}}{dt} = \left(\mathbf{i}'\frac{du}{dt} + \mathbf{j}'\frac{dv}{dt} + \mathbf{k}'\frac{dM_z}{dt}\right) + \left(u\frac{d\mathbf{i}'}{dt} + v\frac{d\mathbf{j}'}{dt} + M_z\frac{d\mathbf{k}'}{dt}\right)$$

$$= \frac{\delta\mathbf{M}'}{\delta t} + \boldsymbol{\omega}' \times \mathbf{M}' \tag{19}$$

We may assume that at a certain time $t = 0$ the axes $xyz$ and $x'y'z'$ coincide, and substitute (19) into (14), obtaining the Bloch equations in the rotating frame:

$$\frac{\delta\mathbf{M}'}{\delta t} = \gamma_N\mathbf{M}' \times (\mathbf{H}'_0 + \boldsymbol{\omega}'/\gamma_N) + \gamma_N(\mathbf{M}' \times \mathbf{H}'_1) - \frac{(\mathbf{i}'u + \mathbf{j}'v)}{T_2}$$

$$- \frac{\mathbf{k}'(M_z - M_0)}{T_1} \tag{20}$$

Separation of the three components of $\mathbf{M}'$ gives

$$\frac{du}{dt} = (\omega_0 - \omega)v - \frac{u}{T_2} \tag{21a}$$

$$\frac{dv}{dt} = -(\omega_0 - \omega)u + \gamma_N H_1 M_z - \frac{v}{T_2} \tag{21b}$$

$$\frac{dM_z}{dt} = -\gamma_N H_1 v - \frac{(M_z - M_0)}{T_1} \tag{21c}$$

After the rf field has been on for a long time the spin precession gets into a steady state, and the stationary solution of Eq. (21) is easily shown to be

$$u = M_0 \frac{\gamma_N H_1 T_2^2(\omega_0 - \omega)}{1 + T_2^2(\omega_0 - \omega)^2 + \gamma_N^2 H_1^2 T_1 T_2} \tag{22a}$$

$$v = M_0 \frac{\gamma_N H_1 T_2}{1 + T_2^2(\omega_0 - \omega)^2 + \gamma_N^2 H_1^2 T_1 T_2} \tag{22b}$$

$$M_z = M_0 \frac{1 + T_2^2(\omega_0 - \omega)^2}{1 + T_2^2(\omega_0 - \omega)^2 + \gamma_N^2 H_1^2 T_1 T_2} \tag{22c}$$

Although we have only solved the equation for the spin in a *rotating* $H_1$ field the solution in an oscillating field is almost the same. This is because a field of strength $2H_1 \cos \omega t$ in the $x$ direction can be regarded as the sum of two counter-rotating fields of strength $H_1$ with cartesian components $(H_1 \cos \omega t, -H_1 \sin \omega t, 0)$ and $(H_1 \cos \omega t, +H_1 \sin \omega t, 0)$. Of these two rotating fields only the clockwise one can be in resonance with the nuclear spins, and the other has very little effect. Thus our single rotating field is practically equivalent to an oscillating field of strength $2H_1$.

$u$ is the component of $M$ which oscillates in phase with $H_1$ while $v$ lags $90°$ out of phase. This can be seen from Fig. 11.2, or by going back to a fixed axis system where

$$H_{1x} = 2H_1 \cos \omega t$$
$$M_x = u \cos \omega t + v \sin \omega t \tag{23}$$

Bloch introduced a complex magnetic susceptibility $\chi(\omega)$ to describe these phase relations. The $x$ components of the oscillating $H_1$ field and the magnetic moment $M$ are written as complex numbers $2H_1 e^{-i\omega t}$ and $2H_1 \chi(\omega) e^{-i\omega t}$, where the real and imaginary parts of $\chi$ are

$$\chi(\omega) = \chi'(\omega) + i\chi''(\omega) \tag{24}$$

The true magnetic moment is then the real part of this expression:

$$M_x = H_1[\chi(\omega)e^{-i\omega t} + \chi^*(\omega)e^{+i\omega t}]$$
$$= 2H_1 \chi'(\omega)\cos \omega t + 2H_1 \chi''(\omega)\sin \omega t \tag{25}$$

$u$ and $v$ are therefore proportional to $\chi'(\omega)$ and $\chi''(\omega)$, while a comparison between Eqs. (22), (23), and (25) shows that the spin susceptibilities are

$$\chi'(\omega) = \frac{1}{2} \chi_0 \omega_0 \frac{T_2^2(\omega_0 - \omega)}{1 + T_2^2(\omega_0 - \omega)^2 + \gamma_N^2 H_1^2 T_1 T_2}$$
$$\chi''(\omega) = \frac{1}{2} \chi_0 \omega_0 \frac{T_2}{1 + T_2^2(\omega_0 - \omega)^2 + \gamma_N^2 H_1^2 T_1 T_2} \tag{26}$$

(Here we have written $\chi_0 H_0$ instead of $M_0$.)

Figure 11.3 shows how the susceptibilities behave as the frequency of $H_1$ passes through resonance. $\chi'(\omega)$ changes sign so that $\mathbf{M}$ is $180°$ out of phase with the rf field at high frequencies. $\chi''$ becomes very large near resonance, and this causes a strong absorption of energy, as we shall now see.

The torque produced by the rotating $H_1$ field acting on the magnetic moments of the spins has a component of strength $vH_1$ about the $z$ axis, and this does work on the precessing spins at a rate $\omega v H_1$. Thus the average power absorbed is

$$\frac{dE}{dt} = 2\omega H_1^2 \chi''(\omega)$$

$$= H_1^2 \chi_0 \omega \omega_0 \frac{T_2}{1 + T_2^2(\omega_0 - \omega)^2 + \gamma_N^2 H_1^2 T_1 T_2} \tag{27}$$

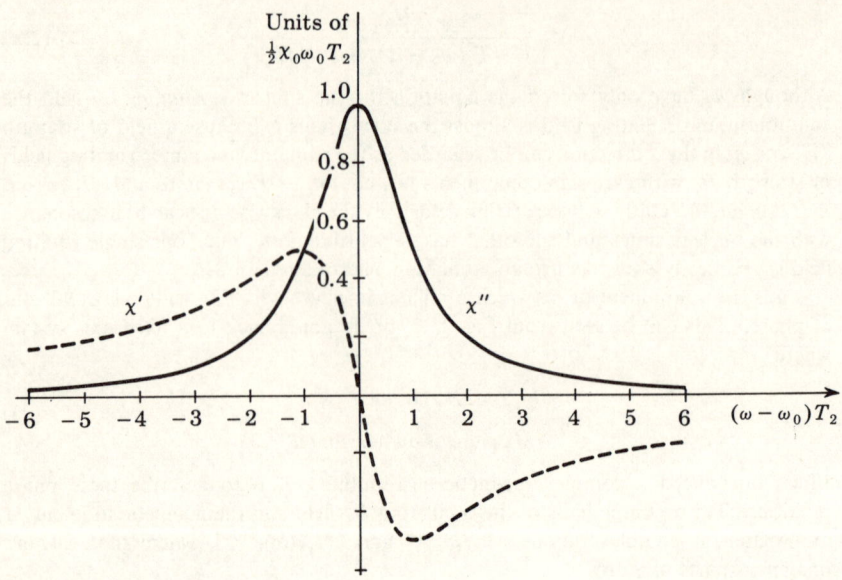

Fig. 11.3. Dispersion and absorption line shapes derived from the Bloch equations.

When the resonance is unsaturated and $\gamma_N^2 H_1^2 T_1 T_2$ is small the absorption has the familiar Lorentz line shape

$$g(\omega) = \frac{T_2}{\pi} \frac{1}{1 + T_2^2(\omega - \omega_0)^2} \tag{28}$$

which we introduced in Chapter 1. On the other hand, in strong rf fields Eq. (27) agrees with the equation

$$\frac{dE}{dt} = n_0 \Delta E \frac{P(\omega)}{1 + 2P(\omega)T_1} \tag{29}$$

which was derived in Sections 1.4 and 2.10 by rather arbitrarily changing the transition probability $P$ to $P(\omega) = (1/2)\pi\gamma_N^2 H_1^2 g(\omega)$. Saturation weakens the central portion of the resonance line relative to the wings and produces an apparent broadening of the spectrum.

The free precession of the spins is often very weakly damped and may continue for several seconds after the $H_1$ field is switched off. Eventually, however, the phase coherence of the individual nuclear spin vectors is lost for various reasons and the oscillations die away. Many beautiful and ingenious experiments have exploited these effects, studying a variety of "spin echoes" which arise from free precession of the spins after a succession of short, carefully timed rf pulses. These methods have not been used much by chemists, although they are a most effective way of measuring relaxation times.

## 11·4  THE ORIGIN OF MAGNETIC RELAXATION

Bloch's equations have taught us that the two relaxation times play quite distinct roles. The spin-lattice relaxation time $T_1$ determines the degree of saturation and $T_2$

determines the unsaturated line width. Both kinds of relaxation are caused by time-dependent magnetic or electric fields at the nucleus (or at the electron) and these fields in turn come from the random thermal motions which are present in any form of matter. A nuclear spin of 1/2, for example, may experience local magnetic fields from the spins of other nuclei moving past it, from unpaired electrons, or from spin-rotational interactions in which the molecular rotation itself generates magnetic fields at the nucleus. There may also be changes in the chemical shielding of the nucleus as a molecule rotates, and these modulate the total effective field which acts on the spin. Nuclei with electric quadrupole moments are further affected by changes in the local electrostatic field gradient when the molecule rotates or vibrates. Electron spins too may find themselves in a variety of fluctuating magnetic fields from other electrons or nuclei. In particular, radicals with an anisotropic $g$ tensor experience fluctuating Zeeman interactions with the external magnetic field, which can be a powerful means of relaxation.

In general, two conditions are necessary for a successful relaxation mechanism. Firstly, there must be some interaction which acts directly on the spins; secondly, it must be time-dependent. Any static interaction can simply be counted as part of the normal spin Hamiltonian. It alters the positions and intensities of the spectral lines without broadening them.

An essential requirement for relaxation is that the molecular motion has a suitable time scale. Interactions which change sign at a rate much faster than the magnetic resonance frequency (typically $10^{10}$ cps for electrons and $10^7$ cps for nuclei) have little effect. Thus electronic motions and molecular vibrations are relatively unimportant. On the other hand, any interaction which causes transitions between the $\alpha$ and $\beta$ spin states and fluctuates strongly at the resonance frequency produces powerful spin-lattice relaxation and line broadening. Another class of interactions contributes to the line width $T_2$ but not to $T_1$. These are random forces which modulate the spin energy levels at low frequencies without causing transitions between them.

The time scale of magnetic resonance is *slow*. Rotation and diffusion motions are very important sources of relaxation in liquids. So are lattice vibrations in solids, collisions in gases, certain slow rotational or torsional motions within molecules, and some chemical exchange processes. It is hard to summarize all the important effects, because every known nuclear interaction coupled with every possible type of motion gives a multitude of relaxation mechanisms. For chemists, the most interesting experiments concern relaxation in liquids. Here most of the effects arise from random Brownian motion of the molecules as they rotate and diffuse through the fluid. The theory is simple and satisfying, and the results give a precise and detailed picture of the rates of molecular motions. The subject of relaxation in liquids is extensive and complicated, so here we shall only cite a few important examples which illustrate the fundamental ideas.

# 1·5 NUCLEAR SPIN RELAXATION IN THE WATER MOLECULE

## 11.5.1 Perturbation Theory

For a start we take one of the proton spins in a drop of water. Here there will be fluctuating local magnetic fields from the other nuclei. The most important one comes from its partner in the same water molecule, which produces a constantly changing dipolar magnetic field as the molecule rotates. There are also smaller fields from nuclear spins in neighboring water molecules. The spin Hamiltonian for the single proton consists of two parts; the Zeeman energy $\mathcal{H}_0 = -\gamma_N \hbar \mathbf{I} \cdot \mathbf{H}_0$ in the steady external magnetic field, and a random perturbation term $V(t)$ due to the local field, which is represented by the matrix

$$\begin{bmatrix} V_{\alpha\alpha}(t) & V_{\alpha\beta}(t) \\ V_{\beta\alpha}(t) & V_{\beta\beta}(t) \end{bmatrix} \tag{30}$$

The off-diagonal matrix elements of $V(t)$ depend on the $x$ and $y$ components of the local field, which can be thought of as containing many fluctuating components oscillating at different frequencies. The parts which oscillate at the nuclear resonance frequency $\omega_0$ induce transitions between the states and cause spin-lattice relaxation. By making a Fourier analysis of $V(t)$ and using time-dependent perturbation theory (see Appendixes E and F) we find that

$$\frac{1}{T_1} = \frac{2}{\hbar^2} \int_{-\infty}^{+\infty} \overline{V_{\alpha\beta}^*(t+\tau)V_{\alpha\beta}(t)} e^{i\omega_0\tau} d\tau \tag{31}$$

Here the integrand is an average of the off-diagonal matrix elements of $V$ taken over all values of the time $t$ and over a large number of samples of the molecular motion. The line width $1/T_2$ depends on the lifetimes of the $\alpha$ and $\beta$ spin states but it is also affected by fluctuations in the energy difference between the two levels, that is, by the $z$ components of the local field, or the diagonal matrix elements of $V(t)$. The expression for $T_2$ is

$$\frac{1}{T_2} = \frac{1}{T'_2} + \frac{1}{2T_1} \tag{32}$$

where $T'_2$ is defined as a time averaged product of the diagonal elements of $V$:

$$\frac{1}{T'_2} = \frac{1}{2\hbar^2} \int_{-\infty}^{+\infty} \overline{[V_{\alpha\alpha}(t+\tau) - V_{\beta\beta}(t+\tau)][V_{\alpha\alpha}(t) - V_{\beta\beta}(t)]} \, d\tau \tag{33}$$

These perturbation formulae apply provided that the local field is reasonably small, so that both relaxation times are long compared with the period of the spin precession $(2\pi/\omega_0)$. It is also necessary that the local fields fluctuate many times during the time $T_1$ or $T_2$.

### 11.5.2 The Power Spectrum of a Random Force

The thermal motions in a liquid cover a wide spectrum and we need some way to measure their frequency distribution and their strength. If $f(t)$ is some random force which fluctuates about a mean value of zero, one measure of its strength is the mean square average $\overline{f^*(t)f(t)}$. To find the frequency variation we can take a sample of the force in a certain time interval from $-T$ to $+T$ and calculate its Fourier transform

$$f_T(\omega) = \int_{-T}^{+T} f(t)e^{i\omega t} \, dt \tag{34}$$

Although $f_T(\omega)$ is itself a random quantity which varies from one sample to another and averages out to zero, its square $|f_T(\omega)|^2$ has a definite average value. For example, if $f(t)$ were the amplitude of some kind of wave motion, $|f_T(\omega)|^2$ would represent the total energy at frequency $\omega$ available in the period from $-T$ to $+T$. If we now allow the time $T$ to increase to infinity the value of $|f_T(\omega)|^2$ increases without limit. However, the quantity

$$J(\omega) = \lim_{T \to \infty} \frac{1}{2T} \overline{f_T^*(\omega) f_T(\omega)} \tag{35}$$

which represents the power or "spectral density" at frequency $\omega$ tends to a definite limit. There is an important theorem in mathematics, the Wiener-Khintchin theorem, which shows that the power spectrum is closely related to the so-called *autocorrelation function* of $f(t)$.

This is defined as the time average value of the product of $f(t)$ and $f^*(t + \tau)$:

$$G(\tau) = \overline{f^*(t + \tau) f(t)} \tag{36}$$

For example, $f(t)$ might be some quantity related to the Brownian motion of a water molecule. Then $f^*(t + \tau)f(t)$ differs for each molecule and each value of $t$, but *on the average* the mean value $G(\tau)$ is the same for all molecules and is independent of $t$ (see Fig. 11.4). The autocorrelation function measures the persistence of the fluctuations; $G(\tau)$ is large for short times, and then dies away to zero as $\tau$ increases. Frequently $G(\tau)$ drops off exponentially with a decay time $\tau_c$, and the autocorrelation

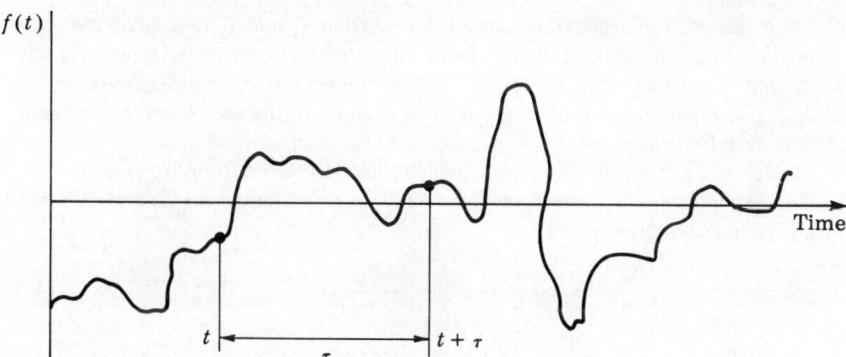

FIG. 11.4. The construction of an autocorrelation function for a random force $f(t)$.

function takes the form

$$G(\tau) = \overline{f^*(t) f(t)} e^{-|\tau|/\tau_c} \tag{37}$$

$\tau_c$ is called the *correlation time*. It is the time taken for a typical fluctuation to die away. In general $G(\tau)$ is a real, even function of $\tau$. We shall show in Appendix G that the power spectrum of $f(t)$ is just the Fourier transform of the autocorrelation function, so that

$$J(\omega) = \int_{-\infty}^{+\infty} G(\tau) e^{i\omega\tau} \, d\tau \tag{38}$$

For example, if there is exponential decay we have

$$J(\omega) = \frac{2\tau_c}{1 + \omega^2 \tau_c^2} \overline{f^*(t) f(t)} \tag{39}$$

This spectrum is illustrated in Fig. 11.5. The power remains fairly constant at frequencies below $1/\tau_c$ and then dies away quite fast at high frequencies.

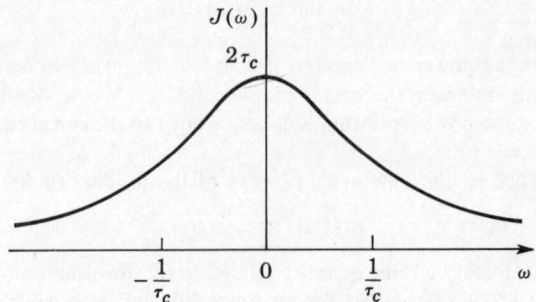

FIG. 11.5. The power spectrum $J(\omega)$ of a random force with an exponentially decaying autocorrelation function and a mean square value equal to one.

### 11.5.3 The Effect of a Local Field

Returning to our relaxation problem, we see that $T_1$ and $T_2$ depend on the autocorrelation functions of the matrix elements of the perturbation $V(t)$. More precisely $T_1$ depends on the spectral density of $V_{\alpha\beta}(t)$ at the nuclear resonance frequency $\omega_0$, while $T'_2$ involves fluctuations in the relative energies of the spin states which occur close to zero frequency.

Let us now suppose that the local magnetic field $\mathbf{H}(t)$ arises from Brownian motion with an exponential autocorrelation function (37). Substitution of (36) and (39) into Eqs. (31) and (33) gives

$$\frac{1}{T_1} = \frac{2}{\hbar^2} \overline{|V_{\alpha\beta}|^2} \frac{2\tau_c}{1 + \omega_0^2 \tau_c^2} \tag{40}$$

$$\frac{1}{T'_2} = \frac{1}{2\hbar^2} \overline{|V_{\alpha\alpha} - V_{\beta\beta}|^2} \cdot 2\tau_c \tag{41}$$

Now in a field $\mathbf{H}$ the matrix elements of $V$ are

$$V_{\alpha\beta} = -\frac{1}{2} g_N \beta_N (H_x - i H_y)$$

$$(V_{\alpha\alpha} - V_{\beta\beta}) = -g_N \beta_N H_z \tag{42}$$

so that the final expressions for the relaxation times are

$$\frac{1}{T_1} = \frac{g_N^2 \beta_N^2}{2\hbar^2} \overline{(H_x^2 + H_y^2)} \frac{2\tau_c}{1 + \omega_0^2 \tau_c^2} \tag{43}$$

$$\frac{1}{T_2} = \frac{g_N^2 \beta_N^2}{2\hbar^2} \left\{ 2\tau_c \overline{H_z^2} + \frac{1}{2} \overline{(H_x^2 + H_y^2)} \frac{2\tau_c}{1 + \omega_0^2 \tau_c^2} \right\} \tag{44}$$

At this point it is interesting to ask what conditions are necessary to make $T_1$ and $T_2$ equal. Since the Brownian motion is completely isotropic we can expect the mean square averages of $H_x$, $H_y$, and $H_z$ to be equal. In that case the ratio of $T_1$ to $T_2$ depends on the relative strengths of the fluctuations at $\omega = \omega_0$ and $\omega = 0$. If the correlation time is long the factors $(1 + \omega_0^2 \tau_c^2)$ in (42) and (43) are large and $T_1 < T_2$ (see Fig. 11.5). However, if the Brownian motions are very rapid $\omega_0 \tau_c$ is small and $T_1 = T_2$.

We can estimate the local field by treating the other proton in the water molecule as a classical bar magnet $g_N \beta_N \mathbf{I}_2$ which is quantized in the steady field $\mathbf{H}_0$ with spin quantum number $m_2 = \pm 1/2$. The local field has a component

$$H_z = \frac{\pm g_N \beta_N}{r^3} \left[ \frac{1}{2} (3 \cos^2 \theta - 1) \right] \tag{45}$$

parallel to $H_0$ and a perpendicular component of magnitude $\frac{3}{2} \sin \theta \cos \theta \, (g_N \beta_N / r^3)$. Here $\theta$ is the angle between the H-H direction and the steady field. The proton-proton distance in water is 1.58 Å and $g_N \beta_N / r^3$ has the value 7.2 gauss. The factor $\frac{1}{2} (3 \cos^2 \theta - 1)$ fluctuates rapidly as the molecule rotates; we must therefore try to form a picture of the Brownian motion and calculate the autocorrelation functions.

### 11.5.4. Rotational Brownian Motion

Water molecules do not so much rotate as "tumble" in a very irregular fashion, because the axis and the direction of motion are continually being altered by collisions. As a result the motion of one proton relative to another is a kind of random walk over the surface of a sphere. The theory is due to Debye.

We picture the situation in Fig. 11.6. The first proton is imagined to be fixed at the center of a sphere, radius $r$, while the other tumbles round it. If the second proton is initially at the pole, $A$, it might pursue a path of the type shown, arriving some time later at the point $B$, with polar coordinates $(\theta, \phi)$. We now need to know the probability $p(\theta, \phi; t)$ that a proton reaches the point $B$ at a time $t$ after starting from $A$. For this purpose we will assume that the mean angular velocity of rotation is $\Omega$ and that the "mean free path," or the mean angle through which the molecule turns between collisions is $\alpha$ ($\alpha$ is assumed small). According to this model the molecule has $\Omega t / \alpha$ collisions in time $t$ and the mean square angle turned through between collisions comes out as $2\alpha^2$. For a random walk this results in a mean square displacement proportional to the time $t$:

$$\overline{\theta^2} = 2\alpha \Omega t = 4D't \tag{46}$$

$D'$ is called the spherical diffusion coefficient, and Debye showed that the probability $p(\theta, \phi; t)$ obeys the differential equation for diffusion over the surface of a sphere

$$\frac{1}{D'} \frac{\partial p}{\partial t} = \nabla^2 p$$

$$= \frac{1}{\sin \theta} \frac{\partial}{\partial \theta} \left( \sin \theta \frac{\partial p}{\partial \theta} \right) + \frac{1}{\sin^2 \theta} \frac{\partial^2 p}{\partial \phi^2} \tag{47}$$

The general solution of this equation is well known. It is an expansion in a series of normalized spherical harmonics $Y_{lm}(\theta, \phi)$:

$$p = \sum_{l,m} C_{lm} Y_{lm}(\theta, \phi) e^{-D'l(l+1)t} \tag{48}$$

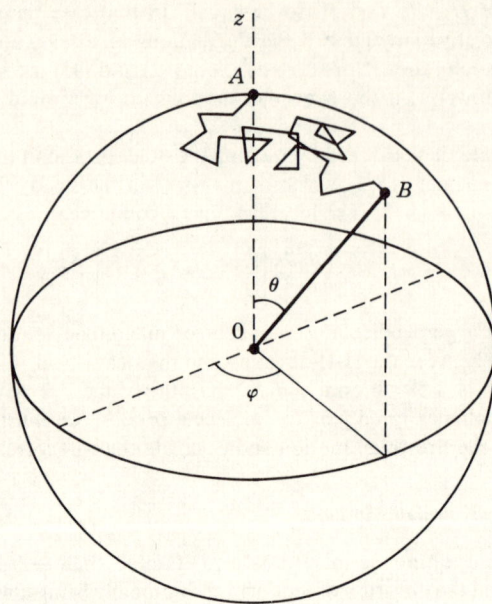

FIG. 11.6. Random motion of one water proton relative to the other.

The coefficients $C_{lm}$ are to be found from the initial conditions by using the expansion theorem for spherical harmonics. One can now show that during the Brownian motion the average value of any angular function $F(\theta, \phi)$ which is a spherical harmonic of order $l$ decays exponentially to zero. The details of the proof are in Appendix H. The result is that

if
$$\overline{F(t)} = \int_0^\pi \int_0^{2\pi} F(\theta, \phi) p(\theta, \phi; t) \sin \theta \, d\theta \, d\phi \tag{49}$$

then
$$\overline{F(t)} = \overline{F(0)} e^{-D'l(l+1)t} \tag{50}$$

Debye treated the problem of dielectric relaxation in water. Here $\theta$ is the angle between the electric dipole moment and the applied field, and the resolved component of the dipole moment is proportional to $\cos \theta(t)$, which has the value

$$\overline{\cos \theta(t)} = e^{-2D't} = e^{-t/\tau_d} \tag{51}$$

The time $\tau_d$ is called the Debye correlation time of the liquid. In magnetic resonance we are concerned with tensors which are spherical harmonics of the second order ($l = 2$). These relax faster, with a characteristic time $\tau_c$ called the rotational correlation time. For the local field in Eq. (45) we find

$$\overline{\tfrac{1}{2}[3 \cos^2 \theta(t) - 1]} = e^{-6D't} = e^{-t/\tau_c} \tag{52}$$

The value of $\tau_c$ depends on complicated details of the molecular motion, but it is easy to estimate it roughly. The molecule is pictured as a small sphere of radius $a$ embedded in a viscous fluid (viscosity $\eta$). When the sphere rotates, it experiences a viscous torque equal to $8\pi\eta a^3(d\theta/dt)$ and the diffusion constant $D'$ is directly related to the frictional constant $8\pi\eta a^3$ by Einstein's famous relation

$$D' = \frac{kT}{8\pi\eta a^3} \tag{53}$$

or

$$\tau_c = \frac{4\pi\eta a^3}{3kT} \tag{54}$$

In water at 20°C the viscosity is $\eta = 0.01$ poise, and, assuming a molecular radius of 1.5 Ångstroms, Eq. (54) gives $\tau_c = 3.5 \times 10^{-12}$ sec. Dielectric relaxation experiments lead to a value of $2.7 \times 10^{-12}$ which agrees rather well with this theory.

The critical reader may well object that the Brownian motion in liquids cannot possibly be so simple as it is pictured in Debye's theory. This is certainly true, but Debye's simple ideas have proved most useful for interpreting resonance line widths, and the more elaborate theories often lead to practically the same conclusions. All relaxation effects depend on the frequency spectrum of the local fields at the nucleus and there is almost always a definite characteristic time scale for the fluctuations, i.e., a *correlation time*.

### 11.5.5 The Calculation of $T_1$ and $T_2$

The picture of the protons in a water molecule being relaxed by a local magnetic field due to the dipolar interaction between the two nuclear moments is a classical one and fails to describe the relaxation effects properly. One reason for this is that the spins obey quantum mechanics rather than classical mechanics. Another is that the two spins are correlated and the total spin angular momentum $\mathbf{I} = (\mathbf{I}_1 + \mathbf{I}_2)$ is conserved during the tumbling. Water molecules can exist either in a singlet spin state ($I = 0$) or a triplet ($I = 1$). The singlet state has the wave function $|s\rangle$ described in Section 3.2 and gives no n.m.r. spectrum, while the three triplets $|t_1\rangle$, $|t_0\rangle$ and $|t_{-1}\rangle$ are responsible for all magnetic effects. Relaxation occurs when the dipolar interaction

$$\mathcal{H}_D(t) = g_N^2 \beta_N^2 \mathbf{I}_1 \cdot \mathbf{D}(t) \cdot \mathbf{I}_2 \tag{55}$$

perturbs the three triplet levels. The Hamiltonian $\mathcal{H}_D(t)$ was defined in Eq. (8) of Section 3.2. It varies with time because the elements of the dipolar coupling tensor alter during the tumbling motion.

Our immediate task is to calculate the effects of $\mathcal{H}_D(t)$ in triplet water molecules. The spins behave as a three-level system with $I = 1$, and following the method of Section 8.2 we could even consider the coupled protons as a single particle of spin 1, having a Hamiltonian

$$\mathcal{H}_D(t) = \tfrac{1}{2} g_N^2 \beta_N^2 \mathbf{I} \cdot \mathbf{D}(t) \cdot \mathbf{I} \tag{56}$$

We now follow Section 3.3 and split $\mathcal{H}_D$ into its various parts $A$, $B$, $C$, $D$, $E$, $F$. Their effects are illustrated in Fig. 11.7. $A$ and $B$ have diagonal matrix elements and modulate the energy levels, contributing to $T_2$; $C$ and $D$ have off-diagonal elements between adjacent energy levels, while $E$ and $F$ connect $|t_{+1}\rangle$ with $|t_{-1}\rangle$. $C$, $D$, $E$, and $F$ determine the transition probabilities for relaxation between the various levels, and we shall use these probabilities to calculate the spin-lattice relaxation time $T_1$.

Let us call the probabilities that the spin quantum number $m = (m_1 + m_2)$ changes by $+1$ or $-1$ $W_{\pm 1}$, and the corresponding probabilities for $\Delta m = \pm 2$ transitions $W_{\pm 2}$, as shown in Fig. 11.8. We suppose that there are $N$ water molecules and

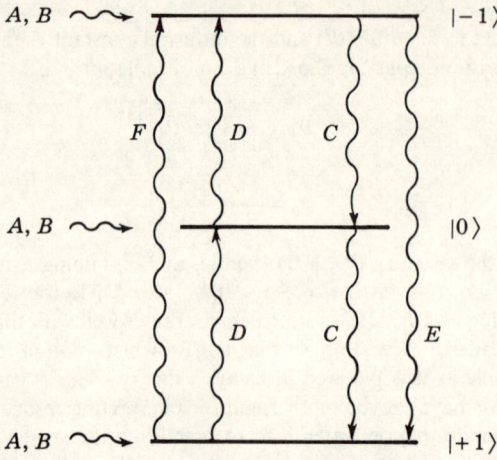

FIG. 11.7. Transitions induced by $\mathcal{H}_D(t)$ within the triplet spin levels of a water molecule.

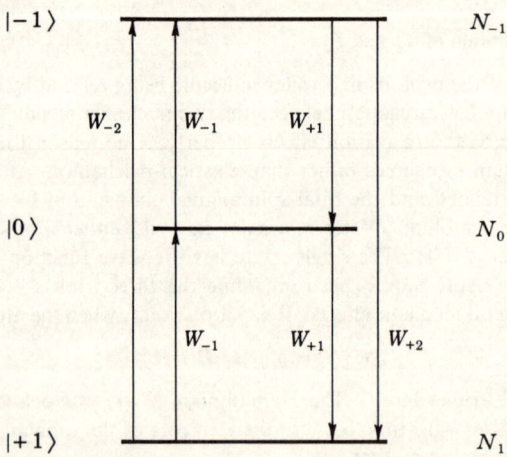

FIG. 11.8. Transition probabilities.

that the populations of the three spin states are $N_1$, $N_0$, $N_{-1}$. In thermal equilibrium the steady state populations obey Boltzmann's law, and this requires that the probabilities for transitions up and down are in the ratios

$$W_{+1}/W_{-1} = e^{+g_N\beta_N H/kT}$$

$$W_{+2}/W_{-2} = e^{+2g_N\beta_N H/kT}$$

(57)

Since $g_N\beta_N H \ll kT$ we can satisfy these conditions by taking

$$W_{\pm 1} = W_1[1 \pm \tfrac{1}{2}(g_N\beta_N H/kT)]$$

$$W_{\pm 2} = W_2[1 \pm (g_N\beta_N H/kT)]$$

(58)

The time evolution of the populations now obeys the differential equations

$$\frac{dN_{-1}}{dt} = -(W_{+1} + W_{+2})N_{-1} + W_{-1}N_0 + W_{-2}N_1$$

$$\frac{dN_0}{dt} = -(W_{+1} + W_{-1})N_0 + W_{+1}N_{-1} + W_{-1}N_1 \qquad (59)$$

$$\frac{dN_1}{dt} = -(W_{-1} + W_{-2})N_1 + W_{+1}N_0 + W_{+2}N_{-1}$$

An arbitrary initial distribution does not relax uniformly to thermal equilibrium because the general solution of (59) involves two different relaxation times. However, a simple solution does obtain provided that a "spin temperature" exists; that is, the populations of all adjacent pairs of spin levels are in a constant ratio, $N_1 : N_0 = N_0 : N_{-1}$. Thus we consider the initial conditions

$$N_{-1} = \tfrac{1}{3}(N - n), \qquad N_0 = \tfrac{1}{3}N, \qquad N_1 = \tfrac{1}{3}(N + n) \qquad (60)$$

with $n$ small, and look for the solution of (59). The magnetization is

$$M_z = \tfrac{2}{3}g_N \beta_N n \qquad (61)$$

and so the relaxation time found for $n$ will be the same as $T_1$. Substitution into (59) gives

$$-\frac{dn}{dt} = n[W_{+1} + (W_{+2} + W_{-2})] - N[(W_{+1} - W_{-1}) + (W_{+2} - W_{-2})]$$

$$0 = -n(W_{+1} - W_{-1}) \qquad (62)$$

$$\frac{dn}{dt} = -n[W_{-1} + (W_{+2} + W_{-2})] + N[(W_{+1} - W_{-1}) + (W_{+2} - W_{-2})]$$

These equations are mutually consistent if we ignore the quantity $n(W_{+1} - W_{-1})$ which is of the second order because both $(n/N)$ and $(W_{+1} - W_{-1})$ are small. Substituting the probabilities from (58) we find that

$$\frac{dn}{dt} = -(W_1 + 2W_2)(n - n_0) \qquad (63)$$

where $n_0 = Ng_N \beta_N H / kT$ is the equilibrium value of $n$. The spin-lattice relaxation time is clearly

$$\frac{1}{T_1} = (W_1 + 2W_2) \qquad (64)$$

and we must now calculate the two transition probabilities.

Time-dependent perturbation theory shows that $W_1$ and $W_2$ are respectively proportional to the mean square fluctuations of the matrix element $\langle t_0 | \mathcal{H}_D | t_1 \rangle$ at frequency $\omega_0$ and $\langle t_{-1} | \mathcal{H}_D | t_1 \rangle$ at $2\omega_0$. In fact, referring to Appendix E, we have

$$W_1 = \frac{1}{\hbar^2} |\langle t_0 | \mathcal{H}_D | t_1 \rangle|^2 \frac{2\tau_c}{1 + \omega_0^2 \tau_c^2}$$

$$\qquad (65)$$

$$W_2 = \frac{1}{\hbar^2} |\langle t_{-1} | \mathcal{H}_D | t_1 \rangle|^2 \frac{2\tau_c}{1 + 4\omega_0^2 \tau_c^2}$$

The matrix of $\mathcal{H}_D$ is found at once by using the operators $A$, $B$, $C$, $D$, $E$, $F$ of Section 3.3. It is

$$\mathcal{H}_D = \frac{g_N^2 \beta_N^2}{r^3} \times$$

$$\begin{bmatrix} \frac{1}{4}(1 - 3\cos^2\theta) & -\frac{3}{2\sqrt{2}}\sin\theta\cos\theta e^{-i\phi} & -\frac{3}{4}\sin^2\theta e^{-2i\phi} \\[2ex] -\frac{3}{2\sqrt{2}}\sin\theta\cos\theta e^{+i\phi} & -\frac{1}{2}(1 - 3\cos^2\theta) & \frac{3}{2\sqrt{2}}\sin\theta\cos\theta e^{-i\phi} \\[2ex] -\frac{3}{4}\sin^2\theta e^{+2i\phi} & \frac{3}{2\sqrt{2}}\sin\theta\cos\theta e^{+i\phi} & \frac{1}{4}(1 - 3\cos^2\theta) \end{bmatrix} \tag{66}$$

The averages of the angular functions over all directions in space are

$$\overline{(1 - 3\cos^2\theta)^2} = \frac{4}{5}$$

$$\overline{(\sin\theta\cos\theta)^2} = \frac{2}{15} \tag{67}$$

$$\overline{(\sin^2\theta)^2} = \frac{8}{15}$$

and it is now straightforward to compute the mean square values of the matrix elements in (65). They are, respectively, $(3g_N^4\beta_N^4/20r^6)$ and $(6g_N^4\beta_N^4/20r^6)$, so that finally, using (64) and (65), the spin-lattice relaxation time is

$$\frac{1}{T_1} = \frac{3}{20}\left(\frac{g_N^4\beta_N^4}{\hbar^2 r^6}\right)\left\{\frac{2\tau_c}{1 + \omega_0^2\tau_c^2} + \frac{8\tau_c}{1 + 4\omega_0^2\tau_c^2}\right\} \tag{68}$$

The calculation of $T_2$ is considerably more complicated than the one we have just performed for $T_1$. It is not possible to use the relation $1/T_2 = 1/T'_2 + 1/2T_1$ which holds for a single spin of 1/2, and the theory needs a more advanced mathematical approach. We merely quote the result

$$\frac{1}{T_2} = \frac{3}{40}\left(\frac{g_N^4\beta_N^4}{\hbar^2 r^6}\right)\left\{6\tau_c + \frac{10\tau_c}{1 + \omega_0^2\tau_c^2} + \frac{4\tau_c}{1 + 4\omega_0^2\tau_c^2}\right\} \tag{69}$$

A typical proton resonance experiment at 60 Mc/s has $\omega_0 = 1.9 \times 10^8$, whereas the correlation time for water at room temperature is $2.7 \times 10^{-12}$ sec. This means that $\omega_0\tau_c$ is extremely small and can be neglected in Eqs. (68) and (69). The two relaxation times should be equal, with

$$T_1 = T_2 \tag{70}$$

$$\frac{1}{T_1} = \frac{3}{2}\left(\frac{g_N^4\beta_N^4}{\hbar^2 r^6}\right)\tau_c \tag{71}$$

and taking $r = 1.58$ Å, $\tau_c = 2.7 \times 10^{-12}$ sec we estimate $T_1 = 6.7$ sec. The observed relaxation times are indeed equal, the experimental value being 3.6 sec.

We have neglected the magnetic fields of the protons on other water molecules which diffuse by, but these also produce relaxation, and again $T_1 = T_2$ under normal conditions. The theoretical expression (71) must therefore be supplemented by a further term, which is

$$\left(\frac{1}{T_1}\right)_{\text{trans}} = \frac{\pi}{5}\left(\frac{g_N^4 \beta_N^4}{\hbar^2}\right)\frac{N}{Db} \tag{72}$$

Here $D$ is the coefficient of self diffusion, $N$ the concentration of spins, and $b$ the distance of closest approach between spins on different molecules. Taking $D = 1.85 \times 10^{-5}$, $N = 6.75 \times 10^{22}$, $b = 1.74$ Å for water we estimate $(T_1)_{\text{trans}} = 12.6$ sec. The total relaxation rate is the sum of the rotational and translational parts

$$\frac{1}{T_1} = \left(\frac{1}{T_1}\right)_{\text{rot}} + \left(\frac{1}{T_1}\right)_{\text{trans}} \tag{73}$$

$$T_1 = 4.4 \text{ sec} \tag{74}$$

The agreement with experiment is fairly satisfactory and gives us confidence that the theory is mainly correct.

### 11.5.6 Short and Long Correlation Times

As long as $\tau_c$ is short, so that $\omega_0 \tau_c \ll 1$, $T_1$ and $T_2$ remain equal. Both contributions to (73) should be proportional to $(\eta/T)$, since the first depends on the correlation time $\tau_c = 4\pi a^3 \eta/3kT$ and the second on the diffusion coefficient, whose value calculated for a spherical molecule is $1/D = 6\pi\eta a/kT$. This prediction is well tested by the experiments of Bloembergen on proton relaxation times in glycerin, a liquid whose viscosity varies very strongly with temperature. The experimental results are shown in Fig. 11.9 where the logarithms of $T_1$ and $T_2$ are plotted against $\log(\eta/T)$. The left side of the figure, where $\eta/T < 10^{-1}$, shows the expected straight lines with $T_1 = T_2$. However, at higher viscosities the curves for $T_1$ and $T_2$ diverge, and this requires some explanation.

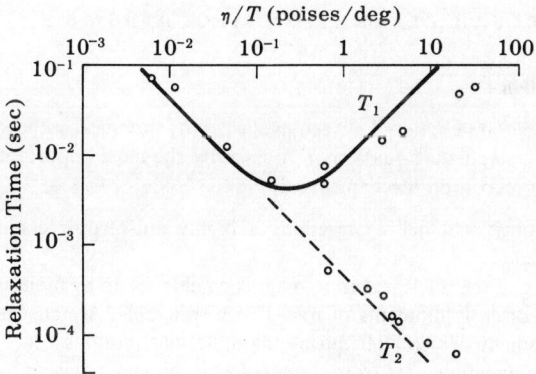

FIG. 11.9. Proton relaxation times in glycerine between 60°C and 35°C. Measured at 29 Mc/s.

As the liquid becomes more viscous and the correlation time increases, the component of the local field fluctuations at the resonance frequency $\omega_0$ decreases, tending to zero when $\omega_0\tau_c \gg 1$. The result is that the value of $T'_2$ decreases, being still proportional to $1/2\tau_c$, but the spin-lattice relaxation time now becomes very long, as it is proportional to $2\tau_c/(1 + \omega_0^2\tau_c^2)$. Hence in the limit of long correlation times $T_1$ is proportional to $\tau_c$ instead of $1/\tau_c$ and begins to increase again. In the region between long and short correlation times $T_1$ goes through a minimum at the point $\omega_0\tau_c = 1$, as we see in Fig. 11.10. The line width $1/T_2$ continues to increase linearly with $\tau_c$ until the molecular motion slows down so much that it is completely frozen. At this stage we have a randomly oriented solid and the line broadening is of the type discussed in Chapter 3.

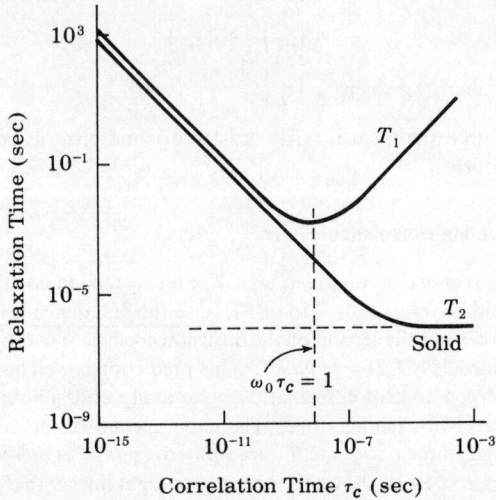

FIG. 11.10. Theoretical behavior of $T_1$ and $T_2$ as a function of the correlation time. Calculated for protons in water at 29 Mc/s ($\omega_0 = 1.9 \times 10^8$ radians/sec.)

# 11·6 OTHER NUCLEAR RELAXATION MECHANISMS

### 11.6.1 Introduction

Dipole-dipole interaction between nuclei is by no means the only source of relaxation in diamagnetic liquids, nor necessarily the most important. Some of the other important relaxation mechanisms are:

1. Fluctuating local fields caused by a highly anisotropic chemical shift in a tumbling molecule.
2. Coupling of the nuclear spin to magnetic fields set up by molecular rotation.
3. The quadrupole moments of nuclei with spin $>1/2$ which can interact with electric field gradients which alter during the molecular motions.
4. Strong magnetic forces from unpaired electron spins present as paramagnetic impurities.

We shall now give examples of these effects.

### 11.6.2 Anisotropic Chemical Shift

The chemical shielding tensor $\boldsymbol{\sigma}$ generally has three distinct principal values $\sigma_1, \sigma_2, \sigma_3$, of which only the average value $\sigma$ can be measured directly in solution. The tensor $\boldsymbol{\sigma}$ can always be divided into an isotropic part $\sigma\mathbf{1}$ and an anisotropic part $\boldsymbol{\sigma}'$ with principal values $\sigma'_1, \sigma'_2, \sigma'_3$ such that $\sigma_1 = (\sigma + \sigma'_1)$, $\sigma_2 = (\sigma + \sigma'_2)$ and so on; the Zeeman energy in a steady external field will then be written

$$\mathscr{H} = -g_N\beta_N H_0(1 - \sigma)I_z + g_N\beta_N \mathbf{H}_0 \cdot \boldsymbol{\sigma}' \cdot \mathbf{I} \tag{75}$$

As the molecule tumbles, the second term in (75) is equivalent to a fluctuating local field at the nucleus with components

$$(H'_x, H'_y, H'_z) = (H_0\sigma'_{zx}, H_0\sigma'_{zy}, H_0\sigma'_{zz}) \tag{76}$$

and we can apply the theory of Section 11.5.3, calculating the mean square values of $H'$ (see Appendix I). When $\tau_c$ is short the result is

$$\frac{1}{T_1} = \frac{g_N^2\beta_N^2 H_0^2}{10\hbar^2}(\sigma'^2_1 + \sigma'^2_2 + \sigma'^2_3)2\tau_c \tag{77}$$

The two relaxation times are not equal, however, and $T_1/T_2 = 7/6$.

The characteristic feature of this mode of relaxation is that $1/T_1$ is proportional to $H_0^2$. Fluorine chemical shifts are sufficiently anisotropic to give observable line broadening; in fluorobenzene, for example $(\sigma_\parallel - \sigma_\perp)$, the anisotropy about the CF bond direction, is about 120 ppm, and the n.m.r. line width in 1, 3, 5-trifluorobenzene shows a definite increase at higher fields in agreement with the theory.

### 11.6.3 Nuclear Spin-Rotational Coupling

When a molecule rotates the motion of the nuclei and electrons sets up magnetic fields proportional to the rotational angular momentum $\mathbf{J}$, and the energy of interaction between the nuclear spin and this field may be written

$$\mathscr{H}_{IJ} = -\mathbf{I} \cdot \mathbf{C} \cdot \mathbf{J} \tag{78}$$

$\mathbf{C}$ is, in general, a tensor, called the spin-rotational tensor. Brownian motion causes continual alterations in the magnitude and direction of $\mathbf{J}$ which have a characteristic correlation time $\tau_J$, and the fluctuating local fields due to $\mathscr{H}_{IJ}$ relax the spin. For short correlation times $T_1$ and $T_2$ are equal and the theoretical expression is

$$\frac{1}{T_1} = \frac{2\tau_J}{9\hbar^2}(C_1^2 + C_2^2 + C_3^2)\langle J(J + 1)\rangle \tag{79}$$

Here $C_1, C_2, C_3$ are the principal values of the $\mathbf{C}$ tensor, and $\langle J(J + 1)\rangle$ is the mean square rotational angular momentum of the molecule. Spin-rotational couplings are particularly large in fluorine compounds. For example, in fluorobenzene the principal values of $\mathbf{C}$ are $-2.0$, $-2.5$, and $-1.9$ Kc/s, referred to axes along the CF bond, across the bond, and perpendicular to the ring. The angular momentum increases with increasing temperature because the rotational energy $\hbar^2 J(J + 1)/2I'$ of a molecule with moment of inertia $I'$ is equal to $\frac{1}{2}kT$. Also, at higher temperatures, the molecules in a liquid rotate more freely and so $\tau_J$ increases as well. Thus, in fluorobenzene, below 0°C the value of $T_1$ is mainly determined by dipolar coupling with the protons and increases with temperature, just as the proton $T_1$ does. At higher temperatures from 0° to 160°C, however, spin-rotational coupling becomes important and the fluorine $T_1$ falls again. The protons have little or no spin-rotational interaction and for them the relaxation time continues to increase.

### 11.6.4 Electric Quadrupole Couplings

The interaction between the nuclear quadrupole moment and the local electric field gradient has the form

$$\mathcal{H}_Q = \mathbf{I} \cdot \mathbf{Q}(t) \cdot \mathbf{I} \tag{80}$$

where $\mathbf{Q}$ is a certain tensor with elements proportional to the quadrupole moment $Q$ and the field gradients $\partial^2 V/\partial x^2$, $\partial^2 V/\partial x \partial y$ etc. Molecular rotation causes changes in the $\mathbf{Q}$ tensor, and in nuclei with spin greater than 1/2 this is often the dominant cause of relaxation. The quadrupole Hamiltonian for a spin of 1 is formally equivalent to the dipole-dipole interaction Hamiltonian (66) for the two protons in water, which we have already treated, provided that $g_N^2 \beta_N^2/r^3$ is replaced by $\frac{1}{2}e^2 qQ$, and the angles $(\theta, \phi)$ now describe the direction of the electric field gradient. Thus we can use the entire theory of Section 11.5.5, recalculating the transition probabilities $W_1$ and $W_2$ and obtaining the relaxation time

$$\frac{1}{T_1} = \frac{3e^4 q^2 Q^2}{80\hbar^2} \left\{ \frac{2\tau_c}{1 + \omega_0^2 \tau_c^2} + \frac{8\tau_c}{1 + 5\omega_0^2 \tau_c^2} \right\} \tag{81}$$

Quadrupole broadening is particularly large and troublesome with $N^{14}$, where it makes the analysis of high resolution n.m.r. spectra almost impossible. Not only is the $N^{14}$ resonance broadened, but so are the lines from other nuclei such as $H^1$ or $C^{13}$ which have spin-spin couplings to the nitrogen. For example, the proton resonance of $N^{14}H_3$ is broadened because of the coupling with $N^{14}$ which is being relaxed rapidly by quadrupole effects. On the other hand, $NH_4^+$ which is tetrahedral has no asymmetry at the nitrogen, so now the spin lifetime is long and the proton resonance is much sharper. (The position here is not really quite so simple as this, because proton exchange also contributes to the line widths. For example, $N^{14}$ splittings are observed in *dry* ammonia in spite of the quadrupole effect. However, in moist ammonia proton exchange washes out the finer details of the n.m.r. spectrum.)

$N^{14}$ line widths have been used to study molecular motions in solution or to estimate quadrupole coupling constants. For example, in acetonitrile, the line width is proportional to the viscosity, and the temperature variation of $T_1$ has been used to estimate an activation energy of 1.9 kcal/mole for molecular reorientation perpendicular to the CN axis.

### 11.6.5 Unpaired Electron Spins

Paramagnetic ions in solution have an extremely strong effect on the relaxation time, even when they are present in low concentrations, because the mean square magnetic fields at the nuclei are proportional to the square of the electron magnetic moment and are about $10^6$ times stronger than the fields produced by other nuclei. For ions whose magnetism comes purely from spin angular momentum, the squared moment is $\mu^2 = g^2 \beta^2 S(S + 1)$, while for others $\mu^2$ has to be replaced by an effective mean square value $\mu_{\text{eff}}^2$. As the ions diffuse past the nuclei they cause line broadening in the same way as the protons of one water molecule moving past another. The contribution to $1/T_1$ is then calculated by analogy with (72):

$$\frac{1}{T_1} = \frac{16\pi^2}{15} \left( \frac{g_N^2 \beta_N^2 g^2 \beta^2}{\hbar^2} \right) \frac{NS(S+1)\eta}{kT} \tag{82}$$

As an example, the $Fe^{+++}$ ion with a magnetic moment $\mu_{eff} = 5.9$ Bohr magnetons lowers the spin-lattice relaxation time of water to $0.1$ sec when the concentration is as low as $10^{18}$ spins/cm$^3$ (or $10^{-3}$ molar). Dissolved oxygen also broadens nuclear resonances in solution. The relaxation effects produced by paramagnetic ions are many and varied, and we shall learn more about them in Chapter 13.

# 1·7 SPIN RELAXATION OF RADICALS IN SOLUTION

## 11.7.1 Relaxation Mechanisms

For an electron spin of 1/2 attached to a radical in solution there are two important anisotropic magnetic interactions within the molecule. One comes from anisotropy of the $g$ tensor, due to spin orbit interaction, the other from dipolar hyperfine coupling with magnetic nuclei. Both interactions fluctuate in a tumbling molecule in solution, but they are usually fairly weak and do not produce much broadening. Therefore organic radicals in solution usually have very well resolved e.s.r. spectra with line widths as low as 50 milligauss. Electron spin-rotational couplings are sometimes important, but we shall not discuss them here.

The isotropic hyperfine interaction $a\mathbf{I} \cdot \mathbf{S}$ does not produce any relaxation unless the value of $a$ varies with time as the result of molecular motion. We shall defer discussion of this effect until Chapter 12.

If the concentration of radicals in solution becomes too high, adjacent pairs of electron spins begin to interact and the lines are broadened by several different processes. In addition to the direct dipolar coupling of the magnetic moments there are much stronger exchange forces of the type $hJ\mathbf{S}_1 \cdot \mathbf{S}_2$ between radicals whose electronic wave functions overlap. Therefore it is advisable to use dilute solutions in most e.s.r. measurements.

Paramagnetic ions with $S > 1/2$ usually have large zero-field splittings and these make the e.s.r. spectra very broad.

We now discuss some of these effects in more detail.

## 11.7.2 Anisotropic $g$ Tensor and Hyperfine Tensor

Let us begin with a free radical of spin 1/2 which has an anisotropic $g$ tensor and tumbles rapidly. Following the treatment of anisotropic chemical shielding given in Section 11.6 we divide the electronic Zeeman Hamiltonian into its average $\mathcal{H}_0 = g\beta H_0 S_z$ and its anisotropic part $V(t) = \beta \mathbf{H}_0 \cdot \mathbf{g}'(t) \cdot \mathbf{S}$. This latter term is time-dependent because of the Brownian motion. $\mathbf{g}'(t)$ is the anisotropic part of the $g$ tensor, and the perturbation $V$ which is responsible for relaxation can be written

$$V(t) = \beta H_0 (g'_{zx} S_x + g'_{zy} S_y + g'_{zz} S_z) \tag{83}$$

The relaxation rates are proportional to the mean squared matrix elements of $V$, and hence to the squared elements of the $g'$ tensor averaged over all orientations of the molecule. As we have shown in Appendix I, these quantities depend on the so-called inner product of the tensor with itself

$$(g':g') = \sum_{ik} g'_{ik} g'_{ki} \tag{84}$$

which is independent of orientation. The required averages are

$$\overline{(g'_{zz})^2} = \frac{2}{15}(g':g'); \quad \overline{(g'_{zx})^2} = \overline{(g'_{zy})^2} = \frac{1}{10}(g':g') \tag{85}$$

and substitution of them into the general perturbation formulae for $T_1$ and $T_2$ gives

$$\frac{1}{T_1} = \frac{(g':g')\beta^2 H_0^2}{60\hbar^2} \left\{ \frac{12\tau_c}{1 + \omega_0^2 \tau_c^2} \right\} \tag{86}$$

$$\frac{1}{T_2} = \frac{(g':g')\beta^2 H_0^2}{60\hbar^2} \left\{ 8\tau_c + \frac{6\tau_c}{1 + \omega_0^2 \tau_c^2} \right\} \tag{87}$$

In the limit of fast tumbling we find the ratio $T_1/T_2 = 7/6$, just as for the nuclear relaxation. The effects of the $g$ tensor anisotropy are usually quite small and become even smaller as the tumbling rate increases. For this reason it is best to use solvents of low viscosity if sharp spectra are desired.

We now turn to the hyperfine interaction. The usual spin Hamiltonian for a radical with a single nuclear spin $I$ splits into the two parts

$$\begin{aligned} \mathcal{H}_0 &= g\beta H_0 S_z + a I_z S_z \\ V(t) &= \beta \mathbf{H}_0 \cdot \mathbf{g}'(t) \cdot \mathbf{S} + \mathbf{S} \cdot \mathbf{T}'(t) \cdot \mathbf{I} \end{aligned} \tag{88}$$

where $a$ and $\mathbf{T}'$, respectively, give the isotropic and anisotropic hyperfine interactions (we have here ignored the nuclear Zeeman energy and the off-diagonal parts of $\mathbf{I} \cdot \mathbf{S}$).

The electron spin therefore sees two fluctuating random local fields, one from the variations of $\mathbf{g}'$ and the other from $\mathbf{T}'$. However, these fluctuations are correlated because both tensors vary as the result of the same molecular motion, and the line width depends on the relative orientations of the $g$ and hyperfine tensors. To see this more clearly let us assume that the nucleus is quantized in the external field with a spin quantum number $m_I$. Then the perturbation $V(t)$ can be written in the form $(f_x S_x + f_y S_y + f_z S_z)$ with the three time-dependent operators

$$\begin{aligned} f_x(t) &= (\beta H_0 g'_{zx} + m_I T'_{zx}) \\ f_y(t) &= (\beta H_0 g'_{zy} + m_I T'_{zy}) \\ f_z(t) &= (\beta H_0 g'_{zz} + m_I T'_{zz}) \end{aligned} \tag{89}$$

The relaxation rate depends on the mean squared values of $f_x, f_y,$ and $f_z$, so that, for example, one contribution to $1/T_2$ is proportional to

$$\overline{|f_z(t)|^2} = \beta^2 H_0^2 \overline{(g'_{zz})^2} + 2\beta H_0 \overline{(g'_{zz} T'_{zz})} m_I + \overline{(T'_{zz})^2} m_I^2 \tag{90}$$

The combined effects of all these terms give a line width which varies systematically from one hyperfine line to another according to the formula

$$\frac{1}{T_2} = A + B m_I + C m_I^2 \tag{91}$$

The linear term $Bm_I$ causes asymmetric broadening, the outer lines being broader on one side of the spectrum than the other, while the $Cm_I^2$ part makes the outer lines broader than the inner ones. In the limit of fast tumbling the values of $B$ and $C$ are

$$B = \frac{7\beta H_0 \tau_c}{15\hbar^2}(g':T'), \qquad C = \frac{\tau_c}{15\hbar^2}(T':T') \tag{92}$$

where $(g':T')$ is the inner product of the $g'$ and $T'$ tensors and $(T':T')$ is the square of the hyperfine tensor.

One of the simplest examples to illustrate these effects is the vanadyl ion $VO^{++}$ which has axially symmetric $g$ and hyperfine tensors. The hyperfine structure from the V nucleus ($I = 7/2$) is shown in Fig. 11.11 and the line widths fit Eq. (91) quite well. More complicated line width variations occur in aromatic hydrocarbon radicals, and in some cases it has been possible to determine the anisotropic part of the $g$ tensor or the signs of the nuclear hyperfine interactions.

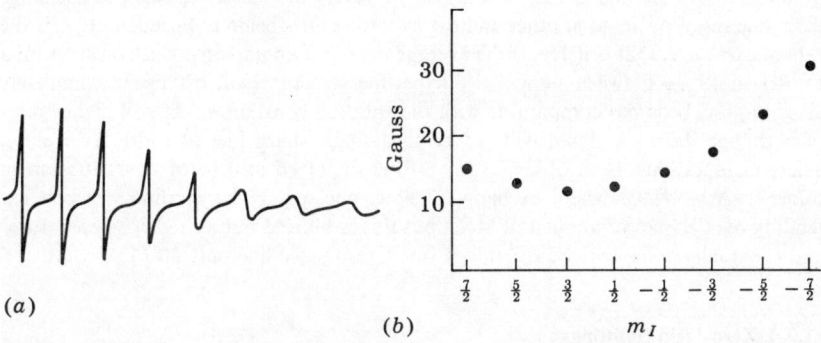

(a)

(b)

$m_I$

FIG. 11.11. (a) The hyperfine spectrum of $VO^{++}$ in solution at 24,300 Mc/s. If the line widths were uniform all lines would be the same height. (b) Line widths plotted against $m_I$.

### 11.7.3 Electron Spin Exchange

Exchange effects become important as the concentration of electron spins in solution rises, for radical-radical collisions occur more frequently, and there is more chance for the electronic wave functions to overlap. During such a collision the electrostatic coupling energy $hJ\mathbf{S}_1 \cdot \mathbf{S}_2$ may interchange the spins of the two radicals so that $\alpha\beta$ becomes $\beta\alpha$, and this can cause either broadening or narrowing of the electron resonance spectrum, depending on the circumstances. In the ideal case that there is no nuclear hyperfine structure and no magnetic dipolar force between different electron spins, exchange has no effect on the resonance spectrum. An interchange of two electron spins does not alter the total magnetic moment of the sample, which is the quantity one measures in an e.s.r. experiment. However, exchange forces do allow an electron *of a definite spin*, say $\alpha$, to jump rapidly from one molecule to another. They have the effect of averaging out electron dipole-dipole forces which would otherwise give a very broad spectral line. This is called "exchange narrowing" and is very important in paramagnetic crystals. For example, the radical diphenyl picryl hydrazyl, or DPPH, has a line width of only 3 gauss in the solid state

$$N-\dot{N}-\underset{NO_2}{\overset{NO_2}{\bigcirc}}-NO_2$$

compared with the value of 100 gauss expected from dipole-dipole broadening. Rapid exchange at a frequency of about $2 \times 10^{10}$ cps is responsible for the difference.

In solution, exchange is limited by the number of collisions between unpaired electrons and depends on the viscosity, so the effect on the dipole-dipole line width will be much smaller, but there are interesting changes in the hyperfine structure. If DPPH is dissolved in tetrahydrofuran from which all oxygen has been removed, at a concentration below $10^{-3}$ M, it shows a well resolved hyperfine structure with over a hundred lines. At higher concentrations, however, the electrons begin to exchange from one set of nuclei to another and the hyperfine lines begin to broaden. If $v_e$ is the rate of exchange, that is if $1/v_e$ is the average time that an unpaired electron stays on a DPPH molecule between jumps, the hyperfine structure will disappear completely as soon as $v_e$ becomes comparable with the splitting constant $a$. At still higher rates of exchange the e.s.r. signal will reduce to a single sharp line of width $\Delta v \approx a^2/v_e$. These changes have been observed for DPPH dissolved at 0.14 M concentration in toluene. At $-41°$C, where exchange is slow, one sees $N^{14}$ hyperfine structure extending over a span of about 150 Mc/s, but this is blurred out at higher temperatures and eventually collapses at 28°C into a single resonance line only 20 Mc/s wide.

### 11.7.4 Zero-Field Splittings

We now turn to consider the factors which affect the e.s.r. line width of transition metal ions in solution. In most cases the ion forms a complex with a definite molecular structure—octahedral, tetrahedral or planar—and the spin energy levels are therefore described by a spin Hamiltonian of the type we discussed in Chapter 10. Typically we expect to find anisotropic $g$ tensors, hyperfine tensors, and zero-field splittings which will all be modulated by the rotational motion and contribute to the line widths.

The anisotropy of the $g$ tensor is the main relaxation mechanism for ions of spin 1/2, and we can divide these compounds into three classes.

1. Ions without low-lying excited states, where anisotropy is small and the lines are so narrow that hyperfine structure is visible. Examples are $Mo(CN)_8^{---}$ and the dibenzene chromium ion.

2. Ions like $Ti^{+++}$, $V^{+4}$, and $Mn^{+6}$ where the orbital of the single $d$ electron is almost degenerate. Here distortions from octahedral symmetry, combined with spin-orbit coupling, give very short relaxation times. No resonance is visible in solution and in crystals it only appears at very low temperatures.

3. A few complexes, notably those containing $Cu^{++}$, where there are low-lying excited states and fairly short relaxation times. Resonance is visible in crystals at room temperature, but the lines in solution are often very broad.

For radicals of higher spin, zero-field splittings dominate the picture and often lead to very strong relaxation. The theory of the line shape for a triplet state is mathematically identical with that for *nuclear* spin-spin relaxation in the water molecule, except that the electronic zero-field tensor replaces the nuclear **D** tensor. The theory gives

$$\frac{1}{T_1} = \frac{\Delta^2}{10}\left\{\frac{2\tau_c}{1 + \omega_0^2\tau_c^2} + \frac{8\tau_c}{1 + 4\omega_0^2\tau_c^2}\right\}$$

$$\frac{1}{T_2} = \frac{\Delta^2}{20}\left\{6\tau_c + \frac{10\tau_c}{1 + \omega_0^2\tau_c^2} + \frac{4\tau_c}{1 + 4\omega_0^2\tau_c^2}\right\}$$

(93)

where

$$\Delta^2 = (X^2 + Y^2 + Z^2)/\hbar^2 \tag{94}$$

In practice the zero-field splittings of practically all molecules in triplet states are so large that the resonance is completely broadened away.

Several $Cr^{+++}$ complexes have rather small zero-field splittings of about 0.05 $cm^{-1}$ and do give observable e.s.r. spectra in solution, although the lines are very broad, with widths ranging from 30 to 400 gauss. The width reflects the deviation from regular octahedral symmetry, since the less symmetrical ions have larger values of $\Delta$ and give broader lines. For a quartet state the resonance line is actually composite because the "inner" transition ($m_S = 1/2$ to $m_S = -1/2$) and the two "outer" transitions ($\pm 1/2$ to $\pm 3/2$) have different widths and strengths. The theory shows that the spectrum consists of two superposed Lorentz lines:

$$\left(\frac{1}{T_2}\right)_{\frac{1}{2}-\frac{1}{2}} = \frac{3}{5}\Delta^2\left\{\frac{2\tau_c}{1 + \omega_0^2\tau_c^2} + \frac{2\tau_c}{1 + 4\omega_0^2\tau_c^2}\right\}$$

$$\left(\frac{1}{T_2}\right)_{\frac{1}{2}\frac{3}{2}} = \frac{3}{5}\Delta^2\left\{2\tau_c + \frac{2\tau_c}{1 + \omega_0^2\tau_c^2}\right\}$$

(95)

with strengths in the ratio $2:3$. In viscous solutions with large $\tau_c$, the outer line will be much broader than the inner.

Sextet states behave similarly, and a large number of $Mn^{++}$ and $Fe^{+++}$ complexes give rather broad electron resonance signals in solution. The theoretical treatment is rather complex and will not be given here.

## PROBLEMS

**1.** Show that an oscillating magnetic field $2H_1 \cos \omega t$ along the $x$ axis is equivalent to the superposition of two rotating fields in the $xy$ plane. Use this result to discuss the solution of Bloch's equations when both $H_1$ and $H_0$ are small. Prove that the absorption susceptibility $\chi''$ is proportional to

$$\frac{T_2}{1 + T_2^2(\omega - \omega_0)^2} + \frac{T_2}{1 + T_2^2(\omega + \omega_0)^2}$$

**2.** Use the Bloch equations without relaxation terms to study the precession of a proton spin in a steady field $H_0$ and a perpendicular $H_1$ field which rotates at angular frequency $-\omega$. Show that in the rotating frame the vector $\mathbf{M}'$ precesses about a steady effective field $\mathbf{H}_{eff}$ such that

$$\mathbf{H}_{eff} = (\mathbf{H}_0 + \omega/\gamma) + \mathbf{H}'_1$$

and show that if relaxation effects are small there is a steady state solution with $M'$ parallel to $H_{eff}$. How does the direction of $M'$ change as one sweeps the field $H_0$ slowly through resonance?

3.   Make a careful sketch of the saturated line shape function

$$\frac{T_2}{1 + T_2^2(\omega_0 - \omega)^2 + \gamma^2 H_1^2 T_1 T_2}$$

taking $\gamma^2 H_1^2 T_1 T_2$ equal to 0, 1, and 4 in turn. How does the width between points of maximum slope change with $H_1$? Describe how you might use these changes to measure $T_1$ and $T_2$.

4.* A proton spin is placed in a steady field $H_0$ parallel to the $z$ axis. Initially, at time $t = 0$ the spin is in a nonstationary state with the wave function

$$|\psi(0)\rangle = \frac{1}{\sqrt{2}}\{|\alpha\rangle + |\beta\rangle\}$$

Verify that this state is an eigenstate of $I_x$, with $I_x = 1/2$. Deduce from the time-dependent Schrödinger equation that at a later time the spin state is

$$|\psi(t)\rangle = \frac{1}{\sqrt{2}}\{|\alpha\rangle e^{(1/2)i\omega_0 t} + |\beta\rangle e^{-(1/2)i\omega_0 t}\}$$

The operator

$$I'_x(t) = I_x \cos\omega_0 t - I_y \sin\omega_0 t$$

represents the spin angular momentum $I'_x$ in the rotating frame. Prove that $|\psi(t)\rangle$ is an eigenfunction of $I'_x(t)$ and interpret this result physically.

5.   Each proton in a rotating hydrogen molecule at room temperature experiences fluctuating magnetic fields of two types: $H_1$ due to the rotational magnetic moment of the nuclei and electrons, and $H_2$ due to the dipolar coupling with the other nucleus. The root mean square values of $H_1$ and $H_2$ are 38 and 65 gauss. Use the results of Section 11.5.3 to estimate $T_1$. You should assume that $H_1$ and $H_2$ fluctuate independently, and that the correlation time $\tau_c$ is equal to the time between collisions which is $10^{-11}$ sec. (The observed relaxation time for ortho hydrogen is 0.015 sec.)

6.   Estimate nuclear spin relaxation rates $1/T_1$ due to the following effects:
   (a) Quadrupole relaxation of $D$ in heavy water, taking $e^2 qQ/h = 0.25$ Mc/s and $\tau_c = 2.7 \times 10^{-12}$ sec.
   (b) A fluorine nucleus relaxed by an anisotropic chemical shift in solution, when $(\sigma_\parallel - \sigma_\perp)$ $= 100$ ppm, the steady field $H_0 = 10,000$ gauss, and the correlation time is $10^{-12}$ sec.
   (c) A proton in water to which a $10^{-7}$ molar concentration of $Mn^{++}$ ions have been added. The spin $S = 5/2$, and the viscosity of water at $20°C$ is $\eta = 0.01$ poise.

7.   What is the electron resonance line width in gauss due to $g$ tensor anisotropy in the $Cu^{++}$ ion in aqueous solution? The magnetic field is 3,000 gauss; the $g$ values are $g_\parallel = 2.4$ and $g_\perp = 2.1$. You should estimate the correlation time from the viscosity of water, taking the radius of the hydrated copper ion as 5 Å.

8.   Imagine a ground state triplet molecule in solution, within which the two unpaired electrons are respectively localized at two points $A$ and $B$ which are a fixed distance $d$ apart. The line width of the e.s.r. spectrum in solution is 100 gauss ($g = 2$) when the correlation time is $10^{-12}$ sec. Estimate the zero-field splittings $D$ and $E$ and the distance between the electrons.

9.   How rapidly would a naphthalene molecule in its lowest triplet state need to tumble in solution in order to give a sharp e.s.r. spectrum 10 gauss wide between points of maximum slope? ($D = 0.1012$ cm$^{-1}$, $E = -0.0141$ cm$^{-1}$.)

# SUGGESTIONS FOR FURTHER READING

Slichter: Chapters 2 and 5. Bloch's equations. Density matrix theory of relaxation.

Pake: Chapter 2. Bloch's equations.

Abragam: Chapters 3 and 8. Exhaustive treatment of Bloch's equations and relaxation theory.

Bloembergen, Purcell, and Pound: *Phys. Rev.*, **73**: 679 (1948). Classic paper on spin relaxation in liquids.

Debye: *Polar Molecules* (New York: Dover Publications, Inc., 1945). Chapter 5. Brownian motion of the water molecule.

Gutowsky and Woessner: *Phys. Rev.*, **104**: 843 (1956). Nuclear relaxation by anisotropic chemical shift.

Pople, Schneider, and Bernstein: Chapter 9. Nuclear spin relaxation in solution.

Powles and Green: *Physics Letters*, **3**: 134 (1962). Spin-rotational nuclear relaxation in fluorobenzene.

Herbison-Evans and Richards: *Mol. Phys.*, **7**: 515 (1964). Quadrupole broadening of $N^{14}$ resonances.

Rogers and Pake: *J. Chem. Phys.*, **33**: 1107 (1960). Line width variations in e.s.r. of $VO^{++}$.

Carrington and Longuet-Higgins: *Mol. Phys.*, **5**: 447 (1962). Simplified theory of electron relaxation by anisotropic $g$ and hyperfine tensors.

Pake and Tuttle: *Phys. Rev. Letters*, **3**: 423 (1959). Exchange broadening in e.s.r. of DPPH in solution.

McGarvey: *J. Phys. Chem.*, **61**: 1232 (1957). Line widths for e.s.r. of transition metal ions in solution.

# CHAPTER 12

# THE STUDY
# OF MOLECULAR
# RATE PROCESSES

## 12·1  THE TIME SCALE OF MAGNETIC RESONANCE EXPERIMENTS

In the last chapter we found that the molecular motions which cause relaxation effects belonged to two different frequency ranges. Frequencies close to the resonance frequency $\omega_0$ were important for spin-lattice relaxation while lower frequencies close to zero led only to line broadening. Many of the relaxation mechanisms we described there led to a generally similar broadening of all the lines in the resonance spectrum, and therefore one could speak of a single, longitudinal or transverse, relaxation time for the spin system.

Many of the molecular processes which are of most interest to chemists involve chemical exchange of the nuclei or electrons. That is, the spin finds itself jumping at random from one environment to another one with a different spin Hamiltonian, and this alters the multiplet structure of complex resonance spectra in a selective way. Some lines broaden, but others stay sharp. Other groups of lines coalesce into a single line. These effects are almost always due to motions which have much lower frequencies than the Larmor frequency. Instead they are comparable with the frequency separation between resonance lines in a spectral multiplet; several megacycles for e.s.r. spectra, or a few cycles in n.m.r.

There are two important limiting cases called fast and slow exchange. In the first case the jumping frequency $\omega_e$ is much greater than the separation $\Delta\omega$ between two spectral lines which are affected by the motion. The result is a single line at the average frequency with a width

$$\delta\omega \approx \frac{(\Delta\omega)^2}{\omega_e}, \quad \text{fast exchange} \tag{1}$$

Here the spin is jumping so rapidly compared with the difference in resonance frequencies that it hardly loses phase from one time it is on a particular site to the next, and so it sees an average of the two environments.

At the other extreme the motion is very slow with $\omega_e \ll \Delta\omega$. Now both lines are resolved but each is broadened, with a width

$$\delta\omega \approx \omega_e, \quad \text{slow exchange} \tag{2}$$

204

This is a lifetime broadening, and once a spin leaves one site for another it loses all phase coherence by the time it returns.

A remarkable variety of line broadening effects have been studied. The results yield precise rate constants and activation energies for processes such as internal rotation, isomerization, electron exchange, and so on.

The theoretical analysis is often subtle and complicated, so we shall only treat one case in any detail, and then describe a few other examples to indicate the scope of the subject.

## 2·2 THE LINE SHAPE FOR A JUMPING SPIN

One of the simplest examples of chemical exchange is a nuclear spin which can jump at random between two sites $A$ and $B$ where it has different resonance frequencies $\omega_A$ and $\omega_B$. These sites may be in different molecules, as in a proton exchange reaction; or they may represent two distinct states of the same molecule, different conformations, or cis and trans isomers.

When the jumping rate is slow, the n.m.r. spectrum shows distinct $A$ and $B$ spectra, but if the rate is fast one finds just one spectrum characteristic of the average environment which the nucleus sees.

We shall calculate the line shape by adapting Bloch's equations to include the effects of exchange. At any given moment there are a certain number of nuclei on site $A$, with a total magnetic moment $\mathbf{M}_A$, while the other nuclei have a moment $\mathbf{M}_B$. In the absence of exchange, $\mathbf{M}_A$ and $\mathbf{M}_B$ would precess independently in the magnetic field and relax to equilibrium with their own relaxation times. Thus each vector, referred to rotating axes, obeys equations of the type

$$\frac{du}{dt} + \frac{u}{T_2} - (\omega_0 - \omega)v = 0$$

$$\frac{dv}{dt} + \frac{v}{T_2} + (\omega_0 - \omega)u = \gamma_N H_1 M_z$$

(3)

We are interested in the solution for a weak $H_1$ field, where $M_z$ is practically equal to $M_0$, and we now introduce a more compact notation in terms of a complex magnetic moment $\hat{M}$ and a complex frequency $\hat{\omega}_0$. Thus (3) becomes

$$\hat{M} = u + iv \tag{4}$$

$$\hat{\omega}_0 = \omega_0 - i/T_2 \tag{5}$$

$$\frac{d\hat{M}}{dt} + i(\hat{\omega}_0 - \omega)\hat{M} = i\gamma_N H_1 M_0 \tag{6}$$

Without exchange, then, the Bloch equations for $\hat{M}_A$ and $\hat{M}_B$ would be

$$\frac{d\hat{M}_A}{dt} + i(\hat{\omega}_A - \omega)\hat{M}_A = i\gamma_N H_1 M_{0A}$$

$$\frac{d\hat{M}_B}{dt} + i(\hat{\omega}_B - \omega)\hat{M}_B = i\gamma_N H_1 M_{0B}.$$

(7)

Because of exchange, nuclei from site $A$ will go over to site $B$, and vice versa. We do not know precisely what takes place during the act of exchange, but it is reasonable to assume that the jump happens very suddenly and takes a time much less than the Larmor period. If so, one can neglect any spin precession during the jump, and then the nucleus does not alter its spin direction while it is in transit. Further we suppose that any nucleus which is on site $A$ has a definite time-proportional transition probability $P_{AB}\delta t$ that it jumps to site $B$ in any time interval $\delta t$. Similarly there is a probability $P_{BA}\delta t$ for the reverse jump. The change of magnetization due to jumping is then described by the equations

$$\frac{d\mathbf{M}_A}{dt} = -P_{AB}\mathbf{M}_A + P_{BA}\mathbf{M}_B$$

$$\frac{d\mathbf{M}_B}{dt} = -P_{BA}\mathbf{M}_B + P_{AB}\mathbf{M}_A$$

(8)

$P_{AB}$ and $P_{BA}$ are not necessarily equal, and in equilibrium the fractions of nuclei $f_A$ and $f_B$ on the two sites will be

$$f_A = \frac{P_{BA}}{P_{AB} + P_{BA}}, \qquad f_B = \frac{P_{AB}}{P_{AB} + P_{BA}}$$

(9)

that is,

$$M_{0A} = f_A M_0, \qquad M_{0B} = f_B M_0$$

(10)

We now combine (7), (8), and (10) into a modified set of Bloch equations for the exchange problem.

$$\frac{d\hat{M}_A}{dt} + i(\hat{\omega}_A - \omega)\hat{M}_A + P_{AB}\hat{M}_A - P_{BA}\hat{M}_B = if_A(\gamma_N H_1 M_0)$$

$$\frac{d\hat{M}_B}{dt} + i(\hat{\omega}_B - \omega)\hat{M}_B + P_{BA}\hat{M}_B - P_{AB}\hat{M}_A = if_B(\gamma_N H_1 M_0)$$

(9)

The steady state values of $\hat{M}_A$ and $\hat{M}_B$ clearly satisfy the linear equations

$$[(\omega - \hat{\omega}_A) + iP_{AB}]\hat{M}_A - iP_{BA}\hat{M}_B = -f_A(\gamma_N H_1 M_0)$$

$$-iP_{AB}\hat{M}_A + [(\omega - \hat{\omega}_B) + iP_{BA}]\hat{M}_B = -f_B(\gamma_N H_1 M_0)$$

(10)

and it is easy to calculate the total magnetic moment. This is given by

$$\hat{M} = \hat{M}_A + \hat{M}_B$$

$$= -\gamma_N H_1 M_0 \frac{f_A(\omega - \hat{\omega}_B) + f_B(\omega - \hat{\omega}_A) + i(P_{AB} + P_{BA})}{[(\omega - \hat{\omega}_A) + iP_{AB}][(\omega - \hat{\omega}_B) + iP_{BA}] + P_{AB}P_{BA}}$$

(11)

and the intensity of the resonance absorption is proportional to $v$, or the imaginary part of $\hat{M}$.

Rather than pursue this complicated calculation to its general solution let us simplify the problem by making two new assumptions.

1. The line broadening due to processes other than exchange is negligible, so that $1/T_{2A} = 1/T_{2B} = 0$.
2. The two sites are equally populated, with $P_{AB} = P_{BA} = P$.

Now Eq. (11) becomes

$$\hat{M} = -\gamma_N H_1 M_0 \frac{(\omega - \bar{\omega}) + 2iP}{(\omega - \omega_A)(\omega - \omega_B) + 2iP(\omega - \bar{\omega})} \tag{12}$$

where $\bar{\omega}$ is the average resonance frequency $\frac{1}{2}(\omega_A + \omega_B)$. The intensity of absorption is the imaginary part

$$v = \frac{1}{2} \gamma_N H_1 M_0 \frac{P(\omega_A - \omega_B)^2}{(\omega - \omega_A)^2(\omega - \omega_B)^2 + 4P^2(\omega - \bar{\omega})^2} \tag{13}$$

The form of the line shape depends on just one dimensionless quantity, the ratio of $(\omega_A - \omega_B)$ to the jumping rate, and typical curves are shown in Fig. 12.1. When $P$ is

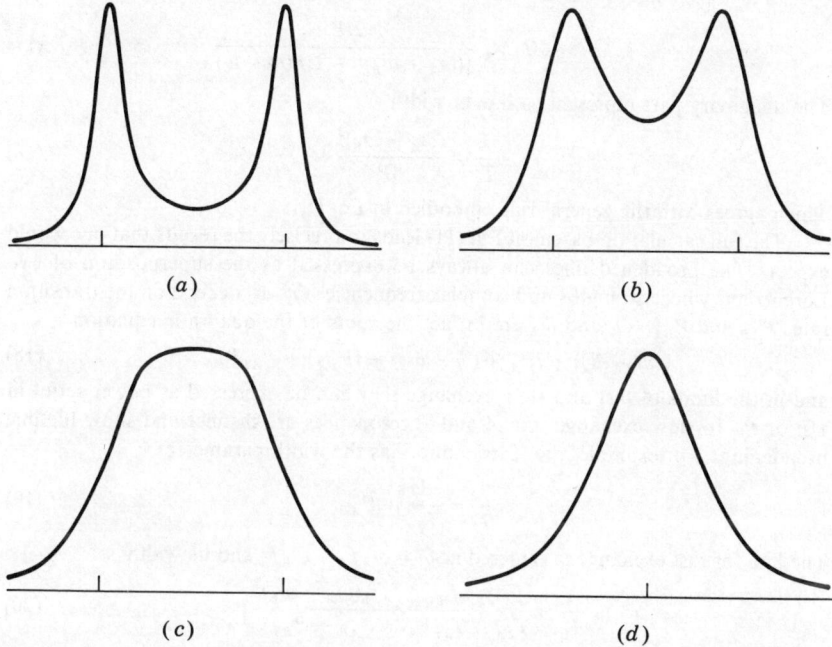

FIG. 12.1. Line shapes for a jumping spin. The ratio of the jumping rate $P$ to the line splitting $(\omega_A - \omega_B)$ takes the values: (a) 1/10; (b) 1/5; (c) $1/2\sqrt{2}$; (d) 1/2 The scale is adjusted so that each spectrum has the same maximum height.

small both lines are sharp and well separated, but as the jumping rate increases the two peaks broaden and move towards one another until they collapse into a single flat-topped line. The shifted positions of the two peaks are found by seeking minima of the denominator of Eq. (13). They occur symmetrically on either side of the center $\bar{\omega}$ and the separation between them is

$$\Delta\omega = \sqrt{(\omega_A - \omega_B)^2 - 8P^2} \tag{14}$$

If $P$ exceeds $(\omega_A - \omega_B)/2\sqrt{2}$ there is just one broad line. Higher exchange rates lead to a sharp *exchanged-narrowed* line centered at $\bar{\omega}$.

The two limiting cases deserve further study. For fast exchange consider the absorption spectrum (13) when $\omega \approx \omega_A$ and $P$ is small

$$v = \frac{1}{2} \gamma_N H_1 M_0 \frac{P}{(\omega - \omega_A)^2 + P^2} \tag{15}$$

This represents a normal Lorentz line with a width given by

$$\frac{1}{T_2} = P \tag{16}$$

The line therefore has "lifetime broadening," since $1/P = T_2$ is the average time a spin stays on site $A$ between jumps. There is also a small frequency shift of amount $2P^2/(\omega_B - \omega_A)$ towards the other line.

In fast exchange we set $\omega \approx \bar{\omega}$ and find from (12) that

$$\hat{M} \approx \gamma_N H_1 M_0 \frac{2iP}{\frac{1}{4}(\omega_A - \omega_B)^2 - 2iP(\omega - \bar{\omega})} \tag{17}$$

The imaginary part represents a line of width

$$\frac{1}{T_2} = \frac{(\omega_A - \omega_B)^2}{8P} \tag{18}$$

which agrees with the general rule embodied in Eq. (1).

The full calculation based on Eq. (11) leads to precisely the results that one would expect. The broadened lines can always be expressed as the superposition of two Lorentzians whose strengths and complex frequencies $\hat{\omega}_1$, $\hat{\omega}_2$ depend on the transition rates $P_{AB}$ and $P_{BA}$. $\hat{\omega}_1$ and $\hat{\omega}_2$ are in fact the roots of the quadratic equation

$$[(\hat{\omega} - \hat{\omega}_A) + iP_{AB}][(\hat{\omega} - \hat{\omega}_B) + iP_{BA}] + P_{AB}P_{BA} = 0 \tag{18}$$

and in the limits of fast and slow exchange they can be expressed as power series in $1/P$ or $P$. In slow exchange, the $A$ and $B$ resonances are distinct and show lifetime broadening. For example, the $A$ resonance has the width parameter

$$\frac{1}{T_2} = \frac{1}{T_{2A}} + P_{AB} \tag{19}$$

The line for fast exchange is centered at $\bar{\omega} = \omega_A f_A + \omega_B f_B$ and its width

$$\frac{1}{T_2} = \frac{f_A}{T_{2A}} + \frac{f_B}{T_{2B}} + \frac{f_A f_B (\omega_A - \omega_B)^2}{P_{AB} + P_{BA}} \tag{20}$$

is the average of the natural widths of $A$ and $B$ plus an extra part due to incomplete averaging of the two resonance frequencies.

## 12·3 CHEMICAL EXCHANGE EFFECTS IN N.M.R. SPECTRA

### 12.3.1 Hindered Internal Rotation

The proton resonance spectrum of N,N-dimethylnitrosamine in the temperature range 25° to 200°C shows striking changes caused by hindered internal rotation about the N—N bond.

The proton spectrum is particularly simple because the $N^{14}$ nuclei are strongly relaxed by electric quadrupole coupling and give no spin-spin splitting, neither is there any splitting from coupling between different methyl groups. At room temperature the n.m.r. spectrum at a frequency of 40 Mc/s shows two discrete lines, 26 cycles apart, from the cis and trans methyl protons. On heating to 140°C internal rotation begins to broaden the lines and draw the maxima together, while at higher temperatures the spectrum passes through the sequence of changes shown in Fig. 12.2. The jumping

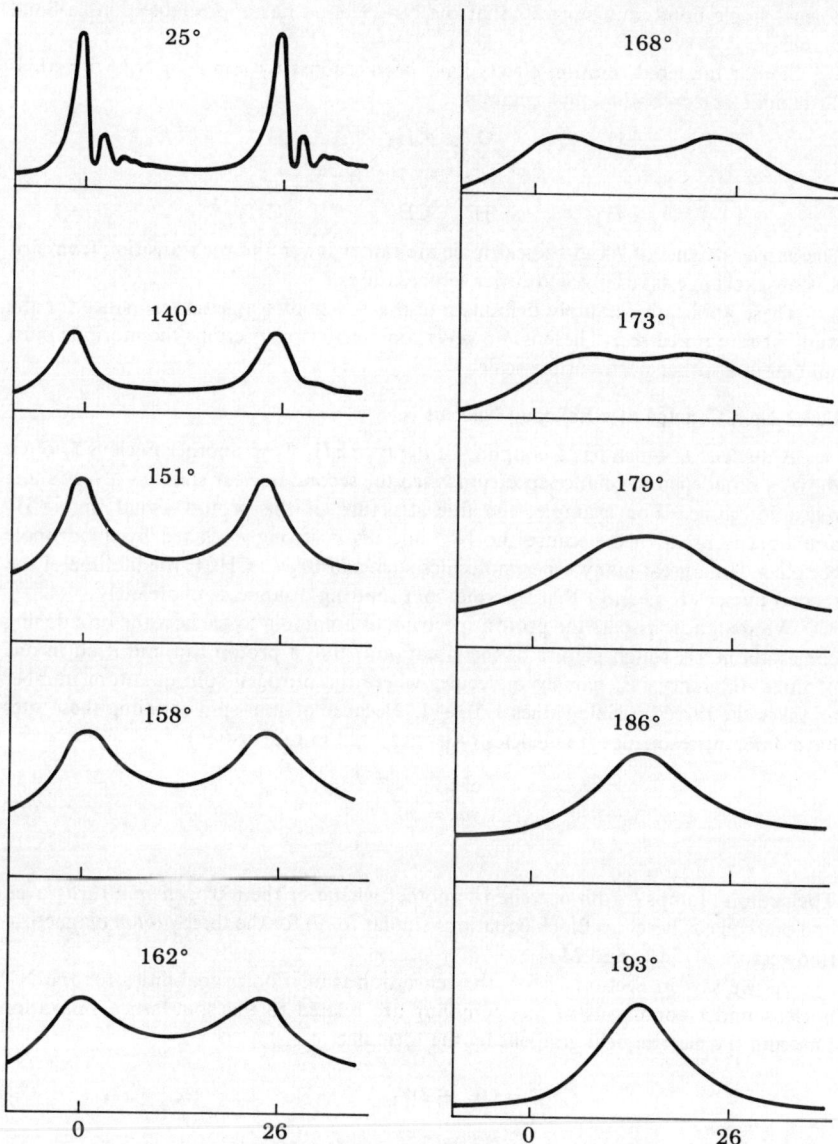

FIG. 12.2. The 40 Mc/s proton resonance spectrum of N,N-dimethylnitrosamine.

rate $P$ can be estimated from Eq. (14), and the critical stage where the two lines merge is reached at about 179°. At this point $P = \pi(v_A - v_B)/\sqrt{2} = 58$ cycles/sec.

The graph of log $P$ against $1/T$ is a good straight line and shows that in this temperature range the jumping rate varies as

$$P = P_0 e^{-E/kT} \tag{21}$$

The effective barrier to internal rotation is $E = 23$ kcal/mole, which is rather high for a normal single bond, and suggests that the N—N bond has appreciable double bond character.

Similar hindered rotation effects have been studied by n.m.r. in N,N-dimethyl-formamide and N,N-dimethylacetamide

The barrier heights of 7 and 12 kcal/mole are rather lower and the transition from fast to slow exchange takes place at lower temperatures.

These applications amply demonstrate the power of magnetic resonance for the study of rate processes. There is, however, one serious restriction: the motions must fall within a rather narrow time scale.

### 12.3.2 Spin Coupled to a Relaxing Nucleus

A nucleus $I_1$ which has a coupling of the type $\hbar J \mathbf{I}_1 \cdot \mathbf{I}_2$ to another nucleus $I_2$ often shows a broadened resonance spectrum when the second nuclear spin has a very short relaxation time. For example, the fine structure of the proton signal in $N^{14}H_3$ ammonia is broadened because the $N^{14}$ nucleus is strongly relaxed by quadrupole coupling. In a great many other molecules such as $PBr_5$ or $CHCl_3$ the lifetime of the second nucleus is so short that the spin-spin splitting disappears completely.

We shall now discuss the proton spectrum of ammonia to see how the broadening comes about. A rough picture of the situation is that a proton can find itself in one of three different sites, namely molecules where the nitrogen spin quantum number $m_I$ takes the three possible values 1, 0, −1. Because of spin-spin coupling these sites have different resonance frequencies ($J$ is measured in radians/sec).

$$\begin{aligned} \omega_1 &= \omega_0 + J \\ \omega_0 &= \omega_0 \\ \omega_{-1} &= \omega_0 - J \end{aligned} \tag{22}$$

The proton "jumps" from one site to another whenever the nitrogen spin turns over, and one can easily set up Bloch equations similar to (9) for the three *proton* magnetization vectors $\hat{M}_1$, $\hat{M}_0$, and $\hat{M}_{-1}$.

As we saw in Section 11.6.4, the relaxation transition probabilities for the $N^{14}$ nucleus under conditions of fast tumbling are related to the spin-lattice relaxation time and the electric field gradient by the formulae

$$\frac{1}{T_1} = (W_1 + 2W_2) \tag{23}$$

$$W_1 = \frac{1}{2} W_2 = 3e^4 q^2 Q^2 \tau_c / 40 \hbar^2 \tag{24}$$

The calculated line shape then depends on a single dimensionless parameter

$$\eta = 5JT_1 \qquad (25)$$

and Fig. 12.3 shows how the spectrum varies. The $1:1:1$ nitrogen splitting of the proton resonance first becomes broadened, the widths of the outer and inner lines being, respectively, $3/5T_1$ and $2/5T_1$. As $T_1$ decreases the three lines collapse into one and finally leave a sharp exchange-narrowed central line.

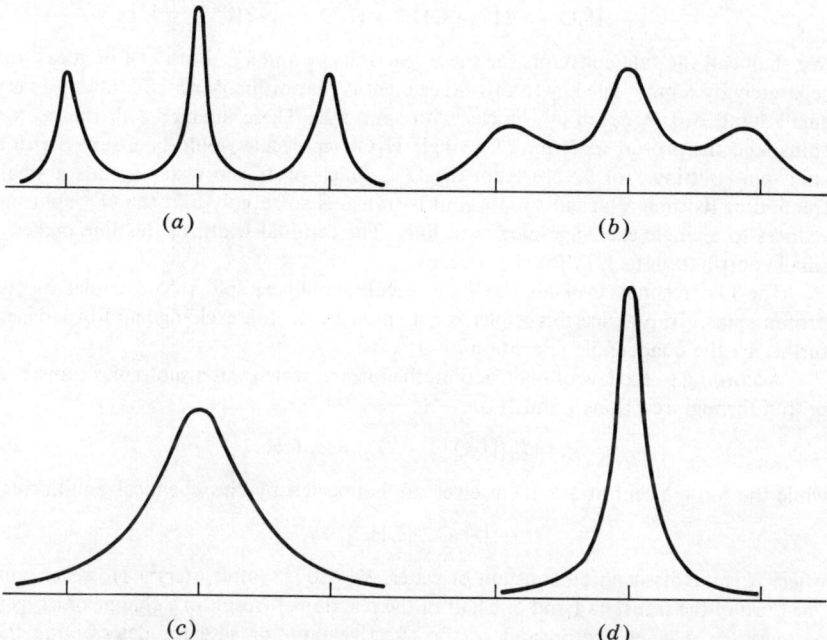

FIG. 12.3. Calculated line shapes for a proton coupled with a relaxing $N^{14}$ nucleus. The curves correspond to (a) $\eta^2 = 1000$; (b) $\eta^2 = 100$; (c) $\eta^2 = 10$; (d) $\eta^2 = 1$.

The spin of the second nucleus also may be altered by exchanging the atom for a fresh one from another molecule, bringing in a new nucleus whose spin state is completely random. As far as the first nucleus is concerned this process only reveals itself as a modulation of the $I_1 \cdot I_2$ coupling and produces the same kind of line broadening. An example is the HF molecule in liquid HF containing a few parts in $10^4$ of water. The HF protons exchange very fast with protons in $H_3O^+$ ions and this gives appreciable changes in the $F^{19}$ relaxation times. When exchange is very fast, comparable to the difference between the proton and fluorine resonance frequencies, the $I_1^+I_2^-$ and $I_1^-I_2^+$ terms in the coupling begin to produce *spin-lattice* relaxation and the line width is twice as large as one would infer from the Bloch theory. A thorough study has been made of relaxation and saturation processes in HF, and the temperature dependence of the proton exchange rate yields an activation energy of about 2 kcal/mole for proton exchange during collisions between HF molecules and $H_3O^+$ ions.

### 12.3.3 Proton Exchange Reactions

The n.m.r. spectra of acids and bases in solution with water often show line width effects due to proton exchange. The rate of transfer usually depends on temperature, pH, and other factors.

Water itself affords a good example of these processes. There are at least two fast proton transfer reactions which contribute to the high mobility of hydrogen and hydroxyl ions.

$$H_2O + H_3O^+ \rightarrow H_3O^+ + H_2O \qquad \text{(I)}$$

$$H_2O + OH^- \rightarrow OH^- + H_2O \qquad \text{(II)}$$

We shall call the rate constants for these reactions $k_1$ and $k_2$. Both can be measured separately by n.m.r. The key to this experiment is that ordinary water contains a very small number (0.037%) of $O^{17}$ nuclei with spin 5/2. These interact with the proton spins, and the proton spectrum of a single $H_2O^{17}$ molecule would be a sextet with a spin-spin splitting $J$ of 92 cps from the $O^{17}$. Each proton in water spends a small fraction of its time attached to $O^{17}$, but it transfers so rapidly that the $O^{17}$ splitting reduces to a single exchange-narrowed line. The residual width of this line makes a small contribution to $1/T_2$ for the protons.

The $O^{17}$ resonance of an $H_2O^{17}$ molecule would be split into a triplet by the proton spins. In practice this triplet is collapsed by proton exchange and broadened further by the quadrupole relaxation of $O^{17}$.

According to the law of mass action, the rates at which water molecules transfer a proton through reactions I and II are

$$P_1 = k_1[H_3O^+] \qquad P_2 = k_2[OH^-] \qquad (26)$$

while the ion concentrations themselves are connected by the chemical equilibrium

$$[H_3O^+][OH^-] = K \qquad (27)$$

where $K$ is the dissociation constant of water, $K = 10^{-14}$ (mole/liter)$^2$. However only one third of the transfers I and one half of the reactions II result in a change of oxygen partners for a *given* proton and so the effective jumping rate for determining the additional proton resonance line width is

$$P_H = \frac{1}{3} P_1 + \frac{1}{2} P_2 = \frac{1}{3} k_1[H^+] + \frac{k_2 K}{2[H^+]} \qquad (28)$$

Similarly the $O^{17}$ line width is determined by the rate

$$P_O = \frac{2}{3} P_1 + P_2 \qquad (29)$$

at which *either* proton transfers out of the water molecule. The line broadening therefore depends strongly on pH and for both nuclei it has a strong maximum near pH = 7. The proton width naturally increases with higher $O^{17}$ concentrations too.

The effects of exchange can be separated from those of normal relaxation because exchange alters $T_2$, but not $T_1$. In the absence of any exchange $T_1$ and $T_2$ should normally be equal, and so the quantity $T^*$ defined as

$$\frac{1}{T^*} = \frac{1}{T_2} - \frac{1}{T_1} \qquad (30)$$

represents a measure of the extra line broadening. The theoretical analysis is complicated, but the experimental curves of $1/T^*$ for $O^{17}$ and H as a function of pH at 25°C eventually lead to the rate constants

$$k_1 = 10.6 \times 10^9$$

$$k_2 = 3.8 \times 10^9$$

Numerous other proton transfer reactions have been studied by n.m.r., including the exchange of hydroxyl protons in methyl or ethyl alcohol, NH protons of methylammonium chloride in water, and many others. In each case the exchange shows itself by selectively broadening out part of the spin-spin splitting in the n.m.r. spectrum, much as $N^{14}$ quadrupole relaxation broadens the proton spectrum of ammonia.

One rather different situation occurs in acetylacetone, which can exist in either *keto* or *enol* forms,

At normal temperatures in pure liquid acetylacetone interconversion is slow, and the n.m.r. spectrum at 40 Mc/s shows two distinct sets of lines from the two forms. The hydrogen bonded OH resonance appears at low field about 10 ppm below the $CH_2$ protons. Since the shift corresponds to a frequency separation of 400 cps, the rate of exchange must be much lower than this. The two methyl groups in the enol form are however equivalent, so that the exchange between O—H⋯O and O⋯H—O *within* the hydrogen bond must be fast. Under normal conditions one can then measure equilibrium concentrations of the two forms, studying the effect of different solvents, and so on.

A dramatic change takes place on adding acetic acid. Now exchange takes place between the hydroxyl protons of the enol form and the acid, and in the temperature range 0° to 100°C the n.m.r. spectrum runs through the whole sequence of stages between slow and fast exchange. Diethylamine also promotes rapid proton exchange. In addition to broadening the OH signals, it speeds up the interconversion between keto and enol forms. These fast proton transfers are presumably due to strong hydrogen bonding between the acetylacetone and diethylamine molecules.

## 12.4 RATE EFFECTS IN E.S.R. SPECTRA

### 12.4.1 Modulation of the Hyperfine Coupling

There are many radicals which show striking line width effects because of some molecular motion which modulates one or more isotropic hyperfine splittings. We can illustrate the principles by considering the vinyl radical, which can exist in two equivalent forms.

(a)                                    (b)

The e.s.r. spectrum of vinyl prepared by photolysis of acetylene in solid argon, at 4°K, shows hyperfine splitting from three inequivalent protons. The trans $\beta$ coupling is 192 Mc/s and the cis is 96 Mc/s. The smallest coupling of 44 Mc/s comes from the $\alpha$ proton. At room temperature the molecule inverts, with the result that each $\beta$ proton coupling is modulated between the extreme values 192 and 96 Mc/s. The effect on the hyperfine spectrum may be understood by reference to Fig. 12.4. If we restrict attention

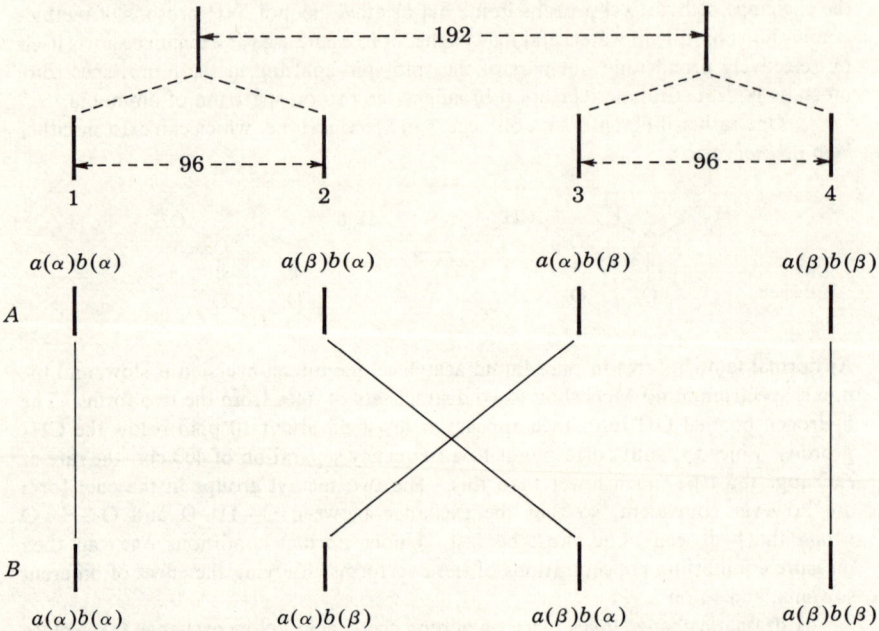

FIG. 12.4. Hyperfine structure of the vinyl radical.

to the $\beta$ protons there are four hyperfine lines from the radical in configuration A. These may be labeled 1, 2, 3, 4 from left to right, and the corresponding proton spin orientations are $a(\alpha)b(\alpha)$, $a(\beta)b(\alpha)$, $a(\alpha)b(\beta)$, $a(\beta)b(\beta)$. On the other hand for configuration B the same spin states correspond to different lines, the order now being 1, 3, 2, 4. When the CH group inverts, the proton spins do not change, and so the two inner lines exchange places, while the outer lines do not shift at all. If the unpaired electron happens to be on a radical with nuclear spins $\alpha\alpha$ or $\beta\beta$, its resonance frequency is the same in both forms and the corresponding hyperfine lines are sharp. However if the spins are $\alpha\beta$ or $\beta\alpha$, the unpaired electron sees different hyperfine fields in forms $A$ and $B$. The two forms behave as different "sites" for the unpaired electron, and the inner hyperfine lines therefore broaden in the manner we have described in Section 12.2. When the inversion rate is comparable to the frequency separation of 96 Mc/s between the two lines they become so broad as to become almost undetectable. This is what is observed in vinyl radicals prepared by electron bombardment of liquid ethylene at $-180°C$. The $\beta$ proton hyperfine spectrum now consists of just two lines, the outer ones of Fig. 12.4, with a separation of $192 + 96 = 288$ Mc/s. At still higher rates of exchange the inner lines would collapse into a single exchange-narrowed sharp line twice as intense as either of the outer ones. We would then describe the $\beta$ protons as equivalent, with an average coupling constant of 144 Mc/s. It is not difficult to calculate the line shape as a function of the inversion rate and to study the temperature dependence. The estimated minimum value for the barrier to inversion in the vinyl radical is 2 kcal/mole.

A rather different example is the naphthazarin semiquinone positive ion in solution.

cis　　　　　　　　　　　trans

One might expect intramolecular hydrogen bonding to be important in this radical, and the electron resonance spectrum suggests that this is indeed the case. The protons between the oxygen atoms give a hyperfine splitting of 0.7 gauss, while the other two hydroxyl proton splittings are 1.8 gauss. The radical can exist in four different hydrogen bonded conformations, two cis and two trans, and each transition from cis to trans or vice versa exchanges the coupling constants of two hydroxyl protons. The electron resonance spectrum shows lines of markedly different widths which are strongly temperature dependent, and the detailed changes in the spectrum are again accounted for most successfully by the modified Bloch theory. The activation energy for switching the hydrogen bond from one side to the other is estimated to be 4 kcal/mole.

### 12.4.2 Ion-Pairing in Solution

When aromatic hydrocarbons are reduced by alkali metals to form their radical ions, the electron resonance spectra sometimes indicate that long-lived ion pairs are formed, since the organic radical shows hyperfine splitting from the metal nucleus. Pyracene anion

$$H_2C\text{---}CH_2$$

$$H_2C\text{---}CH_2$$

is a good illustration, and also shows an unusual type of line broadening due to changes in the structure of the bound ion pairs.

If pyracene is reduced with potassium at $-70°C$ in dimethoxyethane one obtains the spectrum of the free ion. It shows hyperfine splittings of 6.58 gauss from eight equivalent methylene protons, and 1.58 gauss from the four ring protons. Reduction with sodium in methyltetrahydrofuran at $-83°C$ yields quite a different spectrum due to the $(pyracene)^-Na^+$ ion pair. There is now a splitting of 0.17 gauss from the $Na^{23}$ nucleus, and the methylene protons are no longer equivalent, but produce quintet splittings of 6.93 and 6.37 gauss. Presumably the metal ion sits close to the two methylene groups at one side of the molecule. Finally reduction with potassium in tetrahydrofuran at $-30°C$ yields a spectrum in which the nine groups of lines arising from the dominant methylene splitting show strong line width alternation (Fig. 12.5).

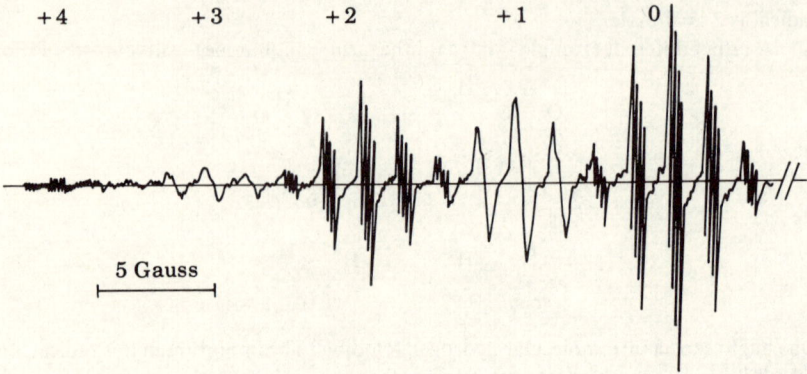

FIG. 12.5. Line-width alternation for (pyracene)$^-$K$^+$ ion pairs in tetrahydrofuran at $-30°C$.

This cannot be due to exchange of the metal ion from one pyracene to another, because the e.s.r. spectrum still shows a small unbroadened $K^{39}$ hyperfine splitting. Instead the metal ion probably jumps from one side of the pyracene ion to the other, modulating the methylene proton splitting constants as it goes.

This motion can produce alternating line widths in the following way. Suppose that the methylene protons divide into two sets of four having hyperfine splittings $a_1(t)$, $a_2(t)$ and total $m_I$ quantum numbers $M_1$ and $M_2$. The frequency of the corresponding hyperfine line is

$$v = v_0 + a_1(t)M_1 + a_2(t)M_2 \tag{31}$$

As the metal ion jumps, it interchanges the values of $a_1$ and $a_2$ in an irregular fashion and the result is a broad line centered at the average frequency $v_0 + a(M_1 + M_2)$ whose width is given by Eq. (18):

$$\frac{1}{T_2} = \frac{\pi^2(a_1 - a_2)^2(M_1 - M_2)^2}{2P} \tag{32}$$

$P$ being the jumping rate. Hyperfine lines with $M_1 = M_2$ are not broadened at all, whereas the remaining ones are too broad to observe. Since $M_1 = M_2$ implies an even value for the total methylene quantum number $M$, the even $M$ lines are sharper than the odd ones, so that the line widths in the e.s.r. spectrum do indeed alternate.

### 12.4.3 Electron Transfer Reactions

One of the earliest rate processes to be studied by electron spin resonance was electron transfer from a radical ion to a neutral molecule in solution. For example, if a small amount of naphthalene is added to a solution containing naphthalene negative ions the unpaired electron can jump from one naphthalene molecule to another, and the hyperfine structure of the electron resonance spectrum is broadened. Consider first an odd electron which has hyperfine coupling to just one proton. At first it is on molecule $A$ where the proton has a definite spin, say $\alpha$. Then the electron jumps to a new molecule $B$ where the spin of the $B$ proton is equally likely to be $\alpha$ or $\beta$. If it is $\alpha$ the hyperfine energy $aI_zS_z$ is unchanged and there is no contribution to the line width; but if it is $\beta$ there is a shift of $\pm a$ in the electron resonance frequency, and line broadening may occur. As usual there are two limiting cases, depending on $\tau_e$, the mean time between electron jumps. In the slow exchange limit each hyperfine line has a lifetime broadening equal to $1/\tau_e$, while fast jumping gives a single exchange narrowed line of width proportional to $a^2\tau_e$.

In the naphthalene ion there are $2^8 = 256$ possible proton spin states which can be grouped into twenty-five distinct hyperfine lines (Fig. 6.5). An electron spin on a given ion therefore finds itself coupled to a definite one of twenty-five proton spin configurations which correspond to different electron resonance frequencies. At each jump the proton spin state is altered at random and so afterwards the electron may see either the same proton configuration or a different one. The average lifetime of a hyperfine energy state is shortest for configurations which belong to the outer lines of the spectrum, and so these lines are more broadened than the inner ones.

The normal line width of the naphthalene negative ion is about 0.1 Mc/s; so transfer rates as low as $10^5$ per sec lead to perceptible broadening. Measurements in the slow exchange region gave bimolecular rate constants for the transfer reaction ranging from $10^7$ to $10^9$ liter mole$^{-1}$ sec$^{-1}$ with different metal ions and solvents.

A much faster electron transfer occurs between benzophenone anions and neutral benzophenone in the presence of sodium. Here the sodium ion plays a particularly interesting part in the reaction. The benzophenone radical forms ion pairs with sodium, and the e.s.r. spectrum shows $Na^{23}$ hyperfine structure in addition to the usual ring proton splitting. If neutral benzophenone is added to the solution the proton hyperfine structure disappears but the sodium splitting stays sharp. This means that the unpaired electron carries the sodium cation along with it while jumping from one benzophenone ion to another.

Electron transfer also sometimes occurs between two halves of the same molecule. For instance it can jump from one benzene ring to another in the paracyclophane anions.

When the rings are separated by more than two methylene groups the electron is effectively trapped on one side or the other and interacts with only four ring protons, but if there are only two the jumping rate is sufficiently fast that the electron interacts with all eight, giving a 9-line hyperfine spectrum with a splitting of 2.7 gauss.

## 12.4.4 Time-Dependent Changes in the Direction of Spin Quantization

So far we have discussed a variety of rate processes in which the motions, whether they are simple changes of conformation or involve actual migration of spins from one molecule to another, produce their main effects by modulating the spin energy levels. We now turn to another kind of process where the changes in the spin Hamiltonian necessitate changes in the direction of spin quantization as well as the energy. The theoretical analysis of these effects is rather difficult, since the modified Bloch equations cannot generally be adapted to describe the motion of the spins, and one must use more advanced density matrix methods. Here we shall describe two examples which illustrate the principles, but make no attempt to calculate the line shape.

The e.s.r. spectrum of copper fluorosilicate present as an impurity in $ZnSiF_6,6H_2O$ shows a striking change at low temperatures. The $Cu^{++}$ ion with nine $d$ electrons is surrounded here by an octahedron of water molecules, but the ion is not stable as a regular octahedron and prefers to take up a distorted shape, elongated along one four-fold axis of the octahedron and compressed along the other two perpendicular directions. At temperatures below 12°K each ion is frozen into a particular distorted form and has an anisotropic spin Hamiltonian of the type

$$\mathscr{H} = g_{\parallel}\beta H_z S_z + g_{\perp}\beta(H_X S_X + H_Y S_Y) + A I_z S_z + B(I_X S_X + I_Y S_Y) \tag{33}$$

The symmetry axis $Z$ may be along either of the $x$, $y$, or $z$ axes of the octahedron, so that three different types of ion are seen in the unit cell. The $g$ values are $g_{\parallel} = 2.46$, $g_{\perp} = 2.10$, and the copper hyperfine splittings are $A = 330$ Mc/s, $B < 90$ Mc/s. Raising the temperature causes the ions to tunnel from one configuration to another at an increasing rate, until at 50°K the spectrum is completely isotropic, being described by an average $g$ value of 2.23 and a hyperfine splitting of about 80 Mc/s. The ion must still be distorted since a regular octahedral complex would have an isotropic $g$ value of 2.00 instead of 2.23.

Another example is provided by the triplet states of benzene or triptycene which we have discussed in Chapter 8. In these molecules a triplet exciton jumps between three sites which have different zero-field splitting tensors. The three tensors have identical principal values $X$, $Y$, $Z$, but the $X$ and $Y$ principal directions differ by rotations through 120° (Fig. 12.6). Referred to a common axis of quantization the three zero-field Hamiltonians are represented by the matrices

$$\mathscr{H}_1 = \begin{bmatrix} X & 0 & 0 \\ 0 & Y & 0 \\ 0 & 0 & Z \end{bmatrix} \qquad \mathscr{H}_2 = \begin{bmatrix} U & -W & 0 \\ -W & V & 0 \\ 0 & 0 & Z \end{bmatrix} \qquad \mathscr{H}_3 = \begin{bmatrix} U & W & 0 \\ W & V & 0 \\ 0 & 0 & Z \end{bmatrix}$$

$$\tag{34}$$

where

$$U = \tfrac{1}{4}(X + 3Y), \qquad V = \tfrac{1}{4}(3X + Y), \qquad W = \tfrac{1}{4}\sqrt{3}(X - Y) \tag{35}$$

When the jumping rate is slow, distinct spectra will be seen, corresponding to the three separate sites, with a small lifetime broadening. However fast exchange leads to a completely different spectrum characteristic of the average spin Hamiltonian $\overline{\mathscr{H}} = \tfrac{1}{3}(\mathscr{H}_1 + \mathscr{H}_2 + \mathscr{H}_3)$. The new Hamiltonian is represented by the matrix

$$\overline{\mathscr{H}} = \begin{bmatrix} \tfrac{1}{2}(X + Y) & 0 & 0 \\ 0 & \tfrac{1}{2}(X + Y) & 0 \\ 0 & 0 & Z \end{bmatrix} \tag{36}$$

and has cylindrical symmetry about the $Z$ axis.

We have dealt with sufficient examples to indicate the wide range of rate processes which can be studied by magnetic resonance. The resonance method has two great

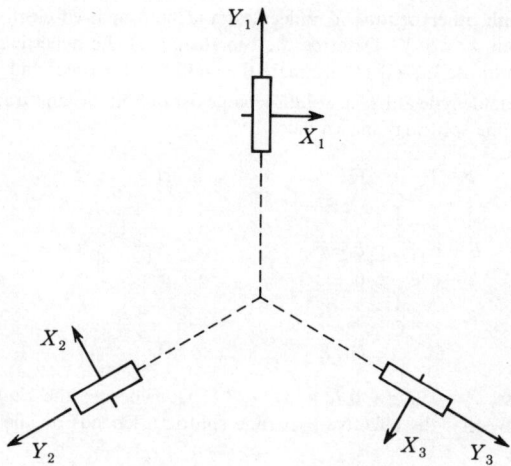

FIG. 12.6. Principal axes of the zero-field tensors for the three modes of excitation of trip-tycene.

advantages. First it measures the rate directly and continuously instead of relying on analysis of the end products of the molecular change. Secondly it does not interfere with the motion being observed. In spite of the obvious restriction of electron reso-nance to molecules containing unpaired electrons, the method is likely to find more and more applications as the theoretical principles become better understood by chemists.

## PROBLEMS

**1.** The line shape in Eq. (13) can be written in the form

$$I(x) = \frac{4P\delta^2}{(x^2 - \delta^2)^2 + 4P^2x^2}$$

where $I(x)$ is the relative intensity, $x = (\omega - \bar{\omega})$, and $(\omega_A - \omega_B) = 2\delta$. Find the positions of the two maxima when exchange is slow, and verify Eq. (14).

**2.*** See if you can deduce the line-width in fast exchange, Eq. (20), by starting from either (11) or (18). Do you have to make any assumptions about the values of $T_{2A}$ and $T_{2B}$?

**3.** Estimate the rates of chemical exchange in the following examples:
  (a) The n.m.r. spectrum of N,N-dimethylnitrosamine shows distinct cis and trans methyl lines with a width $1/T_2 = 1$ cps.
  (b) The same, but now the spectrum has a single line with $1/T_2 = 5$ cps.
  (c) The OH protons of ethanol, and the water protons, in a 1 : 1 molar mixture of ethanol and water at room temperature and 40 Mc/s show a single line ($\delta_{H_2O} - \delta_{OH} = 0.8$ ppm).

**4.** Using the data given in Section 12.3.1, estimate the jumping rate $P$ in N,N-dimethylnitro-samine at 240° and 120° C. What would the line width be at 240° and the peak-to-peak separation at 120°?

**5.** How fast would the vinyl radical have to invert in order for the two inner hyperfine lines (a) to appear resolved, but with a width of 1 gauss; (b) to just coalesce?

**6.** A radical ion $X^-$ with hyperfine splitting from two equivalent protons (splitting constant $a$) is in solution with other neutral $X$ molecules, and the unpaired electron can jump at a rather low rate from $X^-$ to $X$. Describe the broadening of the hyperfine lines in as much detail as you can, and use Eq. (18) to estimate the widths of the inner and outer lines.

**7.\*** The terephthalaldehyde anion in solution can exist in both cis and trans forms, with the approximate hyperfine splittings shown below.

Here $b = 1.4$ gauss, $c = 0.2$, $d = 0.7$, and $e = 3.8$. Considering the ring proton hyperfine structure only, show that the effective hyperfine splitting depends on the nuclear quantum numbers as follows:

$$a = b(m_1 + m_2 + m_3 + m_4) - c(m_1 + m_3 - m_2 - m_4) \qquad trans$$

$$a = b(m_1 + m_2 + m_3 + m_4) - d(m_1 + m_2 - m_3 - m_4) \qquad cis$$

When interconversion takes place the change in hyperfine energy is therefore

$$a = (c - d)(m_4 - m_1) + (c + d)(m_2 - m_3)$$

Show that the $M = \pm 2$ hyperfine lines are not broadened at all by chemical exchange; all the $M = \pm 1$ lines are broadened; out of the six hyperfine states $\alpha\alpha\beta\beta$, $\alpha\beta\alpha\beta$, ... with $M = 0$ two give rise to sharp lines and four to broadened ones.

## SUGGESTIONS FOR FURTHER READING

Pople, Schneider, and Bernstein: Chapter 10. Theory of chemical exchange effects in n.m.r.

McConnell: *J. Chem. Phys.*, **28**: 430 (1958). Modified Bloch equations.

Looney, Phillips, and Reilly: *J. Am. Chem. Soc.*, **79**: 6136 (1957). Internal rotation in $(CH_3)_2NNO$.

Ogg and Ray: *J. Chem. Phys.*, **26**: 1515 (1957). Quadrupole broadening in proton resonance of $NH_3$.

Pople: *Mol. Phys.*, **1**: 168 (1958). Theory of broadening of proton resonance by a coupled $N^{14}$ nucleus.

Meiboom: *J. Chem. Phys.*, **34**: 375 (1961). Proton transfer rates in water.

Reeves and Schneider: *Can. J. Chem.*, **36**: 793 (1958). Keto-enol interconversion.

Fessenden and Schuler: *J. Chem. Phys.*, **39**: 2147 (1963). E.S.R. of the vinyl radical.

Bolton, Carrington, and Todd: *Mol. Phys.*, **6**: 169 (1963). Line width alternation in naphthazarin semiquinone.

De Boer and Mackor: *J. Am. Chem. Soc.*, **86**: 1513 (1964). Ion pairing of pyracene ion with alkali metals.

Weissman: *Z. Elektrochem.*, **64**: 47 (1964). Electron exchange in radical ion solutions.

Abragam and Pryce: *Proc. Phys. Soc.* (*London*), A63: 409 (1953) and Bleaney, Bowers, and Trenam: *Proc. Roy. Soc.* (*London*), A228: 157 (1955). Temperature effect on e.s.r. of copper fluorosilicate.

De Groot and Van Der Waals: *Mol. Phys.*, **6**: 545 (1963). Changes in the axis of quantization in triplet states.

# CHAPTER 13

# NUCLEAR RESONANCE IN PARAMAGNETIC SYSTEMS —DOUBLE RESONANCE

## 13·1 INTRODUCTION

Nuclear resonance in paramagnetic systems presents several qualitatively new aspects which can give chemists new kinds of information about molecular processes. First of all the unpaired electrons produce chemical shifts which are many times larger than the ordinary shifts seen in diamagnetic molecules and depend on temperature in a characteristic way. Then paramagnetic substances can also produce many different kinds of line broadening and relaxation effect. Finally there is the possibility of doing double resonance experiments where radiofrequency fields are applied simultaneously at the electron and nuclear resonance frequencies. Each of these effects involves important new principles which we shall now describe.

## 13·2 THE KNIGHT SHIFT

The distinctive effects produced by the electron spin in nuclear resonance spectra are due to the very strong local magnetic fields which result from hyperfine interactions. If we consider a radical in solution with an isotropic hyperfine splitting $a$, the spin energy levels in a strong external field are given by the expression

$$\mathcal{H} = g\beta H S_z - g_N \beta_N H I_z + a I_z S_z \qquad (1)$$

and the part of the energy which depends on the nuclear spin orientation can be written

$$\mathcal{H} = -g_N \beta_N I_z \left( H - \frac{a S_z}{g_N \beta_N} \right)$$

$$= -g_N \beta_N I_z (H + H_e) \qquad (2)$$

The quantity

$$H_e = -a S_z / g_N \beta_N \qquad (3)$$

represents the effective local field produced by the unpaired electron at the nucleus (see Section 7.3) and may be very large. For example, the proton resonance frequency in a field of 10,000 gauss is 42 Mc/s, and so a moderately large hyperfine splitting

$a = 84$ Mc/s corresponds to a local field $H_e$ of $\pm 10,000$ gauss, depending on the electron spin direction. Thus the n.m.r. spectrum of such a radical would consist of two lines with a separation of 20,000 gauss! This, of course, is totally unrealistic in practice, the point being that in any paramagnetic molecule where nuclear resonance signals are sharp enough to be observable the electron spin relaxes exceedingly fast. In these molecules the *electron* spin relaxation time $\tau_s$ satisfies the condition

$$|a\tau_s| \ll 1 \tag{4}$$

and the nucleus sees only a time averaged local field proportional to the mean value $\langle S_z \rangle$ of the electron spin component. This does not vanish in the magnetic field because the $\alpha$ and $\beta$ electron spin states have significantly different populations. The effective field is $H_e = -a\langle S_z \rangle / g_N \beta_N$ and the nuclear resonance signal shifts to high field by an amount

$$\Delta H = a\langle S_z \rangle / g_N \beta_N \tag{5}$$

Now the average of $\langle S_z \rangle$ is easily calculated from the bulk magnetic susceptibility $\chi$ of the electrons, for the quantity $M_0 = -Ng\beta\langle S_z \rangle$ is simply the equilibrium magnetic moment of the electrons in the external field, being equal to $\chi H$. Thus (5) becomes

$$\Delta H = -\frac{\chi a H}{Ng\beta g_N \beta_N} \tag{6}$$

and the shift is proportional to the applied field $H$. Substituting the standard expression for the spin susceptibility of a free electron

$$\chi = \frac{Ng^2\beta^2 S(S + 1)}{3kT} \tag{7}$$

the shift reduces to

$$\frac{\Delta H}{H} = -\frac{g\beta}{g_N \beta_N} \frac{aS(S + 1)}{3kT} \tag{8}$$

and we see that it is strongly temperature dependent. As the temperature falls the difference between the populations of the $\alpha$ and $\beta$ electron spin states grows larger and the local field $H_e$ therefore becomes stronger. Finally we can relate the shift directly to the unpaired electron spin density at the nucleus. Fermi's formula gives

$$a = \frac{8\pi}{3} g\beta g_N \beta_N \rho(N) \tag{9}$$

and so we find

$$\frac{\Delta H}{H} = -\frac{8\pi}{3N} \chi\rho(N) \tag{10}$$

This nuclear resonance shift is called the Knight shift and it was first observed in metals, where the paramagnetism comes from conduction electrons. Knight shifts also occur in solutions containing paramagnetic ions, in free radical solids, and in individual paramagnetic molecules.

A typical example of Knight shifts in solution is provided by the n.m.r. spectrum of $n$-propanol in the presence of paramagnetic $Co^{++}$ ions at 23°C. Chemical shifts are shown in Fig. 13.1 and are many times larger than normal proton shifts. The alcohol probably forms a loosely bound complex with the cobalt ion and this allows the unpaired spin to penetrate onto the protons. The observed shift should be proportional

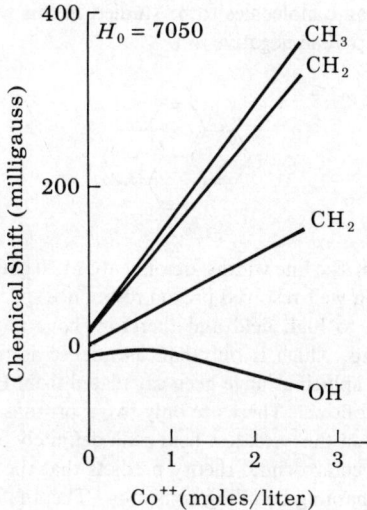

FIG. 13.1. Knight shift produced by $Co^{++}$ ions in $n$-propanol.

to the fraction of time each alcohol molecule spends attached to the ion, and is therefore proportional to the cobalt concentration.

Contact hyperfine interaction is by no means the only interaction which can produce chemical shifts. Apart from an increased bulk susceptibility correction due to the paramagnetism there is also another contribution called the *pseudocontact* hyperfine interaction. This arises from the combined action of an anisotropic $g$ tensor and a dipolar hyperfine coupling and takes its simplest form in a paramagnetic ion with a symmetry axis, tumbling in solution. If a nucleus is attached to the ion at a distance $r$ from the unpaired electron and the vector $\mathbf{r}$ makes an angle $\psi$ with the axis one has to use an effective hyperfine interaction constant

$$A = a + \frac{(3 \cos^2 \psi - 1)}{3r^3} (g_{\parallel} - g_{\perp}) \beta g_N \beta_N \qquad (11)$$

instead of $a$ in calculating the paramagnetic shift. As a rule pseudocontact shifts are only important for nuclei which come very close to the paramagnetic center. Otherwise the factor $1/r^3$ is very small.

## 13·3   UNPAIRED ELECTRON DISTRIBUTIONS BY N.M.R

The Knight shift can be used to measure unpaired spin densities in paramagnetic molecules. The method has several advantages over analysis of electron resonance hyperfine structures. Not only can one determine the *signs* of the spin densities directly from the direction of the shift, but the spectra are much easier to analyze since each nucleus gives rise to only a single n.m.r. line instead of a complicated hyperfine multiplet. Finally, by raising the temperature and extrapolating the observed shifts back to infinite temperature, where the Knight shift disappears, one can identify the separate lines by comparison with the n.m.r. spectra of related diamagnetic molecules. The chief difficulty is that it is hard to find paramagnetic molecules with sufficiently narrow nuclear resonance lines.

One of the first organic molecules to be studied in this way was solid potassium pyrenide containing the pyrene negative ion.

At a temperature of 4.2°K the line widths are only about 20 gauss, and the Knight shifts are large enough to give a well-resolved proton resonance spectrum with three distinct peaks. Two are shifted to high field and therefore have negative proton coupling constants, while the third, which is only half as intense as the first two, appears at low field. The hyperfine splittings have been calculated from Eq. (8), and are assigned to the ring positions as follows. There are only two $\alpha$ protons, compared with four at the $\beta$ or $\gamma$ positions, so that the weak low field peak definitely belongs to the $\alpha$ protons. For the other two, molecular orbital theory predicts that the $\beta$ position should have the larger of the two remaining coupling constants. The final results then are

$$a_\alpha = +1.24, \qquad a_\beta = -4.98, \qquad a_\gamma = -2.81 \text{ (gauss)}$$

They are in good agreement with the splittings of $(+)$ 1.09, $(-)$ 4.70, and $(-)$ 2.06 gauss obtained directly from the electron resonance spectrum of the pyrene negative ion in solution at room temperature. The proton resonance measures the signs of the coupling constants, and these signs demonstrate two important points: namely that the constant $Q$ in McConnell's relation (Section 6.4.3) is negative, and that the $\pi$-*electron* spin density at the $\alpha$ position in the pyrene negative ion is also negative. Both results had previously been predicted theoretically.

A remarkable series of chelated nickel complexes known as the nickel aminotroponeiminates have been synthesized. In them the nickel ion interconverts rapidly between a square planar diamagnetic form and a tetrahedral paramagnetic form with two unpaired electrons. This peculiarity gives the paramagnetic triplet state an unusually short electron spin relaxation time, and the nickel complexes in solution at room temperature show very well resolved proton resonance spectra from the ligands with line widths of only a few cps. Unpaired electron spin density from the metal ion penetrates through the ligand skeleton and transmits itself through as many as ten chemical bonds to produce appreciable Knight shifts. Analysis of the spectra leads to an interesting spin density map of the ligands. For example, nickel N,N'-diphenyl-aminotroponeiminate

+.0095
−.0070
+.0068
+.036
−.019
+.050

has the $\pi$-electron spin distribution shown above. The sign of the spin density alternates from one atom to the next and a theoretical analysis of the results indicates that about one-tenth of an unpaired electron goes into each of the seven-membered ring systems. The benzene rings attached to the nitrogen atoms can be replaced by many other groups such as

and the resonance shifts give a new insight into the transmission of spin effects through conjugated systems.

Knight shifts have also been used to determine the signs and magnitudes of spin densities in paramagnetic sandwich molecules such as nickelocene $Ni(C_5H_5)_2$ or chromocene $Cr(C_5H_5)_2$.

## 13·4 RELAXATION BY PARAMAGNETIC IONS IN SOLUTION

The discovery that small traces of paramagnetic ions could dramatically increase the rate of nuclear spin relaxation was one of the landmarks in the history of magnetic resonance and led to a much sounder understanding of relaxation processes in solids and liquids. The powerful effect of these ions is due mainly to the large local fields produced at the nucleus by the electron spin. Since the electron has a magnetic moment about a thousand times larger than most nuclei, the local field $H_e$ may easily be as high as 10,000 gauss, as we saw in Section 13.2. The other important factor is the short *electron* spin relaxation time of many paramagnetic ions, which causes $H_e$ to fluctuate rapidly and induce fast transitions between the nuclear spin states. Brownian motion will also modulate the anisotropic magnetic interactions in the usual way and contribute to the relaxation, whether the nuclei are attached to the ions themselves, or to other nuclei in solution.

If the nuclear spins in a solid or liquid were unable to move, the paramagnetic centers could produce little effect on the n.m.r. spectrum; most of the nuclei, being out of the sphere of influence of the unpaired electron, would be unaffected, while the few near neighbors would be relaxed so strongly that their n.m.r. lines would be completely washed out. In practice, however, spin exchange and diffusion processes often allow all the nuclei in the material to have frequent encounters with the unpaired electron, so that all nuclei have the same relaxation behavior. The observed relaxation rate is a weighted average over each of the different local nuclear environments.

As a rule the relaxation rate $1/T_1$ or $1/T_2$ is directly proportional to $N$, the concentration of the paramagnetic species, and roughly proportional to $\mu_{eff}^2$, its mean square magnetic moment. However, the detailed interpretation of line widths involves many different factors, and it is usually necessary to study both $T_1$ and $T_2$ over reasonable ranges of concentration and temperature before reaching any firm conclusions.

To illustrate some of the factors involved we shall describe the relaxation behavior of the protons in water which contains added $Mn^{++}$ ions. Some typical experimental results are shown in Fig. 13.2. In every case $1/T_1$ and $1/T_2$ are proportional to $N$, the concentration of manganese ions in moles per liter.

The $Mn^{++}$ ion probably exists in solution as a hydrated octahedral complex $Mn^{++}(H_2O)_6$ surrounded by twelve protons, and one can picture the situation crudely in the following way.

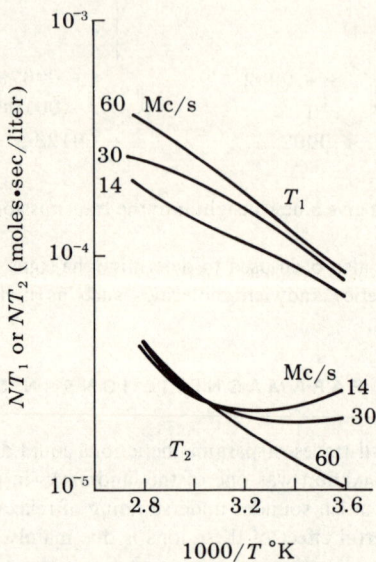

FIG. 13.2. Temperature dependence of proton relaxation times in $Mn^{++}$ solutions at three different resonance frequencies. Upper curves give $NT_1$, lower curves $NT_2$.

Protons in the $Mn^{++}$ coordination shell experience enhanced relaxation due to the unpaired electron, while all the other protons in the solution have the normal relaxation behavior of protons in pure water. Thus if $T_{1M}$, $T_{2M}$ denote the relaxation times of the coordinated protons and $T_{1W}$, $T_{2W}$ represent those in water, we expect to find

$$\frac{1}{T_1} = \frac{(1-f)}{T_{1W}} + \frac{f}{T_{1M}}$$

$$\frac{1}{T_2} = \frac{(1-f)}{T_{2W}} + \frac{f}{T_{2M}} \tag{12}$$

where $f$ is the fraction of time that each proton spends in the coordination shell. If $n(=12)$ is the coordination number and $N_H$ is the molar concentration of protons we have $f = (nN/N_H)$. The line broadening is dominated by the paramagnetic ion, so we may neglect $1/T_{1W}$ and $1/T_{2W}$ in (12) and write

$$\frac{1}{T_1} = \left(\frac{N}{N_H}\right)\frac{n}{T_{1M}}, \qquad \frac{1}{T_2} = \left(\frac{N}{N_H}\right)\frac{n}{T_{2M}} \tag{13}$$

The interpretation of the relaxation times therefore leads us to consider the relaxation of protons in the hydration shell of the $Mn^{++}$ ion in solution.

There are two magnetic interactions which are important. One is the direct dipole-dipole coupling between electron and nuclear spins, which is proportional to $1/r^3$, the inverse cube of the distance between them. The other is the contact hyperfine energy $a\mathbf{I}\cdot\mathbf{S}$. These interactions are modulated by several different time-dependent processes with characteristic correlation times.

1. The electron spin relaxation time $\tau_s$, which is about $3 \times 10^{-9}$ sec at room temperature. The $Mn^{++}$ spin relaxation is dominated by time-dependent zero-field splittings arising from distortions of the water octahedron.

2. The rotational correlation time $\tau_r$ for Brownian motion of the complex, $\tau_r \approx 10^{-11}$ sec.

3. The mean time $\tau_h$ for which a proton remains in the hydration sphere, $\tau_h \approx 2 \times 10^{-8}$ sec.

To begin with, we consider the relaxation effects which are caused by modulation of the contact hyperfine interaction $a\mathbf{I}\cdot\mathbf{S}$. For simplicity we shall concentrate attention on the $I_z S_z$ part, writing

$$\mathscr{H}_C(t) = a(t)S_z(t)I_z \tag{14}$$

to emphasize that the electron spin undergoes fast relaxation. We shall now see that $\mathscr{H}_C(t)$ fluctuates with a characteristic correlation time $\tau_e$, where

$$\frac{1}{\tau_e} = \frac{1}{\tau_h} + \frac{1}{\tau_s} \tag{15}$$

In the first place we know that for the spin *of a particular manganese ion* the auto-correlation function of $S_z$ behaves as

$$\overline{\langle S_z(t+\tau)S_z(t)\rangle} = \langle \overline{S_z^2}\rangle e^{-\tau/\tau_s}$$
$$= \frac{1}{3}S(S+1)e^{-\tau/\tau_s} \tag{16}$$

where $\tau_s$ is the electron spin relaxation time. However a proton stays near a *given electron* only for a limited time $\tau_h$, and so the autocorrelation function decays away with a further exponential factor $e^{-\tau/\tau_h}$. The effective lifetime of the electron spin is therefore given by the expression

$$\overline{\langle S_z(t+\tau)S_z(t)\rangle} = \frac{1}{3}S(S+1)e^{-\tau/\tau_s}\cdot e^{-\tau/\tau_h}$$
$$= \frac{1}{3}S(S+1)e^{-\tau/\tau_e} \tag{17}$$

In a similar fashion the effective correlation time $\tau_c$ for the dipolar interaction is determined by the interplay of rotation, electron relaxation, and proton exchange:

$$\frac{1}{\tau_c} = \frac{1}{\tau_r} + \frac{1}{\tau_h} + \frac{1}{\tau_s} \tag{18}$$

Since each of these times depends on the temperature and differs from one metal ion to another, it is clear that each case must be treated separately. In general however the behavior of $\tau_e$ and $\tau_h$ is dominated by the fastest molecular process which affects it. For instance, in $Mn^{++}$ we find

$$\frac{1}{\tau_e} \approx \frac{1}{\tau_s} = 3 \times 10^8 \ sec^{-1}$$

$$\frac{1}{\tau_c} \approx \frac{1}{\tau_r} = 10^{11} \ sec^{-1} \tag{19}$$

We are now ready to look at the detailed theoretical expressions for the relaxation times. They are

$$\frac{1}{T_{1M}} = \frac{4}{30} S(S+1) \left(\frac{g^2\beta^2 g_N^2\beta_N^2}{\hbar^2 r^6}\right) \left(3\tau_c + \frac{7\tau_c}{1+\omega_s^2\tau_c^2}\right) + \frac{2}{3} S(S+1) \frac{a^2}{\hbar^2} \left(\frac{\tau_e}{1+\omega_s^2\tau_e^2}\right) \tag{20}$$

$$\frac{1}{T_{2M}} = \frac{4}{60} S(S+1) \left(\frac{g^2\beta^2 g_N^2\beta_N^2}{\hbar^2 r^6}\right) \left(7\tau_c + \frac{13\tau_c}{1+\omega_s^2\tau_c^2}\right) + \frac{1}{3} S(S+1) \frac{a^2}{\hbar^2} \left(\tau_e + \frac{\tau_e}{1+\omega_s^2\tau_e^2}\right) \tag{21}$$

Here $\omega_s$ is the electron spin resonance frequency measured in radians/sec and $S = 5/2$. The first terms in (20) or (21) represent the effect of dipolar hyperfine interactions while the second come from the contact term.

The chief point of interest is that the correlation times $\tau_c$ and $\tau_e$ for the dipolar and contact terms lie in quite different ranges as compared with the electron spin resonance frequency $\omega_s$. For instance, let us consider the situation when the *proton* resonance frequency is 30 Mc/s. The corresponding electron resonance frequency $\omega_s = 1.2 \times 10^{10} \ sec^{-1}$ has the property that

$$\omega_s\tau_e \gg 1, \qquad \omega_s\tau_c \ll 1 \tag{22}$$

so that the contact correlation time is "short" and the scalar one is "long." The dipolar terms in (20) and (21) then give equal contributions to $T_1$ and $T_2$, while the contact one contributes only to $T_2$.

$$\frac{1}{T_{1M}} = \frac{4}{3} S(S+1) \left(\frac{g^2\beta^2 g_N^2\beta_N^2}{\hbar^2 r^6}\right) \tau_c$$

$$\frac{1}{T_{2M}} = \frac{3}{4} S(S+1) \left(\frac{g^2\beta^2 g_N^2\beta_N^2}{\hbar^2 r^6}\right) \tau_c + \frac{1}{3} S(S+1) \frac{a^2}{\hbar^2} \tau_e \tag{23}$$

A glance at Fig. 13.2 shows that $T_{1M}$ and $T_{2M}$ are not equal; evidently the contact term does contribute appreciably to the line width, since the ratio $T_1/T_2$ is quite large, being about 7:1. The observed relaxation times for $Mn^{++}$ solutions agree well with Eqs. (13) and (23) if one takes $r = 2,8$ Å, $a/h = 1.0$ Mc/sec and uses the correlation times given in (19). The Knight shift of the water protons has been measured and gives $a/H = 0.62$ Mc/sec, which also fits in with the relaxation measurements.

The effect of paramagnetic ions on nuclear relaxation times has found important applications in chemistry and biology. The $O^{17}$ relaxation in water containing ions of $Mn^{++}$, $Fe^{++}$ and other transition metals has been studied in order to determine the rate of exchange of water molecules between the ion and the solution. The experiments yield rate constants and enthalpies of activation for the reaction, the values for $Mn^{++}$ being $k = 3 \times 10^7 \ sec^{-1}$ and $\Delta H^\ddagger = 8$ kcal/mole.

Biological applications include studies of the binding of metal ions to DNA, enzymes, and adenosine triphosphate.

1. It is found that when some paramagnetic ions are bound to DNA the spin-lattice relaxation time $T_1$ of water protons is much shorter than it is if the ions are simply dissolved in pure water. The rotation of the water molecules bound to the metal ion is slowed down when the ion is attached to a large molecule, and this results in more efficient proton relaxation, due to a larger correlation time $\tau_c$. The observed relaxation times have been used to estimate the number of sites at which a metal ion can bind to DNA.

2. Several enzymes only function with the help of a divalent metal ion, and in many of them the proton relaxation is again strongly enhanced. The degree of enhancement can be correlated with chemical changes at the active site of the protein, and one may also determine how many metal ions can bind to a single enzyme molecule.

3. One of the most satisfying investigations has been the detailed study of the complex formed in solution between adenosine triphosphate (ATP) and $Mn^{++}$ ions. It is well known that the biological activity of ATP requires the presence of a divalent metal ion, but the precise nature of the interaction between the two molecules has been unknown. Nuclear magnetic resonance studies of proton and $P^{31}$ relaxation in strong solutions of ATP (0.35 molar) show that manganese binds simultaneously to the three phosphate groups and part of the adenine ring. Distances between the $Mn^{++}$ ion and the ring protons have been estimated from the observed relaxation times, using Eq. (23).

## 13·5    ELECTRON NUCLEAR DOUBLE RESONANCE

### 13.5.1 The Overhauser Effect

It often happens that two spins $A$ and $B$ are coupled together in such a way that the relaxation of one spin affects the relaxation of the other. One example is the ensemble of electron and proton spin $\mathbf{S}$ and $\mathbf{I}$ in the hydration shell of a $Mn^{++}$ ion in solution, but we have also noted cross-relaxation effects for a proton with spin-spin coupling to $N^{14}$ in ammonia, and for the H and F nuclei in HF in solution. Whenever cross-relaxation occurs one can expect that any change in the populations of the spin states of $A$ will cause secondary changes in the populations for spin $B$. Double resonance or "dynamic polarization" experiments are designed to take advantage of these changes. In a typical arrangement two resonance lines are chosen. The first transition is saturated, thereby equalizing the populations of two spin states, and then one looks for changes in the intensity of resonance absorption at the second line.

One of the most striking changes is called the Overhauser effect. It occurs if one saturates the electron spin resonance of a suitable paramagnetic sample while watching a nuclear resonance absorption line. The intensity of the n.m.r. signal may increase several hundredfold, and in some cases the n.m.r. absorption reverses sign and becomes an emission of energy.

At first sight this appears very mysterious, but the origin of the Overhauser effect only depends on some simple relations between relaxation transition probabilities and on the Boltzmann distribution of energy levels. The essential requirement is that there must be relaxation processes in which both electron and nuclear spins change simultaneously; either $\alpha_e\beta_N \leftrightarrow \beta_e\alpha_N$ or $\alpha_e\alpha_N \leftrightarrow \beta_e\beta_N$.

As an example let us consider an electron spin $S$ and a proton spin $I$, coupled together by a time-dependent contact hyperfine interaction $V(t) = a(t)I \cdot S$ which fluctuates rapidly about an average value of zero. There will be no resolved hyperfine structure in the electron resonance spectrum, and the energy levels will correspond to the Zeeman Hamiltonian

$$\mathcal{H}_0 = g\beta H S_z - g_N \beta_N H I_z \tag{24}$$

In thermal equilibrium the probabilities for the electron and nuclear spins to be $\alpha$ or $\beta$ will obey Boltzmann's law, being $e^{\mp p}$ and $e^{\pm q}$, where

$$p = \frac{g\beta H}{2kT}$$

$$q = \frac{g_N \beta_N H}{2kT} \tag{25}$$

Thus the spin energy levels and their relative populations will be as shown in Fig. 13.3. These populations are maintained by various relaxation processes with different transition probabilities, and we recall the general principle that if $a$ and $b$ are any two

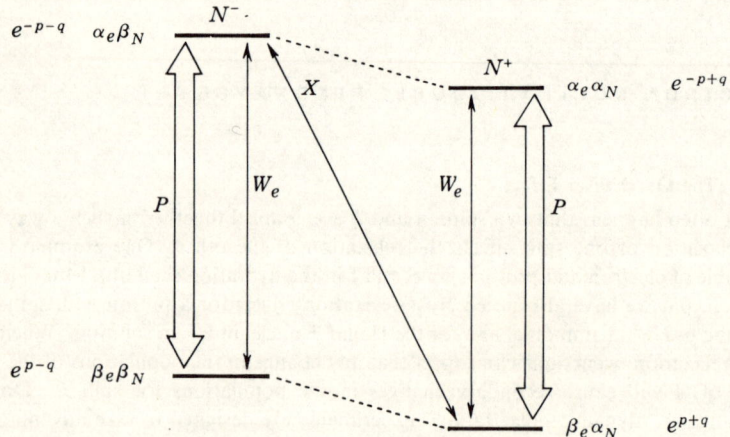

FIG. 13.3. The Overhauser effect with scalar relaxation $a(t)I \cdot S$.

energy levels of a quantum system *in thermal equilibrium* the rates of transitions $a \to b$ and $b \to a$ must balance:

$$N_a W_{ab} = N_b W_{ba} \tag{26}$$

$$\frac{W_{ab}}{W_{ba}} = \frac{N_b}{N_a} = e^{(E_a - E_b)/kT} \tag{27}$$

Here $W_{ab}$ and $W_{ba}$ are the two transition probabilities; $N_a$ and $N_b$ are the populations of the states. Let us now restrict attention to two relaxation processes, electron spin-lattice relaxation and electron-nuclear relaxation by the $I \cdot S$ term. We introduce two pairs of transition probabilities according to the following scheme:

$$
\begin{aligned}
(\alpha_e \to \beta_e), && W_e e^p \\
(\beta_e \to \alpha_e), && W_e e^{-p}
\end{aligned}
\tag{28}
$$

$$
\begin{aligned}
(\alpha_e \beta_N \to \beta_e \alpha_N), && X e^{p+q} \\
(\beta_e \alpha_N \to \alpha_e \beta_N), && X e^{-p-q}
\end{aligned}
\tag{29}
$$

The probability $X$ arises from the time-dependent off-diagonal terms $I^+ S^-$ and $I^- S^+$ in the hyperfine energy, and provides a special form of nuclear relaxation where the nuclear spin can only change from $\alpha$ to $\beta$ if the electron spin makes the reverse transition from $\beta$ to $\alpha$. The effective nuclear relaxation rate therefore depends on the populations of the *electron* spin states. We shall now see how this leads to an Overhauser effect.

In thermal equilibrium the relative populations of the $\alpha$ and $\beta$ nuclear spin states are governed by the Boltzmann law:

$$
\frac{N_0^+}{N_0^-} = \frac{N_{\alpha\alpha} + N_{\beta\alpha}}{N_{\alpha\beta} + N_{\beta\beta}} = e^{2q} = e^{g_N \beta_N H / kT}
\tag{30}
$$

Let us saturate the electron spin resonance, using so much power that the rf transition probability $P$ is much larger than either $W_e$ or $X$. The result is to equalize the populations of the electron spin states

$$
\begin{aligned}
N_{\alpha\alpha} = N_{\beta\alpha} = \tfrac{1}{2} N^+ \\
N_{\alpha\beta} = N_{\beta\beta} = \tfrac{1}{2} N^-
\end{aligned}
\tag{31}
$$

leaving $N^+$ nuclei with spin $\alpha$ and $N^-$ with spin $\beta$. The balance between $N^+$ and $N^-$ is maintained by the $\alpha\beta \leftrightarrow \beta\alpha$ transitions, which must have equal rates in either direction

$$
N_{\alpha\beta} X e^{p+q} = N_{\beta\alpha} X e^{-p-q}
\tag{32}
$$

$$
N^+ / N^- = e^{2p+2q} = e^{(g_N \beta_N + g\beta)H/kT}
\tag{33}
$$

The ratio of the nuclear spin populations in contact with the saturated electron spins is effectively the same as if the nuclear magnetic moment were increased from $g_N \beta_N \mathbf{I}$ to $(g_N \beta_N + g\beta)\mathbf{I}$ and the nuclei were aligned in the steady field $H$. The difference between the populations of the $\alpha$ and $\beta$ spin states is therefore increased by a factor of

$$
f = \left(1 + \frac{g\beta}{g_N \beta_N}\right) = 659
\tag{34}
$$

The intensity of the nuclear resonance signal is proportional to the population difference $(N^+ - N^-)$ and so it too increases by the same factor $f$. One example where Overhauser effects occur in the manner we have just described is the proton resonance in the paramagnetic solutions of sodium in liquid ammonia. Here the electron spins migrate so rapidly from one ammonia molecule to another that there is no resolved hyperfine structure. The contact hyperfine interaction is, however, the main mechanism for nuclear relaxation. Saturation of the electron resonance here increases the strength of the proton resonance up to 400 fold. Similar effects also occur in the solid DPPH radical and in many metals.

The expression (34) represents an ultimate ideal. In practice the Overhauser effect is often spoiled by incomplete saturation and by other relaxation processes, so that a more realistic equation is

$$f = \left[1 + s\xi\left(\frac{g\beta}{g_N\beta_N}\right)\right] \tag{35}$$

Here $s$ measures the degree of saturation of the electron resonance and varies between 0 and 1. It is defined by the equation

$$n = (1 - s)n_0 \tag{36}$$

where $n$ is the population difference between the $\alpha$ and $\beta$ electron spin states. $\xi$ is another numerical factor, and depends on the nature of the relaxation processes.

One very important case is the Overhauser effect in free radicals in solution, where the relaxation is caused by anisotropic hyperfine interaction. We can analyze the populations of the spin states by reference to Fig. 13.4, again assuming that there is no resolved hyperfine structure. For example we might consider a highly concentrated solution of naphthalene negative ions, where fast electron exchange averages out the

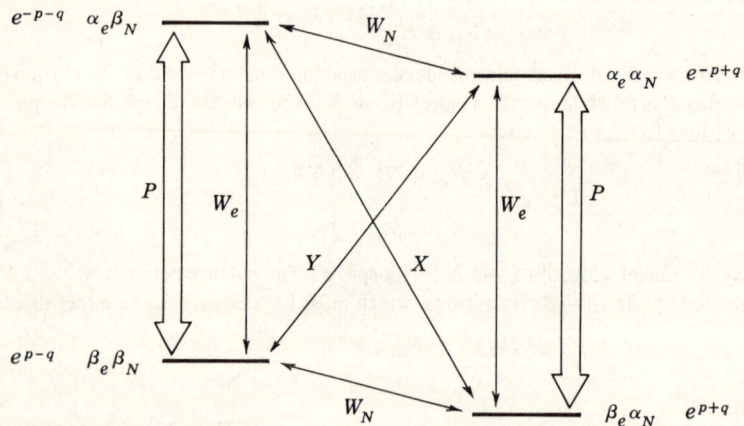

Fig. 13.4. The Overhauser effect in a radical with dipole-dipole relaxation.

isotropic hyperfine splitting. It is necessary to consider two further relaxation processes. Nuclear relaxations $\beta_N \leftrightarrow \alpha_N$ with probabilities $W_N e^{\pm q}$, and double transitions $\alpha_e\alpha_N \leftrightarrow \beta_e\beta_N$ with probabilities $Y e^{\pm(p-q)}$. Upon saturation of the electron resonance transitions we now find that the ratio of the populations of the nuclear spin states is

$$\frac{N^+}{N^-} = e^{2q}\left(\frac{2W_N + Xe^p + Ye^{-p}}{2W_N + Xe^{-p} + Ye^p}\right) \tag{37}$$

To interpret this let us assume that the temperature is high, so that $p$ is small and $e^p = (1 + p)$. Then the bracketed part of (37) becomes

$$\left(\frac{2W_N + Xe^p + Ye^{-p}}{2W_N + Xe^{-p} + Ye^p}\right) = 1 + \frac{2p(X - Y)}{2W_N + X + Y} = 1 + 2\xi p$$

$$= e^{2\xi p} = e^{2\xi q(p/q)}, \tag{38}$$

where

$$\zeta = \frac{(X - Y)}{2W_N + X + Y} \tag{39}$$

That is, $\zeta$ represents the factor in Eq. (35) and its value may be positive or negative, depending on the relative values of $X$ and $Y$. If $\zeta$ is negative the nuclei emit rf energy instead of absorbing. For a radical in solution the anisotropic hyperfine interaction may be written in a form analogous to Eq. (10) of Section 3.3.

$$V(t) = A(t)S_z I_z + B(t)[S^+ I^- + S^- I^+] + C(t)[S_z I^+ + S^+ I_z]$$

$$+ D(t)[S_z I^- + S^- I_z] + E(t)S^+ I^+ + F(t)S^- I^- \tag{40}$$

The transition probabilities $X$, $Y$, and $W_N$ are, respectively, proportional to the fluctuations in $B(t)$, $E(t)$ or $F(t)$, and $C(t)$ or $D(t)$. In the limit of fast tumbling one can show that the probabilities are in the ratio $X : Y : W_N = 2 : 12 : 3$. Thus we obtain

$$\xi = -\tfrac{1}{2} \quad \text{(dipolar relaxation)} \tag{41}$$

The conclusion that scalar and dipolar relaxation mechanisms lead to opposite signs of the nuclear spin polarization is of importance to chemists because it opens up the possibility of studying magnetic interactions in paramagnetic solutions through measurements of $\xi$. For instance the enhancement factor $f$ for the nuclear polarization in naphthalene negative ion solutions is $-60$, and in peroxylamine disulphonate $ON(SO_3)_2^{--}$ it is about $-50$. Here dipolar relaxation must clearly be dominant.

### 13.5.2 The Solid-State Effect

Even if the electrons and nuclei relax quite independently it is still possible to alter the nuclear populations by saturating "forbidden" electron resonance transitions. This is known as the *solid-state effect* and it can be applied to an oriented radical where there is anisotropic hyperfine interaction between the electron and nuclear spins. As we have seen in Section 7.4 the nucleus is quantized in an effective field $H'$ or $H''$ which is not in general parallel to the external field $H$, and this leads to the appearance of new lines in the e.s.r. spectrum corresponding to double transitions like $\alpha_e \alpha_N' \to \beta_e \beta_N''$.

Let us now consider a radical where the nuclear relaxation time is extremely slow, but the electron spin has a fast spin-lattice relaxation. For simplicity we neglect all relaxation transition probabilities except $W_e$ and look at the population of the nuclear spin states when the forbidden transition is saturated. Before the rf field is turned on the populations are determined by Boltzmann's law, as shown in Fig. 13.5. Here

$$p = \frac{g\beta H}{2kT} \tag{42}$$

$$q' = \frac{g_N \beta_N |H'|}{2kT}$$

$$q'' = \frac{g_N \beta_N |H''|}{2kT} \tag{43}$$

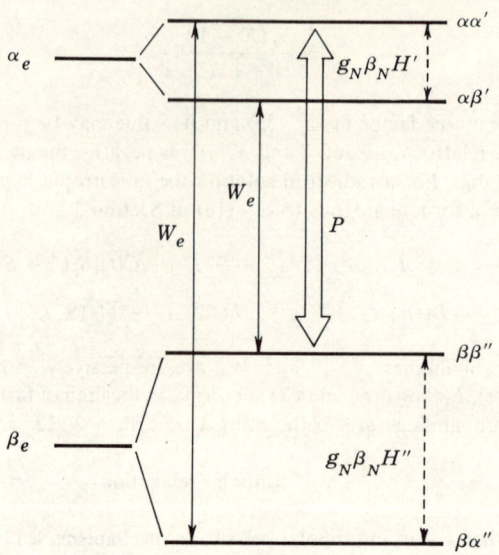

FIG. 13.5. The solid-state nuclear polarization effect.

The energy levels are drawn for the case that the nuclear Zeeman energy is small; that is $H'$ and $H''$ are of opposite sign and approximately equal magnitude. After saturation the relative populations of the states $\alpha_e \alpha_N'$ and $\beta_e \beta_N''$ are equal:

$$N_{\alpha\alpha'} = N_{\beta\beta''} = 1 \tag{44}$$

The electron spin relaxation transition probabilities into and out of these states are however still governed by Boltzmann's law. Thus the rate of relaxation transitions $\alpha_e \alpha_N' \to \beta_e \alpha_N'$ is still greater than the rate of the reverse transition by a factor $e^{2p+q'+q''}$ and so equilibrium requires that

$$N_{\beta\alpha''} = N_{\alpha\alpha'} e^{2p+q'+q''} = e^{2p+q'+q''} \tag{45}$$

Similarly we find by considering the transitions $\alpha_e \beta_N' \leftrightarrow \beta_e \beta_N''$ that

$$N_{\alpha\beta'} = e^{-2p+q'-q''} \tag{46}$$

As a result the nuclear spin polarization is enhanced, with

$$\frac{N_+}{N_-} = \frac{1 + e^{2p+q'+q''}}{e^{-2p+q'+q''} + 1} \approx e^{g\beta H/kT} \tag{47}$$

and the n.m.r. spectrum will show a large increase of intensity.

### 13.5.3 ENDOR

ENDOR stands for *electron-nuclear-double-resonance*, but the term is generally applied to one rather special type of experiment which is used to study the hyperfine structure of free radicals. To illustrate the principle of the experiment we take a radical where the unpaired electron has isotropic hyperfine interaction with a single proton.

We shall further assume that the electron and nuclear spins relax through completely independent processes, so that saturation of the electron resonance does not alter the nuclear spin populations and there can be no Overhauser effects. The populations of the spin states in thermal equilibrium are then as shown in Fig. 13.6, where $r = a/4kT$. Saturation of the hyperfine line which corresponds to the transition $\alpha_e\alpha_N \rightarrow \beta_e\alpha_N$ makes the populations $N_{\alpha\alpha}$ and $N_{\beta\alpha}$ equal, and at high temperatures where $p, q, r \ll 1$ we have

$$N_{\alpha\alpha} = N_{\beta\alpha} = (1 + q) \tag{48}$$

At the same time the electron resonance signal becomes very weak and broad because of saturation. Meanwhile the populations of the states $\alpha_e\beta_N$ and $\beta_e\beta_N$ have not changed:

$$N_{\alpha\beta} = (1 - p + r - q)$$
$$N_{\beta\beta} = (1 + p - r - q) \tag{49}$$

Let us now suddenly apply a strong rf pulse at the frequency

$$h\nu = \tfrac{1}{2}a - g_N\beta_NH \tag{50}$$

which corresponds to the transition $\alpha_e\beta_N \rightarrow \alpha_e\alpha_N$. By special techniques it is possible to reverse the nuclear spins completely and interchange the populations of the two states, so that after the pulse $N_{\alpha\alpha}$ is equal to $(1 - p + r - q)$. The immediate effect is

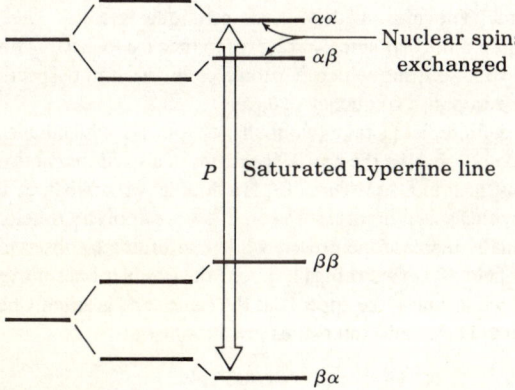

FIG. 13.6. The ENDOR experiment.

that the states $\alpha_e\alpha_N$ and $\beta_e\alpha_N$ no longer have equal populations; the electron resonance signal suddenly ceases to be saturated and absorption rises temporarily until a state of saturation is restored.

The ENDOR experiment provides an alternative method of measuring the hyperfine splittings. To do this it is only necessary to sweep the frequency of the rf pulses slowly through a suitable range and watch the electron resonance. Nothing happens until the n.m.r. frequency hits the value (50). Then the electron resonance signal rises sharply and slowly falls back again to its previous low value. The main advantage of the method is that one can measure very small hyperfine splittings under conditions where the hyperfine structure of the *electron resonance* spectrum contains so many lines that they all overlap.

One of the most remarkable applications is to the hyperfine structure of $F$ centers in alkali halides. If a crystal of, say LiF, is heavily irradiated with $X$ rays some of the fluoride ions are knocked out of place, leaving vacancies in the crystal structure, where a fluoride is missing and the hole is surrounded by lithium ions. Unpaired electrons can occupy these holes, and the resulting paramagnetic centers are called $F$ centers. The point of interest here is that the electron wave function spreads out all around the center and there are hyperfine interactions with all the surrounding $Li^7$ and $F^{19}$ nuclei. The e.s.r. spectrum shows one broad structureless line, but the ENDOR technique allows one to resolve the splittings from up to seven concentric shells of nuclei. The reason why the ENDOR spectrum is so much better resolved is that there is only one line for each distinct group of nuclei with a particular hyperfine splitting constant. The e.s.r. hyperfine structure has approximately $2^n$ lines for $n$ nuclei; as there are far more of them they overlap very badly.

## 13·6 SPIN DECOUPLING IN N.M.R.

There is no difference in principle between the double resonance behavior of an electron-nuclear spin system $I$, $S$ and the behavior of two coupled nuclei $I_1$ and $I_2$. For example, nuclear Overhauser effects occur, and they have been used to study relaxation processes in the HF molecule. However, the most important chemical applications of double resonance in n.m.r. have been as an aid to the analysis of complicated high-resolution spectra. The most widely used technique is called *spin decoupling*. By strong irradiation of a nuclear spin $I_2$ at its resonance frequency $\omega_2$ one can effectively remove the spin-spin splitting which it produces in the n.m.r. spectrum of a second nucleus $I_1$ whose resonance frequency is $\omega_1$.

To be more definite, let us take two nuclei of spin 1/2 which have a large chemical shift ($AX$ system) and observe the n.m.r. spectrum in a fixed magnetic field $H_0$. Under normal conditions the n.m.r. spectrum of $I_1$ is a doublet with two lines at the frequencies $\omega_1 \pm \frac{1}{2}J$ ($J$ is now measured in radians/sec). Now we apply an rf field $H_2$ in resonance with $I_2$ and gradually increase the power, while continuing to observe the spectrum of $I_1$ with a weak rf field $H_1$. Two things happen: the doublet lines move apart and their intensity falls off, while a new line appears at the center and gradually becomes stronger. The resonance frequencies and intensities are as follows:

$$\omega_1 \pm C, \quad \text{strength } \sin^2 \theta,$$
$$\omega_1, \quad \text{strength } 2\cos^2 \theta, \tag{51}$$

where

$$\tan \theta = J/\gamma_2 H_2,$$
$$C = \sqrt{\gamma_2^2 H_2^2 + \tfrac{1}{4}J^2} \tag{52}$$

These effects were first demonstrated in the $F^{19}$ resonance signal of $Na_2PO_3F$ in solution, the nucleus $I_2$ being phosphorus $P^{31}$. The changes in the spectrum are illustrated in Fig. 13.7.

At first sight it may seem that the collapse of the $F^{19}$ spectrum is analogous to the changes which occur in the proton resonance of ammonia, where the $H^1$ nucleus is coupled to a rapidly relaxing $N^{14}$ spin; or again that it resembles the effects of chemical exchange. Both analogies are completely misleading. The spin decoupling effect has

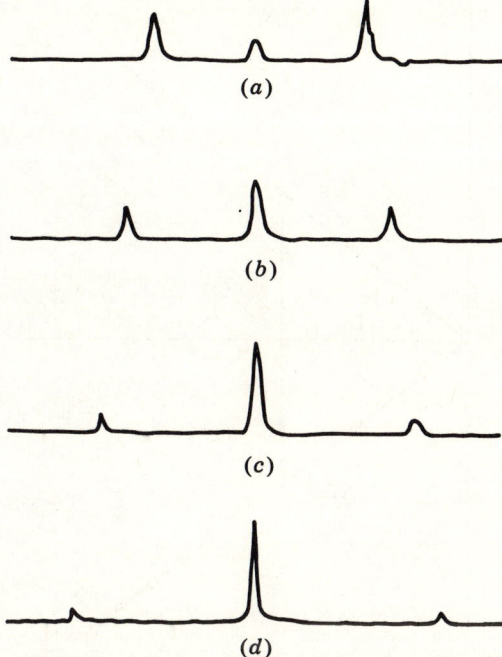

Fig. 13.7. Changes produced in the $F^{19}$ n.m.r. spectrum of $Na_2PO_3F$ by irradiating the $P^{31}$ resonance. The values of $\gamma_2 H_2/J$ are (a) 0.257; (b) 0.463; (c) 0.632; (d) 0.792.

nothing whatsoever to do with the relaxation or saturation of the $P^{31}$ nucleus; there is no broadening of the $F^{19}$ resonance, and the lines remain perfectly sharp. The strong oscillating $H_2$ field is quite different from the random local fields which cause relaxation, and it makes the $P^{31}$ nucleus precess coherently about $H_0$ in a cone making a large angle with the $z$ axis.

We can gain some insight into the motion by looking at the spin precession of nucleus $I_2$ in the rotating coordinate system introduced in Section 11.2, Eq. (20). (See also Fig. 13.8.) We suppose that the frequency of the strong rf field $H_2$ is $\omega$. Later $\omega$ will be set equal to $\omega_2$. The Bloch equation of motion for spin $I_2$, if it were not coupled to $I_1$ is

$$\frac{\delta I'_2}{\delta t} = \gamma_2 I'_2 \times \{(H'_0 + \omega'/\gamma_2) + H_2'\}$$
$$= \gamma_2 I'_2 \times H'_{\text{eff}}$$
(53)

That is, the spin $I_2$ moves relative to the rotating frame precisely as though it was in an effective *steady* magnetic field $H'_{\text{eff}}$ with components

$$H'_{\text{eff}} = (H_0 - \omega/\gamma_2)k' + H_2 i'$$
(54)

The nucleus $I_1$ has a coupling energy $\hbar J I_1 \cdot I_2$ with $I_2$ and this coupling is equivalent to an additional magnetic field of strength $-I_1 J/\gamma_2$ at the second nucleus. Assuming that $I_1$ is quantized along the $z$ axis with $I_{1z} = \pm 1/2$ we see that the second nucleus

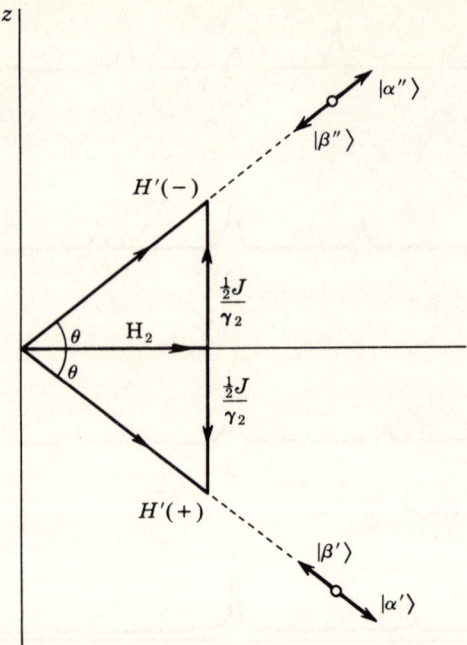

Fig. 13.8. Quantization of spin $I_2$ in the rotating coordinate system.

experiences an additional effective field $\mp J\mathbf{k}'/\gamma_2$. Thus the effective field depends on the orientation of $I_1$, taking either of the values

$$\mathbf{H}'_{\text{eff}}(\pm) = [H_0 - (\omega \pm \tfrac{1}{2} J)/\gamma_2]\mathbf{k}' + H_2\mathbf{i}' \tag{55}$$

and the situation is analogous to the anisotropic hyperfine interaction between an electron and a nucleus which we discussed in Section 7.4. The only difference is that we are now in a rotating coordinate system. It is now possible to introduce the concept that nucleus $\mathbf{I}_2$ is quantized along the direction of $\mathbf{H}_{\text{eff}}$, having a pair of states $|\alpha'\rangle$, $|\beta'\rangle$ in $H_{\text{eff}}(+)$ and $|\alpha''\rangle$, $|\beta''\rangle$ in $H_{\text{eff}}(-)$. The nucleus $\mathbf{I}_1$ also has states $|\alpha\rangle$, $|\beta\rangle$ which are quantized along the $z$ axis, so that the coupled spins possess the four states $|\alpha\alpha'\rangle$, $|\alpha\beta'\rangle$, $|\beta\alpha''\rangle$, and $|\beta\beta''\rangle$. Let us now bring the frequency $\omega$ into resonance at $\omega_2$ and look more closely at the effective fields. They are

$$H'_{\text{eff}}(+) = H_2\mathbf{i}' - \tfrac{1}{2} (J/\gamma_2)\mathbf{k}'$$
$$H'_{\text{eff}}(-) = H_2\mathbf{i}' + \tfrac{1}{2} (J/\gamma_2)\mathbf{k}' \tag{56}$$

and we see from Fig. 13.8 that the directions of $H'_{\text{eff}}(+)$ and $H'_{\text{eff}}(-)$, respectively, make angles $-\theta$ and $+\theta$ with the $\mathbf{i}'$ axis, where $\theta$ is given by (52). Finally, consider the resonance spectrum of the nucleus $I_1$. The transitions $\alpha\alpha' \to \beta\alpha''$ are "allowed" and have strengths proportional to $\cos^2 \theta$. The direction of quantization of $I_2$ turns through an angle $2\theta$, but the Zeeman energy of $I_2$ in $H'_{\text{eff}}$ does not change. Therefore these two transitions occur at the frequency $\omega_1$. The other transitions $\alpha\alpha' \to \beta\beta''$ or $\alpha\beta' \to \beta\alpha''$ are "forbidden." Their relative strengths are $\sin^2 \theta$ and since the Zeeman energy of $I_2$

changes by $\pm \gamma_2 \hbar |H'_{eff}|$ they give rise to a doublet having the frequencies $\omega_1 \pm \frac{1}{2}C$.

The treatment given above is far from rigorous! Nevertheless a more systematic theoretical analysis shows that it is, in essentials, correct.

So far we have assumed that the double resonance experiment is performed in a fixed magnetic field $H_0$ at a fixed frequency $\omega_2$ while one sweeps through the resonance of the first nucleus. In practice it is often more convenient to set both the frequencies $\omega_1$ and $\omega_2$ for resonance in a field $H_0$ and then sweep the field while looking at the $\omega_1$ resonance. The theory then shows that three lines are generally visible, at $H_0$ and $H_0 \pm \Delta H$, where

$$(\gamma_1 \Delta H)^2 = \frac{1}{4} J^2 + \frac{\gamma_2^2 H_2^2}{\gamma_1^2 - \gamma_2^2} \tag{57}$$

However, if $\gamma_1 < \gamma_2$ and $\gamma_2 H_2 > \frac{1}{4}J$, only a single line appears.

The spin decoupling experiment has many applications. It can be used to simplify very complicated spectra and check the analysis, to measure chemical shifts more accurately, and to determine the *relative* sign of spin-spin coupling constants.

## PROBLEMS

1. The $Na^{23}$ Knight shift in sodium metal is almost independent of temperature and has the value $\Delta H/H = -0.113\%$ (resonance shifted to low field). By assuming that each sodium atom has one valence electron in an atomic orbital $\psi(\mathbf{r})$, or otherwise, estimate the spin density $\rho(N)$ at the nucleus. Compare your answer with the value $|\psi(0)|^2 = 5.05 \times 10^{24} \, cm^{-3}$ deduced from the hyperfine structure of sodium atoms. What reasons can you suggest to explain the difference between them? (Spin magnetic susceptibility of sodium metal, per c.c., $\chi = 0.92 \times 10^{-6}$. Density $= 0.971$ gm/c.c. Atomic weight $= 23.00$.)

2. Show that if the nuclear spin relaxation time is very short the averaged hyperfine interaction produces a local field $\Delta H^*$ at the unpaired electron and shifts the electron resonance. This is sometimes called the "Day Shift." What happens to the Day Shift if (a) the nuclear resonance is saturated; (b) the electron resonance is saturated and there is an Overhauser effect?

3. A transition metal complex with $g_\parallel = 2.2$, $g_\perp = 1.98$, and $S = 1/2$ has six nuclei attached to it at the corners of a regular octahedron of radius 4 Å, two of the nuclei being along the direction of $g_\parallel$ and four in the equatorial plane. Calculate the pseudocontact nuclear resonance shifts in solution at room temperature (300°K).

4. The proton n.m.r. spectrum of nickelocene $Ni(C_5H_5)_2$ in toluene at 3,750 gauss and 300°K shows a Knight shift of 1.1 gauss to high field relative to the protons of toluene. Calculate the ring proton hyperfine splitting $a$, and deduce the probability that each of the two unpaired electrons ($S = 1$) is to be found in the $(C_5H_5)_2$ aromatic $\pi$-orbitals.

5. A proton in the hydration shell of a paramagnetic ion experiences purely dipolar relaxation. Within what limits must the ratio $T_1 : T_2$ lie? If relaxation were purely scalar, via the contact term, and $\tau_e = 3 \times 10^{-9}$ sec when the proton resonance frequency $\omega_s = 1.2 \times 10^{10}$, what would the ratio now be?

6. The n.m.r. absorption spectrum of a radical was found to invert when the electron resonance signal was strongly saturated, so that energy was emitted instead of being absorbed. How do you account for this observation?

7. A simple three-level solid state maser can be made from a system which has three unequally spaced energy levels $E_1 > E_2 > E_3$. By saturating the transition $1 \rightarrow 3$ it may be possible to increase the ratio of populations $N_1/N_2$ so that spontaneous emission of energy

occurs. If $X$ and $Y$ represent the average probabilities for the transitions $1 \leftrightarrow 2$ and $2 \leftrightarrow 3$ calculate the saturation value of $N_1/N_2$. At high temperatures, what conditions are necessary to achieve an increase in $(N_1/N_2)$? Compare with Bloembergen, *Phys. Rev.*, **104**: 324 (1956).

**8.** A nucleus $I = 1/2$ is irradiated simultaneously by a strong rf field $H_1$ at a fixed frequency $\omega_1$ and a weak field $H_2$ at a variable frequency $\omega_2$. Show that in a reference frame rotating at angular velocity $\omega_1$ the spin vector $\mathbf{I}'$ precesses about the resultant of a steady field $\mathbf{H}'_{\text{eff}} = (H_0 - \omega_1/\gamma)\mathbf{k}' + H_1\mathbf{i}'$ and a weak field $H'_2$ which rotates at an angular frequency $(\omega_2 - \omega_1)$. Hence deduce that two $H_2$ resonance signals may appear at frequencies

$$\omega_2 = \omega_1 \pm \sqrt{(\omega_0 - \omega_1)^2 + \gamma^2 H_1^2}$$

where $\omega_0 = \gamma H_0$. One of these signals is inverted; can you explain why?

## SUGGESTIONS FOR FURTHER READING

Abragam: Chapters 8, 9, and 12. The full theory of all the effects treated in this chapter, except for ENDOR.

Pake: Chapter 8. Brief descriptions of the Overhauser and solid-state effects, and ENDOR.

Anderson, Zandstra, and Tuttle: *J. Chem. Phys.*, **33**: 1581, 1591 (1960). Knight shifts in the pyrene ion and DPPH crystals.

Eaton, Josey, Phillips, and Benson: *Discussions Faraday Soc.*, **34**: 77 (1962). Unpaired electron distributions from Knight shifts.

Bernheim, Brown, Gutowsky, and Woessner: *J. Chem. Phys.*, **30**: 950 (1959). Proton relaxation by metal ions in water.

Bloembergen and Morgan: *J. Chem. Phys.*, **34**: 841 (1961). Theory of proton relaxation by $Mn^{++}$ ions in solution.

Eisinger, Shulman, and Szymanski: *J. Chem. Phys.*, **36**: 1721 (1962). Binding of metal ions to DNA.

Kowalsky and Cohn: *Ann. Rev. Biochem.*, **33**: 481 (1964). Applications of n.m.r. in biochemistry. A review.

Swift and Connick: *J. Chem. Phys.*, **37**: 307 (1962). Study of hydration to paramagnetic ions by $O^{17}$ n.m.r. line width measurements.

Abragam: *Phys. Rev.*, **98**: 1729 (1955). Theory of the Overhauser effect in solution.

Carver and Slichter: *Phys. Rev.*, **102**: 975 (1956). Overhauser experiment in metal-ammonia solutions.

Richards and White: *Proc. Roy. Soc. (London)*, **A283**: 459 (1965). Chemical applications of the Overhauser effect.

Jeffries: *Phys. Rev.*, **117**: 1056 (1960). The solid-state effect.

Feher: *Phys. Rev.*, **114**: 1219 (1959). ENDOR method described. Experiments on paramagnetic impurity centers in silicon.

Holton, Blum, and Slichter: *Phys. Rev. Letters*, **5**: 197 (1960). ENDOR. Hyperfine structure of F centers in LiF.

Bloom and Shoolery: *Phys. Rev.*, **97**: 1261 (1955). The first spin decoupling experiment.

Baldeschwieler and Randall: *Chem. Revs.*, **63**: 81 (1963). Chemical applications of nuclear magnetic double resonance.

# APPENDIXES

# APPENDIX A: MATRIX ELEMENTS AND EIGENVALUES

The wave functions $\psi_1, \psi_2 \cdots \psi_n \cdots$ for the states of a quantum system are normalized and orthogonal, obeying the relations

$$\int \psi_n^* \psi_n \, dv = 1 \tag{1}$$

$$\int \psi_m^* \psi_n \, dv = 0 \qquad n \neq m \tag{2}$$

In Dirac's notation $\psi_n$ and $\psi_n^*$ are designated $|n\rangle$ and $\langle n|$ while the orthogonality integrals are denoted by $\langle m|n\rangle = \delta_{mn}$. If $A$ is any operator corresponding to a physical quantity the array of integrals

$$A_{mn} = \int \psi_m^* A \psi_n \, dv \equiv \langle m|A|n\rangle \tag{3}$$

are known as the matrix elements of $A$ and give a matrix representation of the operator $A$, of the form

$$\begin{bmatrix} A_{11} & A_{12} \cdots \\ A_{21} & A_{22} \cdots \\ \cdots\cdots\cdots\cdots \end{bmatrix} \tag{4}$$

The matrix is Hermitian, i.e., $A_{mn} = A_{nm}^*$. A state which satisfies the relation $A\psi_n = a_n \psi_n$ or $A|n\rangle = a_n|n\rangle$ is called an eigenstate of $A$ and $a_n$ is the corresponding eigenvalue. Eigenstates having different eigenvalues for $A$ are always orthogonal, and if $\psi_1, \psi_2 \ldots$ are all eigenstates of $A$ the matrix takes the diagonal form

$$\begin{bmatrix} u_1 & 0 \cdots \\ 0 & a_2 \cdots \\ \cdots\cdots\cdots \end{bmatrix} \tag{5}$$

The matrix of the operator $AB$ is obtained from those of $A$ and $B$ by matrix multiplication

$$\langle l|AB|m\rangle = \sum_n \langle l|A|n\rangle\langle n|B|m\rangle \tag{6}$$

This result depends on the property that the states $|n\rangle$ form a complete set, such that

$$\sum_n |n\rangle\langle n| = 1 \tag{7}$$

Equation (7) can be used to expand any wave function $|\psi\rangle$ in terms of the set $|n\rangle$

$$|\psi\rangle = \sum_n |n\rangle\langle n|\psi\rangle \tag{8}$$

In ordinary notation this reads

$$\psi = \sum_n C_n \psi_n, \quad \text{and} \quad C_n = \int \psi_n^* \psi \, dv.$$

Frequently one needs to find the eigenvalues and eigenstates of an operator, usually the energy operator $\mathcal{H}$, and the following procedure is used.

Consider the state $\psi = \sum_n C_n \psi_n$, and let $\mathcal{H}\psi = \sum_n C'_n \psi_n$. Then the vector $C'_n$ is obtained from $C_n$ as $C'_n = \sum_m \mathcal{H}_{nm} C_m$. Now suppose $\mathcal{H}\psi = E\psi$. Then it follows that

$$\sum_m \mathcal{H}_{nm} C_m = EC_n \tag{9}$$

or in matrix notation

$$\begin{bmatrix} \mathcal{H}_{11} - E & \mathcal{H}_{12} & \cdots \\ \mathcal{H}_{21} & \mathcal{H}_{22} - E & \\ \vdots & & \ddots \end{bmatrix} \begin{bmatrix} C_1 \\ C_2 \\ \vdots \end{bmatrix} = 0 \tag{10}$$

The allowed energy values $E$ are the roots of the so-called *secular determinant* $|\mathcal{H}_{mn} - E\delta_{mn}|$ of the matrix (10), while the coefficients $C_n$, which satisfy (10), and are correctly normalized with $\sum_n |C_n|^2 = 1$ are the eigenvectors. The eigenvalues of a Hermitian matrix are always real. In general the matrices of two operators do not commute, $AB \neq BA$; but if they do then both operators have the same eigenvectors.

## PROBLEMS

**1.** Verify by matrix multiplication the relations $\sigma_x \sigma_y = i\sigma_z$ and $\sigma_x^2 = 1$ for the Pauli spin matrices (Appendix C).

**2.** The matrix $\mathbf{I} \cdot \mathbf{S}$ for two spins of $1/2$ is

$$\begin{matrix} \alpha_e \alpha_n \\ \alpha_e \beta_n \\ \beta_e \alpha_n \\ \beta_e \beta_n \end{matrix} \begin{bmatrix} \tfrac{1}{4} & 0 & 0 & 0 \\ 0 & -\tfrac{1}{4} & \tfrac{1}{2} & 0 \\ 0 & \tfrac{1}{2} & -\tfrac{1}{4} & 0 \\ 0 & 0 & 0 & \tfrac{1}{4} \end{bmatrix}$$

Find its eigenvalues and eigenvectors.

## SUGGESTIONS FOR FURTHER READING

Landau and Lifshitz: Chapters 1 and 2.

Schiff: *Quantum Mechanics* (New York: McGraw-Hill Book Co., Inc., 1955). Chapter 6.

Eyring, Walter, and Kimball: Page 31.

Margenau and Murphy: *The Mathematics of Physics and Chemistry*, Vol. I (New York: McGraw-Hill Book Co., Inc., 1956). Chapter 10.

# APPENDIX B: TIME-INDEPENDENT PERTURBATION THEORY

The problem is to find the stationary states of a system with the Hamiltonian $H_0 + V$, where $V$ is a small perturbation. The unperturbed energies and wave functions are $E_1, E_2 \cdots$ and $\psi_1, \psi_2 \cdots$ and we require the corrections to $E_n$ and $\psi_n$ as a series expansion in powers of $V$.

## 1. UNPERTURBED LEVEL NONDEGENERATE

Following Eq. (10) of Appendix A we seek solutions of the linear equations

$$
\begin{bmatrix}
(E_n + V_{nn} - E) & V_{nm} & \cdots \\
V_{mn} & (E_m + V_{mm} - E) & \\
\vdots & & \ddots
\end{bmatrix}
\begin{bmatrix}
C_n \\
C_m \\
\vdots
\end{bmatrix} = 0
\tag{1}
$$

The zeroth approximation is $E = E_n$ and the vector $C$ then has the form $(1,0,0, \ldots)$. Next, the first-order terms from the first row of Eqs. (1) give

$$
(E_n + V_{nn} - E)\cdot 1 + 0 + \cdots = 0
\tag{2}
$$

and the first-order energy correction is $V_{nn}$. The second row, to first order in $V$ becomes

$$
V_{mn}\cdot 1 + (E_m - E_n)C_m = 0
\tag{3}
$$

and yields a correction to $C_m$. Now substitution of the new coefficients back into the first row gives the second-order energy shift. The results then are

$$
\psi = \psi_n - \sum_{n \neq m} \frac{V_{mn}}{E_m - E_n} \psi_m + \cdots
\tag{4}
$$

$$
E = E_n + V_{nn} - \sum_{n \neq m} \frac{V_{mn}V_{nm}}{E_m - E_n} + \cdots
\tag{5}
$$

## 2. DOUBLY DEGENERATE LEVELS COUPLED BY V

We consider two states $\psi_1, \psi_2$ which both have energy $E_0$ initially. Equations (4) and (5) break down because $(E_1 - E_2) = 0$. Instead one must solve the matrix equations directly. Setting $E = (E_0 + \varepsilon)$ they become

$$
\begin{bmatrix}
V_{11} - \varepsilon & V_{12} \\
V_{21} & V_{22} - \varepsilon
\end{bmatrix}
\begin{bmatrix}
C_1 \\
C_2
\end{bmatrix} = 0
\tag{6}
$$

The same procedure is used for any number of degenerate levels.

### 3. DOUBLY DEGENERATE LEVELS COUPLED TO HIGHER STATES

Sometimes the matrix elements of $V$ between $\psi_1$ and $\psi_2$ vanish, but there are elements $V_{1m}$, $V_{2m}$ to higher states, and the equations to be solved are

$$\begin{bmatrix} (E_0 - E) & 0 & V_{1m} \cdots \\ 0 & (E_0 - E) & V_{2m} \cdots \\ \hline V_{m1} & V_{m2} & (E_m + V_{mm} - E) \\ \vdots & \vdots & \ddots \end{bmatrix} \begin{bmatrix} C_1 \\ C_2 \\ \hline C_m \\ \vdots \end{bmatrix} = 0 \tag{7}$$

The perturbed wave functions now take the form

$$\psi = C_1 \left( \psi_1 - \sum_m \frac{V_{m1}}{E_m - E_0} \psi_m \right) + C_2 \left( \psi_2 - \sum_m \frac{V_{m2}}{E_m - E_0} \psi_m \right) \tag{8}$$

while the coefficients and the energy are found by solving the matrix

$$\begin{bmatrix} U_{11} - \varepsilon & U_{12} \\ U_{21} & U_{22} - \varepsilon \end{bmatrix} \begin{bmatrix} C_1 \\ C_2 \end{bmatrix} = 0 \tag{9}$$

The elements $U_{ik}$ $(i, k = 1, 2)$ are defined as

$$U_{ik} = -\sum_m \frac{V_{im} V_{mk}}{E_m - E_0} \tag{10}$$

and constitute an effective coupling between the degenerate states.

If the perturbation has both $V_{12}$ and $V_{1m}$ types of element one must add $U_{ik}$ to $V_{ik}$ and solve the combined $2 \times 2$ matrix.

### SUGGESTIONS FOR FURTHER READING

Landau and Lifshitz: Chapter 6.
Eyring, Walter, and Kimball: Chapter 7.

# APPENDIX C: SPIN ANGULAR MOMENTUM

The same principles apply to both electron and nuclear spins. Here we shall deal with electron spin S.

The *orbital* angular momentum $L\hbar$ of a particle is defined as $\mathbf{r} \times \mathbf{p}$, where $\mathbf{p}$ is the momentum. The quantum mechanical operator for $\mathbf{L}$ is $-i(\mathbf{r} \times \nabla)$, and leads immediately to the conclusion that the different components of $\mathbf{L}$ do not commute, but $L_x L_y - L_y L_x = iL_z$.

The spin angular momentum $S\hbar$ of a group of electrons obeys the same commutation rules as $\mathbf{L}$:

$$S_x S_y - S_y S_x = iS_z \tag{1}$$

and two others derived by rotating the $x, y, z$ suffixes. The squared spin $\mathbf{S}^2 = (S_x^2 + S_y^2 + S_z^2)$ commutes with $S_x$, $S_y$, and $S_z$, and has the value $S(S + 1)$ where $S$ is the total spin. The entire theory of spin follows from (1). First it is useful to introduce the shift operators

$$S^+ = S_x + iS_y$$
$$S^- = S_x - iS_y \tag{2}$$

and remark that (1) leads to the important relations

$$S_z S^\pm = S^\pm (S_z \pm 1) \tag{3}$$

$$(\mathbf{S}^2 - S_z^2) = (S^- S^+ + S_z) = (S^+ S^- - S_z) \tag{4}$$

A single electron has $S = 1/2$, and the two spin states $|\alpha\rangle$, $|\beta\rangle$, are eigenstates of $S_z$ (see Appendix A) with value $\pm 1/2$.

$$S_z|\alpha\rangle = \tfrac{1}{2}|\alpha\rangle \qquad S_z|\beta\rangle = -\tfrac{1}{2}|\beta\rangle \tag{5}$$

We shall now derive important relations for $S^+$ and $S^-$.

$$S^+|\beta\rangle = |\alpha\rangle \qquad S^+|\alpha\rangle = 0$$
$$S^-|\alpha\rangle = |\beta\rangle \qquad S^-|\beta\rangle = 0 \tag{6}$$

The first is proved by showing that $S^+|\beta\rangle$ is an eigenstate of $S_z$ with value $+1/2$. We use (3) and (5).

$$S_z \cdot S^+|\beta\rangle = S^+ \cdot (S_z + 1)|\beta\rangle = S^+ \cdot \tfrac{1}{2}|\beta\rangle$$
$$= \tfrac{1}{2} \cdot S^+|\beta\rangle \tag{7}$$

This only shows that $S^+|\beta\rangle$ is a multiple of $|\alpha\rangle$. It is necessary to check that $S^+|\beta\rangle$ is correctly normalized. The complex conjugate state is $\langle\beta|S^-$, and we obtain the normalization integral $\langle\beta|S^- S^+|\beta\rangle$. By (4) and (5) this reduces to $\langle\beta|\mathbf{S}^2 - S_z^2 - S_z|\beta\rangle = \langle\beta|\mathbf{S}^2 + \tfrac{1}{4}|\beta\rangle$. Finally it is necessary to show that

$$\mathbf{S}^2 = \tfrac{3}{4} \qquad (\text{spin } \tfrac{1}{2}) \tag{8}$$

From (5), $S_z^2$ acting on $|\alpha\rangle$ or $|\beta\rangle$ simply multiplies each by 1/4. Hence $S_z^2 = 1/4$; by symmetry the same is true of $S_x^2$ and $S_y^2$, so (8) holds. Thus we find that $\langle\beta|S^- S^+|\beta\rangle = \langle\beta|\beta\rangle = 1$, and so $S^+|\beta\rangle$ is correctly normalized and may be identified with $|\alpha\rangle$.

To derive the second relation (6) we prove that the normalization $\langle\alpha|S^- S^+|\alpha\rangle$ of $S^+|\alpha\rangle$ vanishes, so the state does not exist. The matrix element is just

$$\langle\alpha|\mathbf{S}^2 - S_z^2 - S_z|\alpha\rangle = \langle\alpha|\mathbf{S}^2 - \tfrac{3}{4}|\alpha\rangle = 0$$

The matrix elements of the operators $S_x$, $S_y$, $S_z$ are usually represented by the Pauli spin matrices $\sigma_x$, $\sigma_y$, $\sigma_z$, with $\mathbf{S} = \tfrac{1}{2}\boldsymbol{\sigma}$. They are

$$\sigma_x = \begin{pmatrix} 0 & 1 \\ 1 & 0 \end{pmatrix} \qquad \sigma_y = \begin{pmatrix} 0 & -i \\ i & 0 \end{pmatrix} \qquad \sigma_z = \begin{pmatrix} 1 & 0 \\ 0 & -1 \end{pmatrix} \tag{9}$$

The states of a general spin $S > 1/2$ are labeled by the eigenvalue $M_S$ or $M$ of $S_z$, and written $|S, M\rangle$ or just $|M\rangle$

$$S_z|S, M\rangle = M|S, M\rangle \tag{10}$$

$M$ takes $(2S + 1)$ possible values $S, (S - 1) \cdots -S$. It can be shown that $S^+$ and $S^-$ only have matrix elements connecting $M$ with $(M \pm 1)$.

$$\langle M + 1|S^+|M\rangle = \langle M|S^-|M + 1\rangle = \sqrt{S(S + 1) - M(M + 1)} \tag{11}$$

Two spin angular momentum vectors *which commute*, such as $\mathbf{I}$ and $\mathbf{S}$, can be combined together to form a resultant

$$\mathbf{F} = \mathbf{I} + \mathbf{S} \tag{12}$$

The quantum states of the coupled system are now eigenstates of $\mathbf{F}^2$ and $F_z$. $\mathbf{F}^2$ has the value $F(F + 1)$, while the total spin $F$ may have one of several values. The spins $I$ and $S$ give a series of multiplets corresponding to the $F$ values $I + S, (I + S - 1) \cdots |I - S|$. Each multiplet consists of $(2F + 1)$ sublevels with different $F_z$ values. Often one needs to know the value of $\mathbf{I} \cdot \mathbf{S}$ (nuclear hyperfine splitting) or $\mathbf{L} \cdot \mathbf{S}$ (spin orbit splitting) in a coupled state. This is found easily:

$$\mathbf{I} \cdot \mathbf{S} = \tfrac{1}{2}[(\mathbf{I} + \mathbf{S})^2 - \mathbf{I}^2 - \mathbf{S}^2] = \tfrac{1}{2}[\mathbf{F}^2 - \mathbf{I}^2 - \mathbf{S}^2]$$

$$= \tfrac{1}{2}[F(F + 1) - I(I + 1) - S(S + 1)]. \tag{13}$$

The spin wave functions for two or more electrons are a little more complicated. The total spin angular momentum vector for two electrons is

$$\mathbf{S} = \mathbf{S}_1 + \mathbf{S}_2 \tag{14}$$

and its components obey the commutation relations (1). The squared spin $\mathbf{S}^2 = (\mathbf{S}_1 + \mathbf{S}_2)^2$ again takes the value $S(S + 1)$ but now the quantum number $S$ may be either 0 (singlet state) or 1 (triplet state) while the resolved component $S_z$ has the possible values $1, 0, -1$. There are four possible spin wave functions

$$\alpha_1\alpha_1 \qquad\qquad S_z = 1$$

$$\alpha_1\beta_2, \beta_1\alpha_2 \qquad\qquad 0$$

$$\beta_1\beta_2 \qquad\qquad -1$$

and we wish to find the wave functions of the singlet and triplet states.

The function $\alpha_1\alpha_2$ has $S_z = 1$ and is clearly a triplet. We check this by using Eq. (4).

$$\mathbf{S}^2\alpha_1\alpha_2 = [S^-S^+ + S_z(S_z + 1)]\alpha_1\alpha_2 \tag{15}$$

But $S^+\alpha_1\alpha_2 = (S_1^+ + S_2^+)\alpha_1\alpha_2 = 0$, and so we find

$$\mathbf{S}^2\alpha_1\alpha_2 = S_z(S_z + 1)\alpha_1\alpha_2 = S(S + 1)\alpha_1\alpha_2 \tag{16}$$

To find a triplet wave function with $S_z = 0$ we use (11) and apply the operator $(1/\sqrt{2})S^-$.

$$\frac{1}{\sqrt{2}} S^-\alpha_1\alpha_2 = \frac{1}{\sqrt{2}}(\alpha_1\beta_2 + \beta_1\alpha_2) \tag{17}$$

A further shift down gives $\beta_1\beta_2$ to complete the set. The only remaining spin function must be a singlet state $(1/\sqrt{2})(\alpha_1\beta_2 - \beta_1\alpha_2)$, since it has $S_z = 0$ and $\mathbf{S}^2 = 0$. To verify this we use (4) again

$$\mathbf{S}^2(\alpha_1\beta_2 - \beta_1\alpha_2) = S^-S^+(\alpha_1\beta_2 - \beta_1\alpha_2) \tag{18}$$

$$S^+(\alpha_1\beta_2 - \beta_1\alpha_2) = (\alpha_1\alpha_2 - \alpha_1\alpha_2) = 0 \tag{19}$$

Similar methods can be used to obtain spin functions for three electrons (or nuclei) which are listed in Section 4.4.4.

APPENDIX D: TENSORS AND VECTORS

## PROBLEMS

**1.** Construct the matrices $S_x$, $S_y$, $S_z$ for spins $S = 1$, $S = 3/2$, and $S = 2$.

**2.** Find the eigenfunctions of $S_x$ and $S_y$ for a spin $S = 1$.

**3.** A magnetic field $H$ makes an angle $\theta$ with the $z$ axis. Find the stationary states for an electron spin $S = 1/2$. Calculate the probabilities that the electron will be found in the states $|\alpha\rangle$ or $|\beta\rangle$.

**4.** A sodium atom ($I = 3/2$, $S = 1/2$) in zero field has the Hamiltonian $\mathscr{H} = a I \cdot S$. Find the energy levels.


## SUGGESTIONS FOR FURTHER READING

Landau and Lifshitz: Chapter 4.

Eyring, Walter, and Kimball: Chapter 9.


# APPENDIX D: TENSORS AND VECTORS

The type of tensor we are concerned with in this book is called a Cartesian tensor. Just as a vector

$$S = S_x \mathbf{i} + S_y \mathbf{j} + S_z \mathbf{k} \tag{1}$$

has various components $S_x$, $S_y$, $S_z$ which are referred to a set of rectangular axes $\mathbf{i}$, $\mathbf{j}$, $\mathbf{k}$, so does a Cartesian tensor. But a tensor T has 9 components

$$\mathbf{T} = \begin{bmatrix} T_{xx} & T_{xy} & T_{xz} \\ T_{yx} & T_{yy} & T_{yz} \\ T_{zx} & T_{zy} & T_{zz} \end{bmatrix} \tag{2}$$

which can be arranged to form a $3 \times 3$ matrix. By multiplying the matrix of **T** on the left with a row vector S one obtains a new row vector $\mathbf{S} \cdot \mathbf{T}$. Similarly matrix multiplication on the right by a column vector I yields a new column vector $\mathbf{T} \cdot \mathbf{I}$. Finally it is possible to form a scalar $\mathbf{S} \cdot \mathbf{T} \cdot \mathbf{I}$ by the matrix multiplication

$$[S_x, S_y, S_z] \begin{bmatrix} T_{xx} & T_{xy} & T_{xz} \\ T_{yx} & T_{yy} & T_{yz} \\ T_{zx} & T_{zy} & T_{zz} \end{bmatrix} \begin{bmatrix} I_x \\ I_y \\ I_z \end{bmatrix} \tag{3}$$

Note that $\mathbf{S} \cdot \mathbf{T} \cdot \mathbf{I}$ may also be regarded as the scalar product of two vectors; either $(\mathbf{S} \cdot \mathbf{T})$ with I, or equally S with $(\mathbf{T} \cdot \mathbf{I})$. In any case the product written in full is

$$S_x T_{xx} I_x + S_x T_{xy} I_y + \cdots S_y T_{yz} I_z + S_z T_{zz} I_z \tag{4}$$

or

$$\sum_{i,k} S_i T_{ik} I_k \qquad (i, k = x, y, z) \tag{5}$$

By analogy with Eq. (1) one sometimes uses the notation

$$\mathbf{T} = \mathbf{i} T_{xx} \mathbf{i} + \mathbf{i} T_{xy} \mathbf{j} + \cdots \mathbf{k} T_{zz} \mathbf{k} \tag{6}$$

and then the scalar products $\mathbf{i \cdot T \cdot i}$, $\mathbf{i \cdot T \cdot j}$ and so on are just the components of the tensor. For example, it follows from the orthogonality of the unit vectors $\mathbf{i}$, $\mathbf{j}$, $\mathbf{k}$ that

$$\mathbf{i \cdot T \cdot j} = (\mathbf{i \cdot i})T_{xx}(\mathbf{i \cdot j}) + (\mathbf{i \cdot i})T_{xy}(\mathbf{j \cdot j}) + \cdots (\mathbf{i \cdot k})T_{zz}(\mathbf{k \cdot j}) \tag{7}$$

$$= 0 + T_{xy} + \cdots 0$$

$$= T_{xy} \tag{8}$$

There are various important types of tensor. *Symmetric tensors* have a symmetric matrix; $T_{xy} = T_{yx}$ etc. *Antisymmetric tensors* have $T_{xy} = -T_{yx}$ and zeros on the diagonal. *Diagonal tensors*, as the name suggests, have the form

$$\begin{bmatrix} t_1 & 0 & 0 \\ 0 & t_2 & 0 \\ 0 & 0 & t_3 \end{bmatrix} \tag{9}$$

$t_1$, $t_2$, $t_3$ are called the principal values of the tensor, and if $\mathbf{T}$ is diagonal in a certain rectangular coordinate system, the coordinate axes are known as the principal axes.

The diagonal sum of a tensor is an important quantity called the *trace* of the tensor denoted by the symbol $\mathrm{Tr}\{\ \}$.

$$\mathrm{Tr}\{\mathbf{T}\} = (T_{xx} + T_{yy} + T_{zz}) \tag{10}$$

Finally a most useful tensor is the unit tensor $\mathbf{1}$, or $(\mathbf{ii + jj + kk})$ whose matrix is the unit matrix

$$\begin{bmatrix} 1 & 0 & 0 \\ 0 & 1 & 0 \\ 0 & 0 & 1 \end{bmatrix} \tag{11}$$

The main reason for using tensor notation is that it is extremely easy to change from one set of rectangular axes to another. The orientation of an axis system $\alpha$, $\beta$, $\gamma$ with respect to another set $x, y, z$ is specified by 9 direction cosines $l_{\alpha x}, l_{\beta x} \cdots$ which form a matrix

$$\mathbf{L} = \begin{bmatrix} l_{\alpha x} & l_{\beta x} & l_{\gamma x} \\ l_{\alpha y} & l_{\beta y} & l_{\gamma y} \\ l_{\alpha z} & l_{\beta z} & l_{\gamma z} \end{bmatrix} \tag{12}$$

Each component is the scalar product of one of the unit vectors $\mathbf{i}$, $\mathbf{j}$, $\mathbf{k}$ with one of the new unit vectors $\mathbf{a}$, $\mathbf{b}$, $\mathbf{c}$; $l_{\alpha z}$ for instance is equal to $(\mathbf{a \cdot k})$, and is the same as $l_{z\alpha}$. The rows (or columns) of (12) form a set of orthogonal normalized vectors.

To calculate the components of vectors and tensors in the new coordinate system we first note that

$$\mathbf{a} = (\mathbf{a \cdot i})\mathbf{i} + (\mathbf{a \cdot j})\mathbf{j} + (\mathbf{a \cdot k})\mathbf{k}$$

$$= l_{\alpha x}\mathbf{i} + l_{\alpha y}\mathbf{j} + l_{\alpha z}\mathbf{k} \tag{13}$$

Hence the $\alpha$ component of a vector $\mathbf{S}$ is

$$S_\alpha = (\mathbf{S \cdot a}) = l_{\alpha x}S_x + l_{\alpha y}S_y + l_{\alpha z}S_z \tag{14}$$

Similarly the $\alpha\beta$ component of a tensor $\mathbf{T}$ becomes

$$T_{\alpha\beta} = \mathbf{a \cdot T \cdot b}$$

$$= (l_{\alpha x}\mathbf{i} + l_{\alpha y}\mathbf{j} + l_{\alpha z}\mathbf{k}) \cdot \mathbf{T} \cdot (\mathbf{i}l_{x\beta} + \mathbf{j}l_{y\beta} + \mathbf{k}l_{z\beta})$$

$$= l_{\alpha x}T_{xx}l_{x\beta} + l_{\alpha x}T_{xy}l_{y\beta} + \cdots \tag{15}$$

The new matrix of **T** can be written as the product of (2) with **L** and its transpose:

$$\begin{bmatrix} l_{\alpha x} & l_{\beta x} & l_{\gamma x} \\ l_{\alpha y} & l_{\beta y} & l_{\gamma y} \\ l_{\alpha z} & l_{\beta z} & l_{\gamma z} \end{bmatrix} \begin{bmatrix} T_{xx} & T_{xy} & T_{xz} \\ T_{yx} & T_{yy.} & T_{yz} \\ T_{zx} & T_{zy} & T_{zz} \end{bmatrix} \begin{bmatrix} l_{x\alpha} & l_{x\beta} & l_{xy} \\ l_{y\alpha} & l_{y\beta} & l_{yy} \\ l_{z\alpha} & l_{z\beta} & l_{zy} \end{bmatrix} \tag{16}$$

The transformation (16) has two important properties. First it follows from the orthogonality of the direction cosines that the trace of **T** is unchanged.

$$(T_{\alpha\alpha} + T_{\beta\beta} + T_{\gamma\gamma}) = (T_{xx} + T_{yy} + T_{zz}) \tag{17}$$

Second, if the tensor is *real* and *symmetric* it is always possible to find a new axis system for which $\alpha$, $\beta$, $\gamma$ are principal axes and the tensor is diagonal. The procedure for finding principal axes is identical with the procedure which was described in Appendix A for finding the eigenvectors of an operator. For instance, to find the direction of the $\alpha$ axis and the corresponding principal value $t_1$ we look for solutions of the equations

$$\begin{bmatrix} (T_{xx} - t_1) & T_{xy} & T_{xz} \\ T_{yx} & (T_{yy} - t_1) & T_{yz} \\ T_{zx} & T_{zy} & (T_{zz} - t_1) \end{bmatrix} \begin{bmatrix} l_{x\alpha} \\ l_{y\alpha} \\ l_{z\alpha} \end{bmatrix} = 0 \tag{18}$$

In connection with these transformations it is noteworthy that the scalar quantity **S·T·I** is invariant, and so is the unit tensor (11). Conversely the only tensor which is invariant under all rotations of axes is the unit tensor **1**, or a multiple of it.

Finally let us consider the average behavior relative to space axes $x, y, z$ of a tensor which belongs to a rapidly rotating molecule in solution. In the molecular axis system $\alpha\beta\gamma$ the components $T_{\alpha\alpha}$, $T_{\alpha\beta}$ etc. are fixed, but the space values $T_{xx}, T_{xy}, \ldots$ fluctuate in a random manner. The time average of **T** is an invariant tensor, so it must be a multiple of **1**. Also the trace of **T** is invariant, so the time average consists of a diagonal tensor with

$$t_1 = t_2 = t_3 = \tfrac{1}{3}(T_{xx} + T_{yy} + T_{zz}) \tag{19}$$

Clearly a tensor with zero trace vanishes on the average.

## PROBLEMS

1.  Write out the vectors **S·T** and **T·I** in matrix notation.

2.  When, if ever, is **S·T·I** equal to **I T S**?

3.  Write out in full the orthogonality conditions for the rows and columns of (12). Hence prove (17).

4.  The tensor **T** is diagonal in the axis system $x, y, z$. The unit vectors **a**, **b**, **c** are obtained by rotating the **i, j, k** axes through an angle $\theta$ in the $xy$ plane. Find the components of **T** in the $\alpha$, $\beta$, $\gamma$ system.

5.  Do Problem 4 again, interchanging the axis systems.

6.  The e.s.r. spectrum of the $CH(COOH)_2$ radical was studied in a single crystal to determine the hyperfine tensor of the CH proton. The components (in Mc/s) referred to crystal axes are

$$\begin{bmatrix} -53 & -7 & -17 \\ -7 & -82 & -15 \\ -17 & -15 & -41 \end{bmatrix}$$

Find the principal values and the principal directions.

7.  The $g$ tensor of the $CH_2(COOH)$ radical in the same crystal as Problem 6 is

$$
\begin{bmatrix}
2.0033 & -0.0005 & 0.0000 \\
-0.0005 & 2.0028 & 0.0010 \\
0.0000 & 0.0010 & 2.0033
\end{bmatrix}
$$

Find its principal values. Are the principal directions the same as in Problem 6? Comment on the result.

## SUGGESTIONS FOR FURTHER READING

Morse and Feshbach: *Methods of Theoretical Physics*, Part I. (New York: McGraw-Hill Book Co., Inc., 1953.) Pages 54–107.

# APPENDIX $E$: TIME-DEPENDENT PERTURBATION THEORY

In this section we consider the behavior of a quantum system which is acted on by a time-dependent perturbation $V(t)$.

It is sufficient to discuss a two-level system with states $\psi_a$, $\psi_b$ having the energies $E_a$, $E_b$. Further, we shall suppose that $V(t)$ takes the form

$$ V(t) = V \cdot f(t) \tag{1} $$

where $V$ is an operator which is independent of time and $f(t)$ is a fluctuating numerical factor which measures the strength of $V$ at different times. The diagonal elements of $V$, $V_{aa}$ and $V_{bb}$, cause a time-dependent modulation of the energy levels $E_a$, $E_b$. The off-diagonal elements $V_{ab} = V_{ba}^*$ induce transitions between them. Here we shall only discuss the transitions.

Suppose $V$ has no diagonal elements. The energy matrix is

$$
\begin{bmatrix}
E_a & V_{ab}f(t) \\
V_{ba}f(t) & E_b
\end{bmatrix}
\tag{2}
$$

and we look for a solution of the Schrödinger equation

$$ i\hbar \frac{\partial \psi}{\partial t} = \mathcal{H}\psi \tag{3} $$

which will have the form

$$ \psi = C_a(t)\psi_a e^{-iE_a t/\hbar} + C_b(t)\psi_b e^{-iE_b t/\hbar} \tag{4} $$

Substituting (4) into (3) and (2) we find that the coefficients satisfy

$$ i\hbar \frac{\partial C_a}{\partial t} = f(t)e^{i(E_a - E_b)t/\hbar} V_{ab}C_b \tag{5} $$

$$ i\hbar \frac{\partial C_b}{\partial t} = f(t)e^{i(E_b - E_a)t/\hbar} V_{ba}C_a \tag{6} $$

Let us now suppose that at $t = 0$ the system starts in $\psi_a$; or $C_a(0) = 1$, $C_b(0) = 0$. Integration of (6) with $C_a = 1$ gives the first-order correction to $C_b$, and after time $T$ we find

$$C_b(T) = -\frac{i}{\hbar} V_{ba} \int_0^T f(t) e^{i(E_b - E_a)t/\hbar} \, dt \tag{7}$$

The probability that the system has made a transition to state $b$ is $|C_b(T)|^2$, or

$$P(a, b) = \frac{1}{\hbar^2} |V_{ab}|^2 \int_0^T dt_1 \int_0^T dt_2 f(t_1) f(t_2) e^{i(E_b - E_a)(t_1 - t_2)/\hbar} \tag{8}$$

The time integral in (8) is simplified by changing to the new variables $t_1 = (t + \tau)$, $t_2 = t$, whence

$$P(a, b) = \frac{1}{\hbar^2} |V_{ab}|^2 \int_0^T dt \int_{-t}^{T-t} f(t + \tau) f(t) e^{i(E_b - E_a)\tau/\hbar} \, d\tau \tag{9}$$

There are two important situations where the probability increases linearly with time, and transitions occur at a definite rate $P_{ab}$, with

$$P(a, b) = P_{ab} T \tag{10}$$

The first is when $f(t)$ is a random force with average value zero, which fluctuates many times in the time interval $T$. The integral (9) is also a random function, but the statistical average value of the expression

$$G(\tau) = \overline{f(t + \tau) f(t)} \tag{11}$$

for fixed $\tau$ generally has a definite value, independent of the time $t$. $G(\tau)$ is known as the *autocorrelation function* of $f(t)$, and we may suppose that $G(0) = \overline{f^2(t)}$ is normalized to unity by a suitable choice of $V$ in (1). Now (9) becomes

$$P(a, b) = \frac{1}{\hbar^2} |V_{ab}|^2 \int_0^T dt \int_{-t}^{T-t} G(\tau) e^{i(E_b - E_a)\tau/\hbar} \, d\tau \tag{12}$$

and $|V_{ab}|^2$ is actually the mean square value of $|V_{ab}(t)|^2$. $G(\tau)$ is generally a rapidly decreasing function of $\tau$, and when $T$ is large the second integral in (12) becomes almost independent of $T$, while integrating with respect to $t$ gives a factor of $T$ in the answer. The transition probability $P_{ab}$ is proportional to the power spectrum $J(\omega)$ of the random fluctuations of $f(t)$ at the resonance frequency. This quantity is defined as

$$J(\omega) = \int_{-\infty}^{+\infty} G(\tau) e^{i\omega\tau} \, d\tau \tag{13}$$

and after a little algebra we obtain from (10), (12), and (13)

$$P_{ab} = \frac{1}{\hbar^2} \overline{|V_{ab}(t)|^2} J(\omega_{ba}) \tag{14}$$

$$\hbar\omega_{ba} = (E_b - E_a)$$

Frequently $G(\tau)$ decays exponentially with a characteristic correlation time $\tau_c$: $G(\tau) = e^{-|\tau|/\tau_c}$. Then the transition probability becomes

$$P_{ab} = \frac{1}{\hbar^2} \overline{|V_{ab}|^2} \frac{2\tau_c}{1 + \omega_{ab}^2 \tau_c^2} \tag{15}$$

The second case of importance is a strictly periodic force

$$f(t) = 2 \cos \omega t \tag{16}$$

We return to Eq. (7) and integrate it directly

$$C_b(t) = \frac{V_{ba}}{\hbar} \left\{ \frac{e^{i(\omega_{ba} - \omega)t} - 1}{(\omega_{ba} - \omega)} + \frac{e^{i(\omega_{ba} + \omega)t} - 1}{\omega_{ba} + \omega} \right\} \tag{17}$$

Let us consider that $E_b > E_a$, then the first term of (17) is very large at resonance and the second may be dropped, so that the transition probability comes out as

$$P_{ab} = \frac{2\pi}{\hbar^2} |V_{ab}|^2 \left\{ \frac{\sin^2 \left[ \frac{1}{2} (\omega_{ba} - \omega)T \right]}{2\pi T \left[ \frac{1}{2} (\omega_{ba} - \omega) \right]^2} \right\} \tag{18}$$

The function in brackets, considered as a function of $\omega$ has a sharp peak at $\omega = \omega_{ba}$ and its area is unity, so it behaves like the $\delta$-function.

$$P_{ab} = \frac{2\pi}{\hbar^2} |V_{ab}|^2 \delta(\omega_{ba} - \omega)$$

$$= \frac{2\pi}{\hbar} |V_{ab}|^2 \delta(E_b - E_a - h\nu) \tag{19}$$

### SUGGESTIONS FOR FURTHER READING

Eyring, Walter, and Kimball: Chapter 8.

Landau and Lifshitz: Chapter 6.

Davidson: *Statistical Mechanics* (New York: McGraw-Hill Book Co., Inc., 1962). Pages 288–301.

# APPENDIX F: CALCULATION OF $T_1$ AND $T_2$ FOR A SPIN OF 1/2

We consider an electron spin $S = 1/2$ in a steady magnetic field $H_0$ which is perturbed by a small time-dependent force consisting of an additional random magnetic field $H^*(t)$. The essential effect here is that $H^*$ disturbs the regular precession of the spin and causes an exponential decay in the magnitude of the spin vector. We shall suppose that $H^*(t)$ has a zero average value and fluctuates with a characteristic correlation time $\tau_c$. The following calculation is only valid if $\tau_c$ is much shorter than the relaxation times.

The Bloch equations for the problem are

$$\frac{dS}{dt} = \gamma[S \times H_0] + \gamma[S \times H^*(t)] \tag{1}$$

(No relaxation terms yet!)

It is simplest to use an axis system $X, Y, Z$ which rotates at the resonance frequency $\omega_0 = \gamma H_0$. Then (1) becomes

$$\frac{dS_X}{dt} = \gamma(S_Y H_Z^* - S_Z H_Y^*)$$

$$\frac{dS_Y}{dt} = \gamma(S_Z H_X^* - S_X H_Z^*) \tag{2}$$

$$\frac{dS_Z}{dt} = \gamma(S_X H_Y^* - S_Y H_X^*)$$

where

$$H_X^* = H_x^* \cos \omega_0 t + H_y^* \sin \omega_0 t$$

$$H_Y^* = -H_x^* \sin \omega_0 t + H_y^* \cos \omega_0 t \tag{3}$$

$$H_Z^* = H_z^*$$

We wish to prove that the *average* behavior of **S** is described by the Bloch equations

$$\frac{dS_Z}{dt} = -\frac{S_Z}{T_1}, \qquad \frac{dS_X}{dt} = -\frac{S_X}{T_2} \tag{4}$$

with relaxation terms included but without **H***.

## 1. MOTION OF $S_Z$

Let us assume initially that $S_Z$ has some nonequilibrium value $S_Z(0)$, while $S_X = S_Y = 0$. After a short time $t_1$ the first-order approximation to the solution of (2) is

$$S_X(t_1) = -\gamma S_Z(0) \int_0^{t_1} H_Y^*(t_2) \, dt_2$$

$$S_Y(t_1) = \gamma S_Z(0) \int_0^{t_1} H_X^*(t_2) \, dt_2 \tag{5}$$

$$S_Z(t_1) = S_Z(0)$$

In the second-order approximation the value of $S_Z$ is

$$S_Z(T) = S_Z(0) + \gamma \int_0^T dt_1 [S_X(t_1) H_Y^*(t_1) - S_Y(t_1) H_X^*(t_1)]$$

$$= S_Z(0)\{1 - \gamma^2 \int_0^T dt_1 \int_0^{t_1} dt_2 [H_X^*(t_1) H_X^*(t_2) + H_Y^*(t_1) H_Y^*(t_2)]\} \tag{6}$$

The integrand of (6) must now be referred back to space-fixed axes, giving

$$\cos \omega_0(t_1 - t_2)[H_x^*(t_1) H_x^*(t_2) + H_y^*(t_1) H_y^*(t_2)]$$

$$- \sin \omega_0(t_1 - t_2)[H_x^*(t_1) H_y^*(t_2) - H_y^*(t_1) H_x^*(t_2)] \tag{7}$$

At this stage we introduce three new assumptions about the random fields.

1. $H_x^*$, $H_y^*$, and $H_z^*$ are not correlated with one another.

2. The mean square values of $H_x^*$ and $H_y^*$ are equal.

3. The autocorrelation function (Appendix E) of each component of $H^*$ takes the typical form

$$\overline{H_x^*(t + \tau)H_x^*(t)} = \overline{H_x^{*2}}e^{-|\tau|/\tau_c} \tag{8}$$

Under these assumptions, and renaming $t_1 = t$, $t_2 = \tau$, Eq. (6) for the value of $S_Z(T)$ becomes

$$[S_Z(T) - S_Z(0)] = -\gamma^2 S_Z(0)\{\overline{H_x^{*2}} + \overline{H_y^{*2}}\} \int_0^T dt \int_0^t d\tau \cos(\omega_0\tau)e^{-\tau/\tau_c} \tag{9}$$

After a long time the time integral in (9) increases linearly with $T$ and the value of $S_Z$ has the form

$$[S_Z(T) - S_Z(0)] = -\frac{TS_Z(0)}{T_1} \tag{10}$$

required by (4), with the *spin-lattice relaxation time*

$$\frac{1}{T_1} = \gamma^2[\overline{H_x^{*2}} + \overline{H_y^{*2}}]\frac{\tau_c}{1 + \omega_0^2\tau_c^2} \tag{11}$$

## 2. MOTION OF $S_X$ AND $S_Y$

If we start with the spin polarized along the $X$-axis with a value $S_X(0)$ and again integrate the Bloch Eqs. (2), the analogue of Eq. (6) is

$$[S_X(T) - S_X(0)] = -\gamma^2 S_X(0) \int_0^T dt_1 \int_0^{t_1} dt_2[H_Y^*(t_1)H_Y^*(t_2) + H_Z^*(t_1)H_Z^*(t_2)] \tag{12}$$

Assuming no correlation between different components of $H^*$, the statistical average of the integrand in (12) is

$$\frac{1}{2}[H_x^*(t_1)H_x^*(t_2) + H_y^*(t_1)H_y^*(t_2)]\cos\omega_0(t_1 - t_2) + H_z^*(t_1)H_z^*(t_2) \tag{13}$$

We now use (8) and follow the proof of Eqs. (9–11) again. The result is

$$\frac{1}{T_2} = \gamma^2\left\{\tau_c\overline{H_z^{*2}} + \frac{1}{2}[\overline{H_x^{*2}} + \overline{H_y^{*2}}]\frac{\tau_c}{1 + \omega_0^2\tau_c^2}\right\} \tag{14}$$

## 3. SECULAR AND NONSECULAR TERMS

Equation (11) tells us that $T_1$ depends on the transverse components of $H^*$ which fluctuate at the resonance frequency $\omega_0$. Indeed the result already bears a remarkable resemblance to Eq. (15) of Appendix E. We recall that $1/T_1$ is defined in Chapter 1 as just twice the transition probability induced by the random force $V(t)$

$$\frac{1}{T_1} = \frac{1}{\hbar^2}|V_{\alpha\beta}|^2\frac{4\tau_c}{1 + \omega_0^2\tau_c^2} \tag{15}$$

When $V(t)$ is due to the magnetic field $H^*$ its matrix element is $V_{\alpha\beta} = \frac{1}{2}\gamma\hbar(H_x^* + iH_y^*)$ and (15) is clearly identical with (11). We could obviously derive both $T_1$ and $T_2$ by using time-dependent perturbation theory. The results naturally agree with (11) and (14), but the proofs are fairly heavy.

Equation (14) demonstrates that $1/T_2$ is the sum of two parts

$$\frac{1}{T_2} = \frac{1}{T'_2} + \frac{1}{2T_1} \tag{16}$$

The second term, known as the *nonsecular* part or the *lifetime broadening* arises from the finite lifetime of the spin states $|\alpha\rangle$ and $|\beta\rangle$. The other is called the secular term

$$\frac{1}{T'_2} = \gamma^2 \tau_c \overline{H_z^{*2}} \tag{17}$$

It arises from modulation of the energy levels by the diagonal matrix elements of $V(t)$ and may also be written

$$\frac{1}{T'_2} = \frac{\tau_c}{\hbar^2} \overline{(V_{\alpha\alpha} - V_{\beta\beta})^2} \tag{18}$$

Finally if $\mathbf{H}^*$ is isotropic with $\overline{H_x^{*2}} = \overline{H_y^{*2}} = \overline{H_z^{*2}}$ and if $\tau_c$ is sufficiently short that $\omega_0\tau_c \ll 1$ the two relaxation times are equal

$$T_1 = T_2 \tag{19}$$

## SUGGESTIONS FOR FURTHER READING

Slichter: Chapter 5.
Abragam: Chapter 8.
Bloembergen: Chapter 4.

# APPENDIX G: THE POWER SPECTRUM OF A RANDOM FUNCTION

The Fourier transform of a function $f(t)$ is defined by the equations

$$f(\omega) = \int_{-\infty}^{+\infty} f(t)e^{i\omega t}\, dt$$

$$f(t) = \frac{1}{2\pi} \int_{-\infty}^{+\infty} f(\omega)e^{-i\omega t}\, d\omega \tag{1}$$

The Fourier transform of $f(t) = 1$ is the Dirac delta function

$$\int_{-\infty}^{+\infty} e^{i\omega t}\, dt = 2\pi\delta(\omega) \tag{2}$$

If $f(t)$ represents the fluctuations of some physical quantity, it does not die away as the time tends to infinity, and its Fourier integral does not converge. However, the power at any frequency tends to a definite limit. In order to have a consistent mathematical description it is useful to look at the Fourier transform of a truncated function $f_T(t)$ which is defined to be

$$f_T(t) = f(t) \qquad \text{if } -T < t < T$$
$$= 0 \qquad \text{otherwise} \tag{3}$$

Its Fourier transform, $f_T(\omega)$, will exist for any finite value of $T$. Later we shall allow $T$ to become very long.

We now introduce a truncated autocorrelation function

$$G_T(\tau) = \frac{1}{2T} \int_{-\infty}^{+\infty} f_T^*(t + \tau) f_T(t) \, dt \tag{4}$$

which tends to a definite limit as $T \to \infty$. The limit is clearly the time average of $f^*(t + \tau)f(t)$, or just the usual autocorrelation function.

$$\lim_{T \to \infty} G_T(\tau) = \overline{f^*(t + \tau)f(t)} = G(\tau) \tag{5}$$

To express $G_T(\tau)$ in terms of the Fourier transform we write

$$f_T(t) = \frac{1}{2\pi} \int_{-\infty}^{+\infty} f_T(\omega) e^{-i\omega t} \, dt \tag{6}$$

substitute into (4), and use (2).

$$G_T(\tau) = \frac{1}{2T} \left( \frac{1}{4\pi^2} \right) \iiint_{-\infty}^{+\infty} d\omega \, d\omega' dt f_T^*(\omega) f_T(\omega') e^{i(\omega - \omega')t} e^{i\omega\tau}$$

$$= \frac{1}{2T} \left( \frac{1}{2\pi} \right) \iint d\omega \, d\omega' \delta(\omega - \omega') f_T^*(\omega) f_T(\omega') e^{i\omega\tau}$$

$$= \frac{1}{2T} \left( \frac{1}{2\pi} \right) \int |f_T(\omega)|^2 e^{i\omega\tau} \, d\omega \tag{7}$$

Now let $T$ tend to $\infty$, and define the quantity

$$J(\omega) = \lim_{T \to \infty} \frac{1}{2T} \overline{f_T^*(\omega) f_T(\omega)} \tag{8}$$

Clearly (7) and (8) demonstrate that $J(\omega)$ is the Fourier transform of the autocorrelation function. The definition of $J(\omega)$ also shows that it represents the average power of the fluctuations at frequency $\omega$ during the long time interval $-T$ to $T$, and the integral of $J(\omega)$ over all frequencies gives the mean square value of $\overline{|f(t)|^2}$. From (4), (7), and (8), with $\tau = 0$ we find

$$\overline{|f(t)|^2} = G(0) = \frac{1}{2\pi} \int_{-\infty}^{+\infty} J(\omega) \, d\omega \tag{9}$$

## SUGGESTIONS FOR FURTHER READING

Kittel: *Elementary Statistical Physics* (New York: John Wiley & Sons, Inc., 1958). Page 133.
Davidson: *Statistical Mechanics* (New York: McGraw-Hill Book Co., Inc., 1962). Page 290.

# APPENDIX H: THE DIFFUSION EQUATION FOR BROWNIAN MOTION

Debye's theory leads to the equation

$$\frac{1}{D'}\frac{\partial p}{\partial t} = \nabla^2 p$$

$$= \frac{1}{\sin\theta}\frac{\partial}{\partial\theta}\left(\sin\theta\frac{\partial p}{\partial\theta}\right) + \frac{1}{\sin^2\theta}\frac{\partial^2 p}{\partial\phi^2} \tag{1}$$

for the time evolution of the probability $p(\theta, \phi; t)$ that the molecular axis points in the direction $(\theta, \phi)$. The spherical harmonics $Y_{lm}(\theta, \phi)$ satisfy the equation

$$\nabla^2 Y = -l(l+1)Y \tag{2}$$

so that a particular solution of (1) is

$$p = Y_{lm}(\theta, \phi)e^{-l(l+1)D't} \tag{3}$$

The general solution is found by expanding $p(\theta, \phi; 0)$ at $t = 0$ in terms of the $Y_{lm}$. Initially

$$p(\theta, \phi; 0) = \sum_{lm} C_{lm} Y_{lm}(\theta, \phi) \tag{4}$$

and so

$$p(\theta, \phi; t) = \sum_{lm} C_{lm} Y_{lm}(\theta, \phi)e^{-l(l+1)D't} \tag{5}$$

The normalized harmonics obey orthogonality relations

$$\int_0^\pi \int_0^{2\pi} Y_{l,m}^*(\theta, \phi)Y_{l'm'}(\theta, \phi)\sin\theta\,d\theta\,d\phi = 1 \quad \text{if } l = l' \text{ and } m = m'$$
$$= 0 \quad \text{otherwise} \tag{6}$$

Multiplying both sides of (4) by $Y_{lm}^*(\theta, \phi)$ and integrating we find the expansion coefficients are

$$C_{lm} = \int_0^\pi \int_0^{2\pi} Y_{lm}^*(\theta, \phi)p(\theta, \phi; 0)\sin\theta\,d\theta\,d\phi \tag{7}$$

We now wish to calculate how the average value of some molecular quantity $F(\theta, \phi)$ approaches equilibrium. The average is defined as

$$\overline{F(t)} = \iint F(\theta, \phi)p(\theta, \phi; t)\sin\theta\,d\theta\,d\phi \tag{8}$$

We show that when $F$ is a spherical harmonic it relaxes exponentially to zero with a characteristic decay time

$$\tau_l = \frac{1}{D'l(l+1)} \tag{9}$$

The required average is

$$\overline{Y_{lm}(t)} = \iint Y_{lm}(\theta, \phi)p(\theta, \phi; t)\sin\theta\,d\theta\,d\phi \tag{10}$$

We note that

$$Y_{lm}(\theta, \phi) = Y^*_{l,-m}(\theta, \phi) \tag{11}$$

and compare (10) with (5) and (7). Clearly $\overline{Y_{lm}(0)} = C_{l,-m}$ and the average is

$$\overline{Y_{lm}(t)} = \overline{Y_{lm}(0)}e^{-l(l+1)D't} \tag{12}$$

For the example quoted in Section 11.5.4 the initial probability distribution is a $\delta$ function concentrated at $\theta = \phi = 0$.

$$p(\theta, \phi; 0) = \frac{1}{2\pi} \delta(\cos \theta - 1) \tag{13}$$

Substitution into (7) gives

$$C_{lm} = Y^*_{lm}(0, 0) \tag{14}$$

and all terms with $m \neq 0$ vanish. We recall that $Y_{l0}$ is a Legendre polynomial $\sqrt{(2l + 1)/4\pi} \, P_l(\cos \theta)$, and $P_l = 1$ when $\theta = 0$. Hence the expansion of (13) gives

$$p(\theta, \phi; t) = \sum_l \left(\frac{2l+1}{4\pi}\right) P_l(\cos \theta)e^{-l(l+1)D't} \tag{15}$$

Finally, the average values we require are

$$\overline{\cos \theta(t)} = e^{-2D't} \qquad (l = 1) \tag{16}$$

$$\overline{\tfrac{1}{2}[3 \cos^2 \theta(t) - 1]} = e^{-6D't} \qquad (l = 2) \tag{17}$$

## SUGGESTIONS FOR FURTHER READING

Debye: *Polar Molecules* (New York: Dover Publications, Inc., 1945). Chapter 5.

# APPENDIX I: TENSOR AVERAGES IN A ROTATING MOLECULE

In the calculation of relaxation rates we require the mean values of products of the elements of a symmetric tensor $T$ referred to axes $x$, $y$, $z$ fixed in space, while the molecule rotates. The components of the tensor are constant in an axis system $\alpha$, $\beta$, $\gamma$ attached to the molecule. The required average is $\overline{T_{ij}T_{kl}}$, where $i, j, k, l$ are any of the $x, y, z$ axes. Using the transformation from molecular axes described in Appendix D this becomes

$$\overline{T_{ij}T_{kl}} = \sum_{\alpha,\beta,\gamma,\delta} \overline{(l_{i\alpha}T_{\alpha\beta}l_{\beta j})(l_{k\gamma}T_{\gamma\delta}l_{\delta l})}$$

$$= \sum_{\alpha,\beta,\gamma,\delta} \overline{l_{i\alpha}l_{\beta j}l_{k\gamma}l_{\delta l}} \, T_{\alpha\beta}T_{\gamma\delta} \tag{1}$$

If the axes $\alpha$, $\beta$, $\gamma$ make the tensor $T$ diagonal, this simplifies to

$$\overline{T_{ij}T_{kl}} = \sum_{\alpha,\beta} \overline{l_{i\alpha}l_{\alpha j}l_{k\beta}l_{\beta l}}\, T_{\alpha\alpha}T_{\beta\beta} \tag{2}$$

and the problem reduces to determining the averaged products of four direction còsines. The average vanishes by symmetry if any of the indices $i, j, k, l$ differs from all the other three. Hence they must be equal in pairs and the average takes the form

$$\overline{l_{i\alpha}l_{\alpha j}l_{k\beta}l_{\beta l}} = a\delta_{ij}\delta_{kl} + b\delta_{ik}\delta_{kl} + c\delta_{il}\delta_{jk} \tag{3}$$

where $a$, $b$, $c$ are numbers. Let us first suppose $\alpha \neq \beta$. Since $l_{i\alpha} = l_{\alpha i}$ one can rewrite the left-hand side of (3) as $l_{i\alpha}l_{\alpha j}l_{l\beta}l_{\beta k}$, and so it follows that $b = c$. The orthogonality of the $\alpha$ and $\beta$ axes implies that

$$\sum_i = l_{i\alpha}l_{\beta i} = 0 \tag{4}$$

Hence

$$\sum_i \overline{l_{i\alpha}l_{\alpha j}l_{k\beta}l_{\beta i}} = \sum_i (a\delta_{ij}\delta_{ki} + b\delta_{ik}\delta_{ji} + c\delta_{jk})$$

$$= (a + b + 3c)\delta_{jk} = 0 \tag{5}$$

and

$$a + b + 3c = a + 4b = 0 \tag{6}$$

Furthermore the normalization condition gives

$$\sum_i l_{i\alpha}l_{\alpha i} = \sum_k l_{k\beta}l_{\beta k} = 1 \tag{7}$$

so that we deduce

$$\sum_{i,k} \overline{l_{i\alpha}l_{\alpha i}l_{k\beta}l_{\beta k}} = \sum_{i,k} (a + b\delta_{ik}\delta_{ik} + c\delta_{ik}\delta_{ik})$$

$$= 9a + 3b + 3c$$

$$= 9a + 6b$$

$$= 1 \tag{8}$$

Thus the result is

$$\overline{l_{i\alpha}l_{\alpha j}l_{k\beta}l_{\beta l}} = \frac{1}{30}(4\delta_{ij}\delta_{kl} - \delta_{ik}\delta_{jl} - \delta_{il}\delta_{jk}) \tag{9}$$

When $\alpha = \beta$ the average must have the symmetrical form

$$\overline{l_{i\alpha}l_{\alpha j}l_{k\alpha}l_{\alpha l}} = p(\delta_{ij}\delta_{kl} + \delta_{ik}\delta_{jl} + \delta_{il}\delta_{jk}) \tag{10}$$

where $p$ is a constant. To find $p$, use Eq. (7) and form the sum

$$\sum_{i,k} \overline{l_{i\alpha}l_{\alpha i}l_{k\alpha}l_{\alpha k}} = p\sum_{i,k}(\delta_{ii}\delta_{kk} + \delta_{ik}\delta_{ik} + \delta_{ik}\delta_{ik})$$

$$= p(9 + 3 + 3)$$

$$= 1 \tag{11}$$

Therefore $p = 1/15$ and

$$\overline{l_{i\alpha}l_{\alpha j}l_{k\alpha}l_{\alpha l}} = \frac{2}{30}(\delta_{ij}\delta_{kl} + \delta_{ik}\delta_{jl} + \delta_{il}\delta_{jk}) \tag{12}$$

Equations (9) and (12) are substituted into (2) and the sums over $\alpha$ and $\beta$ performed. There are just three distinct types of term $\overline{T_{ij}T_{kl}}$ which do not vanish. The first is

$$\overline{T_{zz}T_{zz}} = \frac{1}{30}(4-1-1)\sum_{\alpha \neq \beta} T_{\alpha\alpha}T_{\beta\beta} + \frac{2}{30}(1+1+1)\sum_{\alpha} T_{\alpha\alpha}T_{\alpha\alpha}$$

$$= \frac{1}{15}\sum_{\alpha \neq \beta} T_{\alpha\alpha}T_{\beta\beta} + \frac{1}{5}\sum_{\alpha} T_{\alpha\alpha}T_{\alpha\alpha}$$

$$= \frac{1}{15}\left(\sum_{\alpha} T_{\alpha\alpha}\right)^2 + \frac{2}{15}\sum_{\alpha}(T_{\alpha\alpha})^2 \tag{13}$$

The other two are found similarly.

$$\overline{T_{zz}T_{xx}} = \frac{2}{15}\left(\sum_{\alpha} T_{\alpha\alpha}\right)^2 - \frac{1}{15}\sum_{\alpha}(T_{\alpha\alpha})^2 \tag{14}$$

$$\overline{T_{zx}T_{zx}} = -\frac{1}{30}\left(\sum_{\alpha} T_{\alpha\alpha}\right)^2 + \frac{1}{10}\sum_{\alpha}(T_{\alpha\alpha})^2 \tag{15}$$

Most of the tensors whose averages are required have trace zero, i.e., $\sum_{\alpha} T_{\alpha\alpha} = 0$, and then the result is proportional to $\sum_{\alpha}(T_{\alpha\alpha})^2$. This quantity, the trace of the tensor $T^2$, is invariant under rotations. It is written $(T:T)$ and has the value

$$(T:T) = \sum_{\alpha,\beta}(T_{\alpha\beta})^2 \tag{16}$$

irrespective of whether or not $\alpha$, $\beta$, $\gamma$ are principal axes.

Thus for tensors of trace zero the averages are

$$\overline{T_{zz}T_{zz}} = \frac{2}{15}(T:T)$$

$$\overline{T_{zz}T_{xx}} = -\frac{1}{15}(T:T) \tag{17}$$

$$\overline{T_{zx}T_{zx}} = \frac{1}{10}(T:T)$$

The more complicated problem of taking averages with two different tensors may be solved in similar fashion. Thus if $g$ and $T$ are both anisotropic tensors we have

$$(g+T:g+T) = (g:g) + 2(g:T) + (T:T) \tag{18}$$

and by replacing $T$ in (17) by $(g+T)$ it is easy to prove that

$$\overline{g_{zz}T_{zz}} = \frac{2}{15}(g:T)$$

$$\overline{g_{zz}T_{xx}} = -\frac{1}{15}(g:T) \tag{19}$$

$$\overline{g_{zx}T_{zx}} = \frac{1}{10}(g:T)$$

Now $(g:T)$ stands for the *inner product* of the two tensors

$$(g:T) = \sum_{\alpha\beta} g_{\alpha\beta}T_{\beta\alpha} \tag{20}$$

# INDEX